HUMAN GEOGRAPHY: A SERIOUS INTRODUCTION

Second Edition

By Barney Warf
University of Kansas

Bassim Hamadeh, CEO and Publisher
Carrie Montoya, Manager, Revisions and Author Care
Kaela Martin, Project Editor
Berenice Quirino, Associate Production Editor
Jess Estrella, Senior Graphic Designer
Natalie Lakosil, Licensing Manager
Kathryn Ragudos, Interior Designer
Natalie Piccotti, Senior Marketing Manager
Kassie Graves, Director of Acquisitions and Sales
Jamie Giganti, Senior Managing Editor

Copyright © 2019 by Cognella, Inc. All rights reserved. No part of this publication may be reprinted, reproduced, transmitted, or utilized in any form or by any electronic, mechanical, or other means, now known or hereafter invented, including photocopying, microfilming, and recording, or in any information retrieval system without the written permission of Cognella, Inc. For inquiries regarding permissions, translations, foreign rights, audio rights, and any other forms of reproduction, please contact the Cognella Licensing Department at rights@cognella.com.

Trademark Notice: Product or corporate names may be trademarks or registered trademarks, and are used only for identification and explanation without intent to infringe.

Cover image copyright © 2016 iStockphoto LP/marchello74.

Printed in the United States of America.

ISBN: 978-1-5165-2902-5 (pbk) / 978-1-5165-2903-2 (br)

CONTENTS

Chapter 1: Understanding Human Geography   1
Main Points   1
1.1 What is Space?   7
1.2 Five Themes for Studying Human Geography   13
Key Terms   23
Study Questions   24

Chapter 2: Early Human Cultures   27
Main Points   27
2.1 The Evolution of the Hominids   28
2.2 Hunting and Gathering   34
2.3 The Neolithic Revolution   39
2.4 Preindustrial Agriculture   50
2.5 The Urban Dawn and the Rise of Civilizations   55
Key Terms   90
Study Questions   91

Chapter 3: The Geography of Languages   97
Main Points   97
3.1 What Is Language?   98
3.2 Geography of the World's Major Language Families   100
3.3 Linguistic Conflict   129
3.4 Language Death   131
Key Terms   133
Study Questions   134

Chapter 4: The Geography of Religions   139
Main Points   139
4.1 Major World Religions   140
4.2 Geographies of Secularism   174

Key Terms   178
Study Questions   179

Chapter 5: The Rise of Global Capitalism   185
Main Points   185
5.1 Feudalism and the Birth of Capitalism   187
5.2 The Emergence and Nature of Capitalism   196
5.3 The Industrial Revolution   206
5.4 Colonialism: Capitalism on a Global Scale   217
Key Terms   240
Study Questions   241

Chapter 6: Globalization   245
Main Points   245
6.1 International Trade   247
6.2 International Money and Capital Markets   253
6.3 Transnational Corporations   254
6.4 The World Trade Organization, International Monetary Fund, and The World Bank   256
6.5 Regional Economic Integration   259
6.6 Global Resource and Energy Flows   263
6.7 Tourism   269
6.8 Cultural Globalization   272
6.9 Myths About Globalization   276
6.10 Anti-Globalization   280
Key Terms   287
Study Questions   289

Chapter 7: Geographies of Development   293
Main Points   293
7.1 Measuring Economic Development   295

7.2 Urbanization in Developing Countries	305
7.3 Problems of Developing Countries	316
7.4 Development Strategies	329
Key Terms	333
Study Questions	334

Chapter 8: Population and Resource Consumption — 337

Main Points	337
8.1 The Growth and Distribution of the World's Population	338
8.2 Population Change	345
8.3 Malthusian Theory	347
8.4 Neo-Malthusianism	350
8.5 Other Views of Population Growth	351
8.6 The Demographic Transition	353
8.7 Population Structure	368
8.8 Migration	370
Key Terms	378
Study Questions	379

Chapter 9: Political Geography — 383

Main Points	383
9.1 The Nation-State	384
9.2 Nationalism	390
9.3 Geopolitics	394
9.4 World-Systems Theory	398
9.5 The Geography of Wars and Terrorism	400
9.6 Electoral Geography	408
Key Terms	413
Study Questions	414

Chapter 10: Industrial Agriculture — 417

Main Points	417
10.1 Commercialized Agriculture	420
10.2 U.S. Agricultural Policy	429
10.3 Sustainable Agriculture	432
Key Terms	434
Study Questions	435

Chapter 11: Manufacturing — 437

Main Points	437
11.1 Major Concentrations of World Manufacturing	439
11.2 Deindustrialization	449
11.3 Major Manufacturing Sectors	452
Key Terms	460
Study Questions	461

Chapter 12: Services — 465

Main Points	465
12.1 Defining Services	468
12.2 The Growth of Services	471
12.3 Services Labor Markets	476
12.4 The Geography of Services	483
12.5 Telecommunications and Geography	484
Key Terms	495
Study Questions	496

Chapter 13: Culture, Identity, and Everyday Life — 499

Main Points	499
13.1 Introducing Culture	501
13.2 Folk Culture and Popular Culture	502
13.3 Cultural Landscapes	504
13.4 Race and Ethnicity	505
13.5 Gender, Identity, and Place	517
13.6 Sexuality and Space	524
13.7 Orientalism	525
Key Terms	528
Study Questions	530

Index — 535

Chapter One

UNDERSTANDING HUMAN GEOGRAPHY

MAIN POINTS

1. Geography is the study of the space of Earth's surface, including its physical and human-made characteristics and how regions reflect the combinations of these features.

2. Geographies—the spatial distributions and movements of people, goods, technologies, ideas, and other phenomena on Earth's surface—are human creations that change over time.

3. Space and landscapes have three key aspects: the natural environment; the distribution of social phenomena, such as population and culture; and the way that people think about and represent their worlds.

4. Maps help people represent and understand their worlds by structuring spatial information.

5. Five themes help us understand human geography:

    Theme 1: Historical context: History is the study of people through time, and human geography studies people in space. Geographies are produced historically; histories unfold geographically; and understanding one requires understanding the other.

    Theme 2: Networks and interdependence: Every place is part of a system of places. Places are always interconnected to one another.

Theme 3: People and the environment: Human geographies always reflect how people are affected by and, in turn, affect the natural environment.

Theme 4: Culture and space: To know how people make geographies, we must understand their culture, including the ideologies, languages, religions, and webs of meaning they use to make sense of the world.

Theme 5: Social relations: Societies are structured systems of economic, political, and cultural activity. Geographies reflect the societies that make them.

6. Landscapes are both tangible and intangible expressions of how societies are organized economically, politically, and culturally.
7. The term **region** refers to broad areas with similar characteristics.
8. Geographers use a broad array of analytical techniques, both quantitative and qualitative, to gather data and to interpret the distributions of people and human-produced phenomena.

Think of the **landscape** of your campus: the distribution of its buildings, the open spaces in between, the areas in which people congregate (for example, for lunch) and those that are empty (maybe the library), the sports facilities, the administration offices, and the classrooms. A campus is a geography, a **place**, a locale where faculty, students, staff, administrators, and others come together. It is a system of people gathered in a **location** (specific spot on the planet's surface) for a variety of related purposes. Elements of the natural environment appear in it, but this environment has been extensively modified over time. It both resembles other campuses and differs from them; it combines the general and the unique. A campus is the site of both physical interactions and intangible, symbolic ones (for example, learning or the quest for social status). The activities of its inhabitants follow regular patterns that fluctuate on a daily and weekly basis, as well as over the academic year. The campus's organization reflects many forces, including its history, its planning and architectural codes, and the behavior of its users. For example, wealthy private schools will have different landscapes and different student bodies than underfunded public ones. In all these ways, a campus is a microcosm of what human geographers study: the ways in which different combinations of social forces come together to produce a unique spatial arrangement. A campus is simultaneously physical and cultural, economic and political, local and global, a union of many factors that shape its form, function, and structure over time.

The word "geography" has Greek roots: *geo* means "Earth," *graph* means "to write." Literally, then, what geographers do is Earth-writing. Geography is the study of **space**, how Earth's surface is used, how societies produce places, and how human activities are stretched among different locations. In many respects, geography is the study of space in much the same way that history is the study of people in time. This conception differs vastly from popular stereotypes that often portray geographers as concerned only with making maps or drawing boundaries and obsessed with memorizing the names of obscure capital cities. **Physical geography** addresses the physical world (for example, climates and landforms, rivers and streams, soils, plants, and animals). **Human geography** addresses the human world (for example, agricultural activities, migration, the spread of disease, political conflicts, and settlement patterns) (Figure 1.1).

Essentially, the *discipline of geography examines why things are located where they are.* Simply knowing where things are located is relatively simple; anyone with a good atlas can find out, for example, where bananas are grown or the distribution of petroleum reserves. Geographers are much more interested in explaining the *processes* that give rise to **spatial distributions,** not simply mapping those patterns. For example, rather than merely noting the changing locations of jobs in banking, geographers seek to understand why firms move from one city to another or

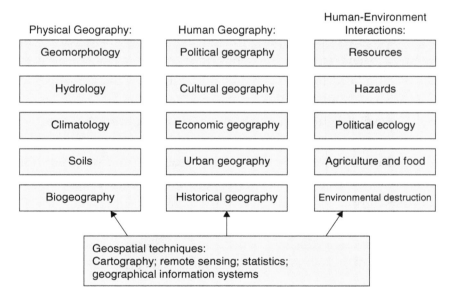

Figure 1.1: Human geography is one part of the broader discipline of geography, which is concerned with how Earth's surface is organized and shaped by physical and social processes. Geographers study a wide array of topics that include both environmental and human phenomena, and they use several geospatial techniques such as maps, statistics, remotely sensed images, and geographical information systems.

from one country to another in an age of **globalization**. Much more interesting than simply finding patterns on Earth's surface is the *explanation* linking the spatial outcomes to the social and environmental processes that give rise to them. Thus, geographers examine not only where people and places are located, but how people understand those places, give them meaning, change them, and are, in turn, changed by them. To understand how geographies are created and change over time is to understand the ways in which societies give rise to new forms of spatial organization. For example, comprehending the shift of the U.S. population from the Northeast and Midwest to the South and West entails appreciating the impacts of migration, birth and death rates, immigration, government spending, favorable climates, water shortages, urban growth, and the changing national distribution of jobs. Because this issue involves both social and environmental topics, geography studies the distribution of both human and natural phenomena. For this reason, it lies at the intersection of the social and physical sciences, as well as the arts and humanities.

Although many geographers have tried to combine the discipline's physical and human components, human geography and physical geography are typically distinct fields with different methods and topics of study. Human geographers do not deny the relevance of physical geography to their work, but they also do not insist that their studies always include a physical component. Because human geography studies human beings, it has close ties with social sciences, such as history, economics, anthropology, sociology, psychology, and political science. One of the strengths of human geography is that it considers the human world in terms of physical and human factors, as necessary. Where appropriate, this book introduces aspects of physical geography. For example, to understand world population distributions, we need to know about global climates. We also need to know what people think of those climates and how different groups have adapted to inhabit different climates. Thus, human geography is related to, but separate from, physical geography in much the same way as it is related to, but separate from, the other disciplines that study human beings.

All social processes and problems are simultaneously spatial processes and problems, for everything social occurs somewhere. More important, *where* something occurs shapes *how* it occurs. Geographers ask questions related to location: Why are there skyscrapers downtown? Why do Asians rely so heavily on rice? Why is the number of world languages declining? Why are there famines in Africa? How do foreign Muslims view the United States, and why do they hold those views? Why is China rapidly becoming a global economic superpower? How does globalization affect different parts of the world?

To view the world geographically is to see space as socially produced (created by people). The landscape and the distribution of various things on it are shaped by people in interaction with one another and with nature (Figure 1.2). They are constructed over time as societies link places together in ever-changing ways. Geographies are created as a product of social relations, a set of

Figure 1.2: Landscapes are a central topic of geographic analysis and reflect both the influences of the natural environment and the political, economic, and cultural forces of societies as they unfold historically over the world. This scene illustrates how landscapes are always mixtures of natural and social features.

patterns and distributions that change over time. Geographic landscapes are social creations produced by people, in the same way that your shirt, computer, school, and family are also social creations. Geographers maintain that the production of space involves different spatial scales, ranging from the smallest and most intimate—the body—to progressively larger areas, including neighborhoods, regions, nations, and the least intimate of all, the entire world.

Because places and spaces are populated—inhabited by people, shaped by them, and given meaning by them—geographers argue that social processes are embodied. This means that places are made by people who live in the world, changing it and being changed by it, through their physical bodies. The body is the most personal of spaces, the "geography closest in," the part of the world that is also part of us. Our bodies are intimately connected to the world we live in, through the air we breathe, the food we eat, and the buildings in which we live and move. Individuals create geographies in their daily lives as they move through time and space in their ordinary routines. In local communities, neighborhoods, and cities—the next larger scale—individuals' movements form regular patterns that reflect a society's organization, division of labor, cultural preferences and traditions, and political opportunities and constraints. Geographies, thus, reflect the class, gender, ethnicity, age, and other categories into which people sort themselves. Spatial patterns reflect the historical legacy of earlier social relations; political and economic organization of resources; the technologies of production, transportation, and communication; the cultures that guide behavior; and legal and regulatory systems.

Geographers study how societies and their landscapes are intertwined. For example, to understand a society like China, which emerged over thousands of years, we must comprehend its vast agricultural landscape suitable for producing large amounts of rice to feed a growing population. More recently, we must understand the cities that are home to half of China's people. China's history, society, economy, and culture cannot be separated from the places in which it has existed. To appreciate this idea, we must recognize that social processes and spatial structures shape each other in many ways. Societies involve complex networks that tie together economic relations of wealth and poverty, political relations of power, cultural relations of meanings, and environmental processes, as well. Geographers examine not only the ways in which people organize themselves spatially, but also how they view their worlds, how they represent and give meaning to space. Divorcing one dimension, say the economic, from another, such as the political or the cultural, is ultimately fruitless.

The distributions of different natural and social phenomena often form regular patterns over Earth's surface, which can be analyzed to determine what causes them and how they change. For example, the location of the world's population can be understood in terms of climatic factors, such as the length of growing seasons and soil fertility, as well as social dynamics, such as the impacts of colonialism, the role of the global and national economies, the formation of cities, and so forth. This distribution is constantly changing through uneven geographies of birth rates, death rates, and migration (Chapter 8). The geography of population unfolds at multiple spatial **scales**, including the global, that within individual countries, the local (e.g., neighborhoods), and the individual body. One way of understanding geographies is to focus on broad patterns of one phenomenon, an approach often called **systematic geography**. For example, geographers might examine the distribution of a particular industry around the world or spatial patterns of fertility. Another approach is to examine how different phenomena combine within the context of a particular area, an approach known as *regional geography*. For example, geographers may study the combination of environmental and social factors that produce the culture and landscape of, for example, Oaxaca, Mexico. Geographers use both approaches to see how the general and the unique intersect to produce patterns over Earth's surface.

1.1 WHAT IS SPACE?

If human geography is the study of people and space, what is this nebulous but crucial thing we call "space?" The space of Earth's surface (as opposed to other kinds of spaces, like outer space) is a complex set of changing phenomena that defies easy explanation. To assist in this task, it is useful to examine three ways in which space and place can be understood: the natural environment; the distribution of human activities; and how people perceive and give meaning to places (Figure 1.3).

Figure 1.3: Space assumes a variety of forms and meanings, ranging from the biophysical processes of the natural environment and cultural landscapes to the perceptions and meanings that people assign to them. For the Asian farmers who created this terraced hillside, the landscape represents a place to live and work and the source of their primary food, rice.

First, there is the physical surface of the planet in all of its geological, atmospheric, and biological complexity. Space, in this meaning, includes the shape of the land and the chemical and biological processes that shape and change it: the topography (shape) of the land, soil types, ecosystems, climate types, and so forth, all of which formed over millennia or longer (Figure 1.4). As we shall explore shortly, this natural environment is vitally important to how people produce and distribute themselves over landscapes. In this respect, human geography intersects and overlaps with physical geography. Yet, because

Figure 1.4: This desert landscape depicts part of the biophysical environment, what we might think of as space that exists "outside" of human activities. Certainly, such physical landscapes preceded the emergence of human beings. But given the extensive human modification of environments around the world, even "natural" landscapes are not so natural anymore. Rather, they reflect the enormous imprint of human cultures and behavior. Even this seemingly pristine landscape has been modified by many years of human action.

human activities have affected almost all of the natural world, to appreciate this dimension of space, we must include the ways in which people change the physical landscape and give meaning to it.

A second meaning of space involves the spatial distribution of social phenomena, such as population (Figure 1.5), different cultures and religions, cities, different types of jobs, and the locations of wealth and poverty. Obviously, to some extent, the location of human phenomena reflects the natural world. Yet, this contributes to only part of the explanation. Cultural, political, and economic landscapes do not simply result from climate or topography, but emerge from social relations of power, culture, and economy as they unfold over time. For example, if we wish to understand why pockets of poverty exist in large cities, such as New York, or on Native American reservations, we must focus on the mechanisms that created these places. Such factors include labor markets (the number and types of available jobs); public policies;

Figure 1.5: Although influenced by the world's climates, soils, topography, and ecosystems, the location of people is also the product of many social forces and processes that play out over the surface of Earth. The geography of the world's population is a reflection of many factors, including the availability of agricultural land, colonialism, urbanization, and migration.

public and private investment (or lack of it); technological changes, migration into and out of a region; and people's cultural beliefs about work, leisure, saving, and spending.

A third dimension of space, one many people do not automatically associate with geography, concerns its symbolic, psychological, and ideological contours. Space is not simply what is "outside" of people's minds. It also exists in how people perceive and give meaning to their world. Geography exists, therefore, simultaneously as a concrete, material world and as an intangible, symbolic one. A rich tradition in human geography studies how people represent their worlds. All human beings develop a symbolic geography, including **mental maps** of where they think places are located, whether accurate or not. Every society represents its spaces and places to itself and to others. Different cultures describe and interpret the parts of the world they inhabit. They do so through narrative descriptions and stories, as well as paintings, photographs, and maps. The history of cartography (mapmaking), for example, is full of examples of how different societies mapped their worlds. How we know the world shapes the world itself; reality and our knowledge of it can never be effectively separated. For example, as Europeans made increasingly accurate world maps during the Renaissance (Figure 1.6) and the period of colonialism that began in the sixteenth century, their ability to control the world's oceans and continents grew correspondingly. Their maps did not simply mirror the world; they helped to create a global economy with Europe at the center (Chapter 5).

Because **maps** summarize vast amounts of information efficiently, **cartography** (the creation and study of maps) has a long history associated with geography. Maps extend our perceptions of the world, allowing us to "see" places and their relations to one another, even if we have never been there. Cartographers have studied maps extensively, including their history, the role of scale, rules for including some information and leaving some out (generalization), computer maps, map symbols, and so forth. No map is perfect. For example, it is impossible to represent a three-dimensional globe on a two-dimensional surface without some distortion of area size or shape. Cartographers have developed several different standard ways of doing this, known as **map projections.** Moreover, no map can show everything about a place; of necessity, it must be selective in what it represents. Maps are thus not simply "objective" statements about the world, but reflections of culture and power.

Figure 1.7 is a simple map showing how the land surface of the world is divided into countries. What do these borders tell us? Have they always existed? Are the divisions changing today—possibly even between the time when I am writing and the time when you read these words? Is the total number of countries increasing or decreasing? What do these political divisions reflect—population distribution, cultures, development? Are there meaningful groupings? Do basic groupings, such as north versus south, or east versus west, have any value? These are the sorts of questions that human geographers ask.

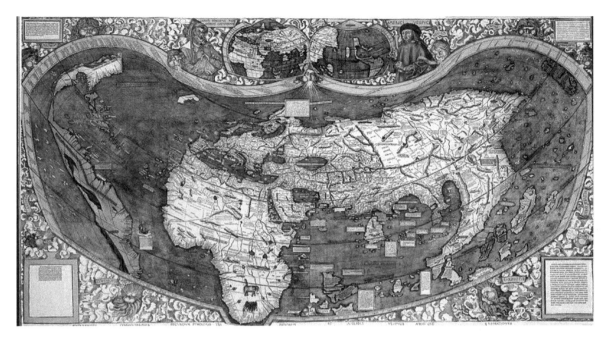

Figure 1.6: Maps testify to geography's concern with how cultures perceive, understand, and give meaning to space, as well as how landscapes and distributions are produced. The Waldseemuller map here, the first to show the Americas, is among the most famous in history.

Political boundaries interest human geographers for at least three reasons. First, the political partitioning of the world exemplifies regionalization, or the organization of space. Second, political units usually play a major role in determining such matters as the number of births and deaths and the availability of paid employment. Third, many political divisions reflect differences in ethnicity and/or culture (for instance, in language or religion). Human geographers identify relationships between culture and political divisions, often with the help of maps. For example, the breakup of the former state of Yugoslavia in the 1990s occurred largely along linguistic and religious lines, particularly between Bosnian Muslims, Orthodox Christian Serbs, and Catholic Croatians.

The world is divided not only into political units but also into larger groups of countries labeled "more developed" and "less developed." These divisions reflect the tendency of capitalist societies to produce uneven distributions of wealth, a theme we will explore at length in later chapters. There are enormous differences among individuals and

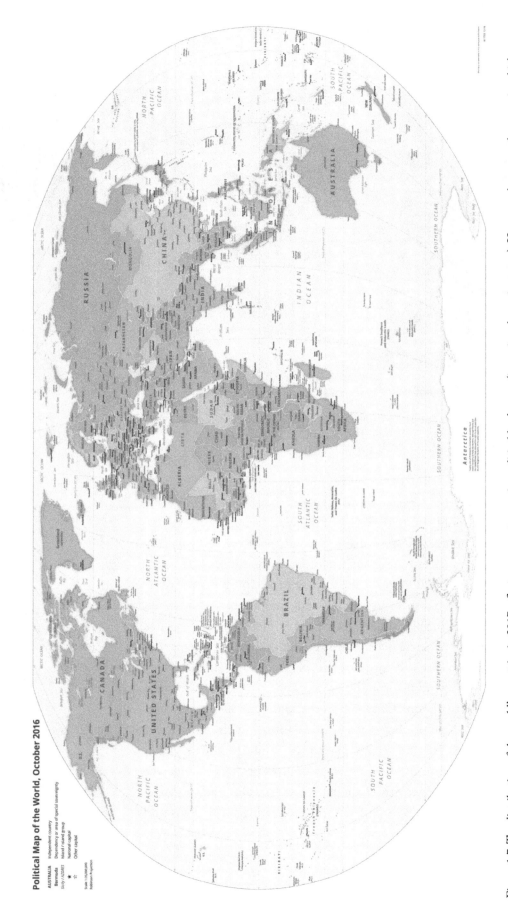

Figure 1.7: The distribution of the world's countries in 2017 reflects one way to organize political landscapes (i.e., using the nation-state). However, there are other ways in which power is manifested geographically, including tribes and globalization, that challenge this seemingly natural system.

more general differences among groups of individuals. Similarly, landscapes vary. Each particular location is unique, of course, different from every other—but it is often more useful to note the more general differences that exist between regions. (We discuss the concept of regions, areas that share certain characteristics, later in this chapter.)

1.2 FIVE THEMES FOR STUDYING HUMAN GEOGRAPHY

One means of understanding human geography is through five themes that will reappear in different ways throughout this book. These are broad generalizations designed to encourage you to think about how societies and places reflect each other. They include (1) the historical context in which geographies are produced; (2) the interconnectedness of regions, particularly with the rise of the global capitalist economy; (3) the interrelatedness of human and natural systems; (4) the importance of culture and everyday life; and (5) the centrality of social structures and relations in human geography.

HISTORICAL CONTEXT

The study of space is inseparable from the study of time. If one accepts geography as the study of human beings in space and history as the study of people over time, then this theme reveals the profound links between geography and history. For example, one cannot understand the contemporary geography of the United States—the locations of its cities and agricultural areas, the distribution of its people and different ethnic and religious groups—without examining the evolution of American society over the past four or five centuries. Because the accumulated decisions of actors in the past—individuals, organizations, governments, and others—created the present, we cannot explain the contemporary world meaningfully without continual reference to their actions. Awareness of the importance of history undercuts the common assumption that the present is "typical" or "normal," for history, as much as geography, teaches us the full range and diversity of human behaviors, cultures, and social systems. The organization of space occurs as societies go through changes over time, and all changes over time occur in places (Figure 1.8). Such an emphasis leads us to question of *how* histories and geographies are produced. Historical geography—and *all* geography is inescapably historical—does not just reconstruct past worlds, but also analyzes how social systems have changed over space and time.

Figure 1.8: Because time and space cannot be separated, all geographies must be understood historically. To know a landscape is to know how it came to be and how it has changed over time. This photo of Istanbul, Turkey, reveals a landscape shaped over many centuries, revealing how history is enfolded in every landscape.

But, there is a broader, deeper implication here. Taking history seriously means acknowledging that geographies are never fixed but are always changing. Change, of course, occurs at many temporal and spatial scales, from gradual, long-term processes, such as population growth, to short-term, rapid ones, such as movements of information through the Internet. History and geography develop through the dynamics of everyday life, the routine interactions through which people create societies, reproduce them, and change them. An historical perspective reveals that social and geographical realities are not predetermined, but made and can be unmade. There is, for example, no need to accept the geography of the world's poverty as inevitable, or to accept the repression of women or mistreatment of children as unchangeable. Views of geography that try to capture a "snapshot in time," therefore, are deceptive; it is the *process* underlying the creation of places that is central, the social dynamics at work, not their appearance at one particular historical moment. Geographies are inherently unstable and never frozen, so we must approach them as processes rather than static states.

NETWORKS AND INTERDEPENDENCE

Every place is part of a system of places (Figure 1.9). All regions are interconnected; they never exist in isolation from one another. Things that tie places together to a greater or lesser extent include the natural environment (e.g., winds and currents, flows of pollution); flows of people (migration); movement of capital and money (investment); exchange of goods (trade); and the **diffusion** (spread) from source areas of information, ideas, innovations, and disease. Places always belong to a network because contemporary social relations stretch across regions. For example, to understand the geography of a city, such as Miami, one must examine how it arose historically. This includes the role of railroads in the early twentieth century, and more recently, migration streams from the northeastern United States, Cuba, and many other places.

Figure 1.9: Geography is not just about places, but is also about the networks and flows that bind them together, as this map of the diffusion of Polynesians across the Pacific illustrates.

Because places are always tied together, what happens in one place must affect events in others. For example, the Internet links people in distant places in many ways; the North American Free Trade Agreement (NAFTA) joins regions from southern Mexico to Quebec; immigration has brought parts of the developing world into wealthy countries; and the jet airplane has made the contemporary world vulnerable to new diseases, such as acquired immune deficiency syndrome (AIDS) and Ebola. Although this theme holds in the study of many places throughout history, it is especially relevant since the rise of global capitalism beginning in the fifteenth century (discussed in Chapter 5). Today's global system of nations and markets has tied places together to an unprecedented degree. For example, as any trip to the grocery store will attest, we can buy products from all over the world. Globalization is just the most recent manifestation of how people have developed relations over ever-larger distances that shape multiple places (Chapter 6). Thus, to understand regions and places as linked is to acknowledge that what happens in one place always reverberates to shape others.

One compelling example of changing geographies can be found in the tragic events of September 11, 2001. The world felt the impact of the terrorist attacks that destroyed the twin towers of the World Trade Center in New York City and damaged the Pentagon near Washington, DC, killing more than 3,000 people. The consequences of those attacks, some immediate, some longer term and still uncertain, have reshaped both local and global geographies. The skyline of New York City will never be the same. Nor will the pattern of global politics, because the United States responded to the attacks by invading Afghanistan, where the ruling Taliban had supported terrorist groups. In the wake of the attacks, there was widespread acknowledgment that fighting terrorism would require in-depth understanding of local geographies, both physical and human.

PEOPLE AND THE ENVIRONMENT

Human action always occurs in a natural environment. Unfortunately, the term "nature" often carries connotations of being separate from human activities. The natural environment includes the climate, topography (shape of the land surface), soils, ecosystems and vegetation, and mineral and water resources of a region. It shapes everything from the length of a growing season of agricultural crops to transport costs to energy supplies. Natural disasters, such as hurricanes, tornadoes, droughts, and earthquakes also shape where and how people live and the risks they face. Although these factors certainly *affect* the creation of histories and geographies, they do not *control* the destiny of societies (as a bankrupt early theory called **environmental determinism** once claimed).

But conversely, the relationship between people and nature is a two-way street. Everywhere, nature has been changed by human beings. Examples include the modification of ecosystems;

annihilation of plant and animal species; soil erosion; air, ground, and water pollution; changed drainage patterns; agriculture; deforestation; desertification; disruptions of natural cycles of nutrients, including carbon and nitrogen, and energy loops in ecosystems; and more recently, global climate change, a process that has had severe ecological impacts ranging from the melting of glaciers and ice caps to the acidification of the oceans. Indeed, human beings cannot live in an ecosystem without modifying it. Human-induced changes within the planet's environments have been so extensive that many geologists talk of a new era, the Anthropocene (the Human Age), in which humans are the major agents changing the world's landscapes. In short, the formation of geographies is neither reducible to the natural environment nor independent of it (Figure 1.10).

Figure 1.10: Landscapes are neither purely "natural" nor purely "cultural," but the product of interactions of the two, revealing how society and nature are shot through with each other. Thus, although nature determines the climate and topography of this scene, humans have extensively modified it.

People produce geographies in the context of particular natural environments, and those environments are always altered through human actions. The spread of human beings accelerated the growth of grasslands worldwide as hunters burned savannahs and prairies. Prior to the Columbian encounter in the sixteenth century, Native Americans produced enormous ecological impacts, including widespread hunting and the burning of prairies and forests. Such facts run counter to the myth that native peoples left their world in an untouched, "natural" state. Indeed, in many ways "nature" is not natural anymore but has been shaped by societies over eons. Humans have redistributed numerous species of plants, animals, viruses, and bacteria over the world, and exterminated many others. The spatial structure of the Industrial Revolution was profoundly influenced by the location of the large coal deposits in Britain and the northern European lowlands. In these examples, the natural environment—the subject of physical geography—was important to understanding how human geographies were created.

CULTURE AND SPACE

Human consciousness is fundamental to understanding how geographies are made. Human beings always possess an awareness of themselves and their world, as manifested in their perceptions, thoughts, language, and ideas. Far more than any other species, humans produce and use symbols to manipulate their world and to communicate. Because people are social animals, the shared medium of **culture** (Chapter 13) organizes their consciousness—their language, assumptions, beliefs, priorities, norms, and values (Figure 1.11). Thus, social science differs fundamentally from analyses of the nonhuman world (except, perhaps, the behavior of some animals).

Traditional definitions of culture portray it as the sum of learned behavior or a "way of life" (religion, language, mores, traditions, roles, and other elements). Thus, culture is not biological and is not inherited through genes. Culture tells us how to behave in the world and relate to one another. We can specify this idea further if we define culture as the knowledge we take for granted, what we know without knowing it, "common sense." Everyone must internalize a culture to survive in it, to know how to act male or female, young or old, wealthy or poor. Within each social context, culture defines what is normal, important, and acceptable—and what is not.

People acquire culture through a lifelong process of **socialization.** Individuals never live in a social vacuum but develop their sense of identity from their interactions with others, from the cradle to grave. The family, peer groups, schools, media, and other social institutions play key roles in this process. As people acquire a culture in daily life, they reproduce the society in

Figure 1.11: Because people are conscious, sentient beings, their thoughts, beliefs, and behaviors are learned through culture. Culture consists of the web of values, norms, mores, ideologies, and symbolic systems that socialize people, let them makes sense of their worlds, and guide their behaviors. These Maori women in New Zealand are illustrative of an ancient traditional culture that has been challenged and modified by years of exposure to Western culture.

which they live: A society can only exist from one day to the next, or from one generation to the next, if it passes down its cultural norms over time. When people go about their daily lives, they reproduce their social world, largely unintentionally. In turn, that world structures their lives. Thus, individuals are both shaped by and shapers of history and geography. Everyday thought and behavior do not simply mirror the world; they create it.

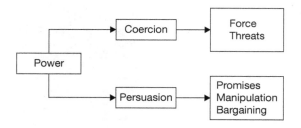

Figure 1.12: Power takes multiple forms and is always unevenly distributed within and among societies. Class, gender, and race are major forms through which power is manifested. Power may involve coercion (force or the threat of force), as well as persuasion and manipulation.

SOCIAL RELATIONS

Social relations describe the structure and organization of societies and places. Every society includes structures of power and wealth that benefit some groups and not others. For example, one cannot hope to understand a society, such as the United States, without looking at how its economy has generated types of jobs (e.g., in finance) that typically favor certain social groups (e.g., white men) over others (e.g., women or ethnic minorities). The analysis of social relations allows us to understand societies as more than just the sum of individual actions (Figure 1.12).

We can define the study of social relations as **political economy**. In this context, the term refers to the organization of societies as systems of power and wealth, including class relations, gender, and ethnicity. A focus on power emphasizes, "who gets what, when, where, and why" in the making of geographies. That is, political economy emphasizes the ways in which social resources are distributed as defining characteristics of everyday life and as part of the fundamental dimensions of political struggle. An economy may be defined as the means by which societies allocate scarce resources (e.g., land, labor, capital), a system of producing and distributing goods and services. Political relations are closely related to economic ones; wealth always brings with it power. Politics is the struggle for power, and power takes many forms, including coercive and persuasive power the possession or withholding of information, wealth, violence, and bureaucratic authority (Figure 1.13). Power is always unevenly distributed and unequally shared by different groups divided by class, gender, ethnicity, and place. As we will see throughout the following chapters, class structures human geographies in many ways.

A social perspective contrasts with the widespread individualism found in Western societies, particularly the United States. Many disciplines, such as economics, take individual decision making as their starting point. In contrast, from a political economy perspective, individuals are the product, not the producer, of social relations

Figure 1.13: From a political economy perspective, it is not individuals but the relations among them that are of paramount importance. Societies are more than the sum of their parts, and individuals are the products, as well as the producers, of social relations. Although we all like to think of ourselves as unique individuals, as indicated by the photo of this group, in reality we are all inevitably products of our social environments.

(Figure 1.13). All individuals are socially produced, and their choices and opportunities are constrained with fields of power that fluctuate over time and space.

Similarly, feminists have shown how uneven power relations between men and women affect all aspects of life, including allocations of resources and modes of thought and behavior. Geographers study spatial differences between men and women in housing, work, and commuting patterns and how such relations have typically favored men over women. For example, manufacturing-based economies generated particular types of jobs dominated by men, giving rise to urban landscapes in which women primarily stayed at home. Conversely, service-based economies have pulled many women into the labor market, changing their role in the household. The rise of the two-income family has altered residential location decisions and commuting patterns of these households.

Finally, as with class and gender, ethnicity is a powerful dimension of social relations. All over the world, ethnic minorities tend to be second-class citizens, with less access to wealth and power than the majorities of their societies. For example, many American cities contain large concentrations of impoverished African Americans. Geographers seek to explain how ethnicity intersects with the dynamics of power, the economy, and cultural attitudes toward race to explain such ghettos. In this book, we will see how class, gender, and ethnicity repeatedly structure the unfolding of human geographies over time.

■ KEY TERMS

Cartography: the science and art of making and understanding maps

Culture: the webs of meaning that people use to interpret the world

Diffusion: the spread of a phenomenon over time and space

Environmental determinism: the belief that the natural environment dictates human affairs

Globalization: the processes that link people together in far-flung places around the world

Human geography: the portion of the discipline of geography concerned with the spatial dimensions of human existence and the distributions of social phenomena

Landscape: Earth's surface at a local or regional scale

Location: position on Earth's surface as measured either in absolute terms of latitude and longitude or in relative terms compared with other places

Map: a graphical device portraying the distribution of phenomena over space or regions

Map projections: various means of converting Earth's three-dimensional surface onto a two-dimensional map

Mental maps: maps of places that people carry in their heads

Place: a location to which people give meaning and experience everyday life

Political economy: the study of social relations, that is, how societies are structured as systematic networks of wealth, power, and culture

Region: a part of the world that shares a broad set of characteristics in common

Scale: the amount of space over which particular processes play out

Socialization: the internalization of social norms and rules by individuals

Space: in geography, the surface of Earth

Spatial distribution: the uneven geographical location of various phenomena

■ STUDY QUESTIONS

1. Define the term *geography*, and explain what the discipline studies.
2. In what ways do people construct space?
3. How does the study of society connect to that of space?
4. How and why do maps selectively portray spatial features?
5. What does the relationship between space and time imply about history and geography?
6. Define *political economy*, and explain its relationship to geography.
7. Is culture comprehensible without referring to how societies and places are structured? Why or why not?
8. Why can it be argued that "nature isn't natural anymore"?
9. Can places be fully understood when considered in isolation? Why or why not?

BIBLIOGRAPHY

Castree, N. (2005). Nature. *London: Routledge.*

De Blij, H., Muller, P., & Nijman, J. (2012). Geography: Realms, Regions and Concepts *(15th ed.). Hoboken, NJ: Wiley.*

Fouberg, E., Murphy, A., & de Blij, H. (2012). Human Geography: People, Place, and Culture. *Hoboken, NJ: Wiley.*

Harvey, D. (1984). On the history and present condition of geography: An historical materialist manifesto. Professional Geographer, 36, *1–10.*

Knox, P., & Marston, S. (2012). Human Geography: Places and Regions in Global Context *(6th ed.). Upper Saddle River, NJ: Prentice Hall.*

Massey, D. (2005). For Space. *London: Sage.*

Rubenstein, J. (2013). The Cultural Landscape: An Introduction to Human Geography *(11th ed.). Upper Saddle River, NJ: Prentice Hall.*

IMAGE CREDITS

- Figure 1.2: U.S. Fish and Wildlife Service, "Navarre Marsh Scenic Aerial," http://commons.wikimedia.org/wiki/File:Navarre_marsh_scenic_aerial.jpg. Copyright in the Public Domain.
- Figure 1.3: Copyright © CEphoto, Uwe Aranas (CC BY-SA 4.0) at https://commons.wikimedia.org/wiki/File:Banaue_Philippines_Batad-Rice-Terraces-02.jpg.
- Figure 1.4: Copyright © Stan Shebs (CC BY-SA 3.0) at http://commons.wikimedia.org/wiki/File:Pahrump_Nevada_aerial.jpg.
- Figure 1.5: http://commons.wikimedia.org/wiki/File:World_population_density_1994.png. Copyright in the Public Domain.
- Figure 1.6: Ptolemy, "Waldseemüller World Map 1508," https://commons.wikimedia.org/wiki/File:Waldseem%C3%BCller_world_map_1508.jpg. Copyright in the Public Domain.
- Figure 1.7: Source: https://www.cia.gov/library/publications/the-world-factbook/docs/refmaps.html.
- Figure 1.8: Copyright © Selda Yildiz and Erol Gülsen (CC BY-SA 3.0) at http://commons.wikimedia.org/wiki/File:Istanbul2010.jpg.
- Figure 1.9: Copyright © Gringer / Bibi Saint-Pol (CC by 3.0) at https://commons.wikimedia.org/wiki/File:Map_Polynesian_migration-fr.svg.
- Figure 1.10: davyb, http://pixabay.com/en/maori-maori-group-kiwi-culture-89317/. Copyright in the Public Domain.
- Figure 1.11: Copyright © chensiyuan (CC BY-SA 4.0) at https://commons.wikimedia.org/wiki/File:1_yuanyang_rice_terrace_duoyishu_sunrise_2012.jpg.
- Figure 1.13: Copyright © Depositphotos/luminastock.

Chapter Two

EARLY HUMAN CULTURES

MAIN POINTS

1. Human beings (*Homo sapiens*), one of several hominid species, evolved in eastern Africa. Our species emerged roughly 300,000 years ago and spread throughout the world over several thousand years. Equipped with a far greater intelligence than any other animal, we became the planet's first superspecies.

2. More than 95% of human existence occurred during the Paleolithic era of hunting and gathering. During this period, small, nomadic groups collected a variety of foods. This lifestyle has almost disappeared today but survives in a few isolated areas.

3. The Neolithic Revolution occurred in several areas around the world, beginning about 10,000 years ago. It included the domestication of plants and animals. People began to engage in agriculture and live in permanent settlements. These changes led to innovations, including writing, metalworking, the wheel, the state, class-based societies, and cities.

4. Early cultures fed themselves through a variety of preindustrial agricultural methods, including slash-and-burn or swidden practices, Asian rice farming, and peasant-based modes.

5. Cities arising from agriculture became centers of power, religion, production, and trade. Several early civilizations shaped the formation of the world's human geographies.

To understand the world's cultural geographies, it is important to trace their origins and development over time. In many cases, the distributions of the planet's peoples and cultures reflect processes and events that can be traced back over thousands of years. Thus, an understanding of early human cultures is imperative to appreciate the trajectory that led to the contemporary world, as geographies are not created overnight. One of the five themes raised in Chapter 1 concerned the historical construction of human geographies. Today's landscapes and settlement patterns have emerged over centuries or longer. Many current institutions, such as agriculture and cities, have origins in events and processes that began far in the past.

This chapter charts the earliest roots of contemporary human geographies in several ways. It begins by examining hunting and gathering. This lifestyle dominated more than 95% of human history. Second, it focuses on the Neolithic Revolution, which included the rise of agriculture and the **domestication** of plants and animals. These innovations led to enormous social, economic, and geographic changes. Third, it explores the nature of various preindustrial agricultural systems. These came from the agricultural revolution and still feed many people today. Fourth, and finally, it summarizes some early civilizations that began in the Neolithic era. These civilizations played key roles in shaping the world's peoples and cultures.

2.1 THE EVOLUTION OF THE HOMINIDS

Human beings are one of a large family of apes that evolved in eastern and southern Africa over many millennia. Although some people mistake evolutionary theory as holding that humans "descended from monkeys" (which have tails, whereas apes do not), in fact, the fossil evidence suggests that humans and other apes all descended from a common ancestor, gradually diverging into distinct biological genera and species. Determining when humans first appeared depends on the definition of "human" that one uses; different forms and variants of humans emerged at different moments in time. For example, the group known as hominids (or hominins) emerged in Africa roughly 6 million years ago (Figure 2.1), when the line that eventually led to humans departed from chimpanzees and bonobos, with whom humans share 98% of their DNA. Many anthropologists held that the rise of hominids reflected changing climatic circumstances on the savannahs of eastern Africa, which experienced long-term drying, leading forests and woodlands to contract, causing grasslands to expand, and resulting in considerable ecological change. For example, hominids are characterized by bipedalism, or walking upright, which was an evolutionary advantage for surviving in the grasslands.

The first-known hominid group is called Australopithecus, a genus of which there were numerous species (e.g., *A. africanus, robustus, ramidus,* and *afarensis*), which lived in eastern and

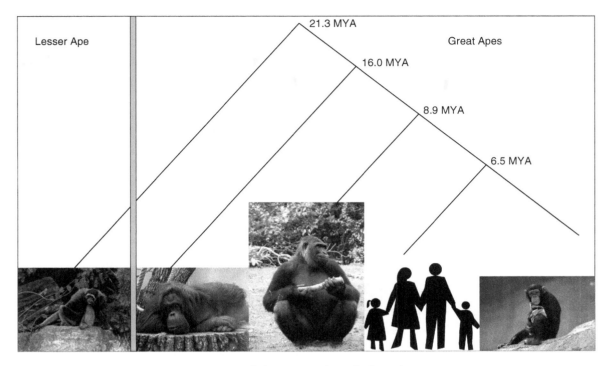

Figure 2.1: The divergence between humans and chimps, our closest biological relatives, began about 6 million years ago. Humans are one of the great apes, and we share most of our genes with our closest cousins, chimpanzees and bonobos.

southern Africa about 3.5 million years ago. Australopithecenes were small, relatively hairy creatures that hunted animals, used stone tools, and may or may not have had fire. The last of this genus disappeared around 1 million years ago, likely the losers in an evolutionary selection process that saw the eventual rise of modern people. Around 2 million years ago, the genus *Homo* appeared, to which all modern humans belong. Thus, if one defines "human" by our genus, 2 million years ago is a feasible starting date. Like Australopithecus, there were several species of *Homo* that emerged and died out over time, including *Homo habilus* ("tool using man") and, around 1.8 million years ago, *Homo erectus*. Fossils in the Turkana Basin of East Africa indicate that *H. erectus* and *H. habilus* overlapped there for roughly 500,000 years. *H. erectus* subsequently spread over much of Africa and into parts of Europe and Asia, becoming the first "humans" to leave Africa in what would be the first wave of migration among many. Indeed, the bulk of *H. erectus* fossils are found in Asia, including northern China and Java.

It is important to avoid simplistic accounts that portray evolution, including that of humans, as a linear process inevitably leading to "higher" or superior life forms. Evolution means adaptation to change, not necessarily progress. In the nineteenth century, when Darwinian ideas were being misused to legitimate inequalities among racial groups, some advocated the view of evolution as a series of stages from primitive apes to modern, typically white, people, a view still remarkably widespread today (Figure 2.2). This account is flawed in many respects. First, it is implicitly racist in that it associates darker-skinned people with animals whereas more "advanced" ones have lighter skin. Second, this view ignores the view of evolution advocated today by biologists, who see the process as contingent, not inevitable, with many dead ends and false starts; evolution resembles a branching tree rather than a straight line.

The species that comprises all human beings today, *Homo sapiens* ("wise man"), emerged roughly 300,000 years ago, giving yet another starting date for "humans." *H. sapiens* was characterized by several

Figure 2.2: Classical depictions of human evolution were simplistic, erroneous, and often racist, such as this portrayal of beings that become whiter as they become more human.

features that gave it considerable advantages over its rival species, including a larger brain size, the ability to throw accurately, and the use of language, which is critical in shaping consciousness and facilitating working in groups. Indeed, despite their numerous conflicts, humans excel at working together in large numbers, something other primates cannot do. Ready access to large quantities of meat, in the form of herbivores on the east African plains, was also important: Large and powerful brains consume a great deal of energy, and meat contains large quantities of calories that can sustain this organ. Indeed, its sentience allowed *H. sapiens* to become the planet's first superspecies, dominating all other forms of life on Earth. Following on the heels of its predecessor *H. erectus*, *H. sapiens* left Africa and began to disperse across the globe (Figure 2.3).

As humans spread across the world, penetrating into a wide variety of climatic and ecological niches, they began to differentiate into slightly different genetic groups. What are often called "races" refer

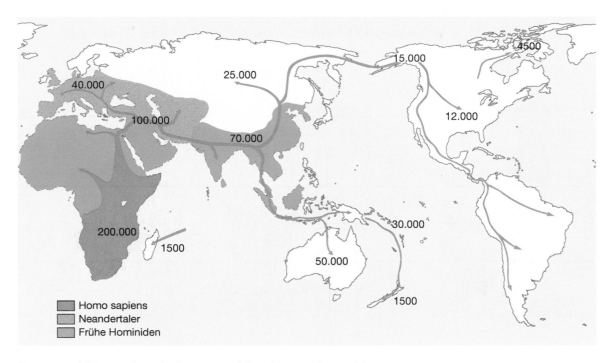

Figure 2.3: All hominids evolved in eastern Africa, but spread out of the continent in several waves, including *Homo erectus* and later, *Homo sapiens*.

to groups of people with similar physical characteristics, such as skin color or hair texture. As we will see in Chapter 13, the idea of "race" actually has little biological data to support it, largely because genetic variations within "racial" groups exceed the variations among them. Genetically, the similarities among different races are vastly greater than the differences, and there are no measurable differences in terms of intelligence or creativity. "Races" as slightly genetically different groups of people likely began to emerge shortly after *H. sapiens* left Africa, giving rise to groups commonly known as black (or Negro), Asian, or Caucasian.

Race is often measured on the basis of skin color, which reflects the amount of melanin in the skin, the same chemical that gives hair its color. At low latitudes, where solar energy is most intense, melanin helps to protect against ultraviolet poisoning, leading to peoples with darker skin; at high latitudes, melanin interferes with vitamin D absorption from sunlight, leading people to have lighter color skin. The geography of skin color, therefore, is very much a function of the intensity of solar energy over different latitudes of various places in which humans have lived for long periods of time (Figure 2.4).

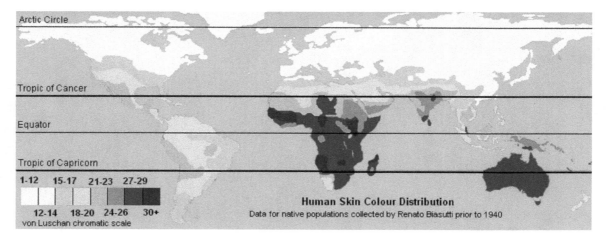

Figure 2.4: The global map of melanin concentration. Melanin is the chemical that gives skin and hair color. At low latitudes it helps to protect against ultraviolet radiation, so skins are darker; at high latitudes, melanin interferes with vitamin D absorption, so skins tend to be lighter.

One variant of *H. sapiens* that has received considerable attention was the Neanderthals, named after the Neander Valley in Germany where the first such skeletons were discovered in the nineteenth century. Neanderthals lived in a broad swath of territory that included the Middle East, parts of Central Asia, and most of southern and eastern Europe from roughly 200,000 to 35,000 years ago (Figure 2.5). They closely resembled modern humans, but were slightly shorter in stature and more heavyset. The relations between Neanderthals and *H. sapiens* have been the subject of considerable controversy. Many claimed that they formed a distinct species, *H. neanderthalensis*, which implied an inability to breed with *H. sapiens*. Others claimed that they were so closely related to modern people that they interbred with them (i.e., they were a subspecies, *H. sapiens neanderthalensis*). The decoding of the Neanderthal genome in 2010 shed much light on this topic, indicating that there did exist considerable breeding between the two; indeed, on average, all humans not of African ancestry derive about 4% of their genes from Neanderthals. Neanderthals and prehistoric *H. sapiens* (known in Europe as Cro-Magnon people) coexisted for several thousand years, until the Neanderthals, likely pushed out of the best hunting grounds into more marginal environments, finally died out around 35,000 BC.

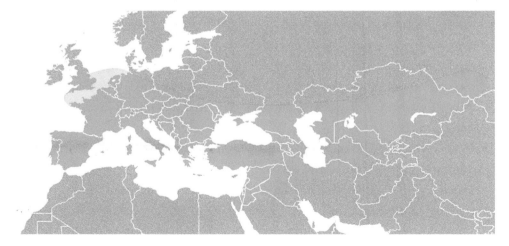

Figure 2.5: Neanderthals were a species closely related to our own that lived in much of Europe, the Middle East, and Central Asia. They died out about 35,000 years ago; roughly 4% of the genes of all people from non-African ancestry are theirs.

2.2 HUNTING AND GATHERING

For the overwhelming majority of time in which human beings have existed—97% to 99% of the period since the first hominids appeared—they lived in a type of society commonly known as **hunting and gathering**. The period of hunting and gathering is synonymous with the **Paleolithic**, or "Old Stone Age," which ended with the Neolithic, or "New Stone Age." (Some anthropologists also refer to a Mesolithic, "Middle Stone Age," although these terms are not universally accepted). Moreover, the line between agricultural and hunting and gathering societies is not always clear. Many societies both farmed and hunted.

Prior to agriculture, humans obtained sustenance through several methods. These included fishing, hunting, and gathering a wide variety of foods. Reliant on the sources in their immediate environment, small groups of 12 to 50 people wandered nomadically. They followed the seasonal rhythms of plants and migratory cycles of animals. The constant movement has led some to call hunting and gathering a "permanent camping trip." Members of hunting and gathering societies acted as food *collectors*, not producers. Stone and bone tools, such as knives, adzes, needles, and scraping implements, were the norm in cultures that lacked metal. Typically, such small bands and tribes exhibited a sexual division of labor (Figure 2.6). Men and boys hunted—often unsuccessfully—and women gathered, often with babies on their backs or small children nearby.

Paleolithic hunters were very skilled and played a critical role in the extinction of very large animals, such as mammoths, all over the world. Everywhere that humans made their first appearance, such as Australia and the Americas, saw the wholesale annihilation of large fauna, including mammoths and the woolly rhinoceros. However, smaller game, such as deer and rabbits, were much more common food sources. Hunting with spears and arrows was the norm, with slings and knives used when possible. The early domestication of the dog also helped hunters; the dog was by far the earliest domesticated animal, likely a descendent of wolves that hung around camp fires looking for scraps. Cats were domesticated much later, when settled societies with large quantities of stored grain used them to reduce the rodents that were attracted to these food supplies.

Despite their prowess in hunting, gathering provided the bulk of nutrients in most Paleolithic societies. Because humans are opportunistic, adaptable improvisers, gathering allowed an exceptionally diverse variety of foods to be collected. Examples include roots, fruits, berries, insects, eggs, grubs, worms, frogs, seeds, nuts, and slow animals such as snails or turtles. This form of resource exploitation, thus, depended on lightly using a wide variety rather than intensively using a few. Because hunting and gathering does not generate large quantities of calories

Figure 2.6: The gender division of labor during the Paleolithic era saw men hunt and women gather. Although prehistoric hunting of megafauna has received the most publicity, most prey were small, such as deer or rabbits.

per unit area, population densities remained low. Numerous variations in hunting and gathering existed in different local climates and ecologies.

Several innovations allowed hunter-gatherers to spread to almost every ecological niche in the world. The controlled use of fire, as well as the needle and thread, enabled people to move into northerly latitudes and withstand the harsh winters of northern Eurasia, eventually migrating into North America via the Bering Strait. In addition to keeping people warm, fire allowed food to be cooked, making what were inedible foods edible and often sanitizing and detoxifying it. Fire played a key role in the Paleolithic modification of many habitats (Figure 2.7), especially midlatitude grasslands and forests, where it was often used to hunt large herbivores. Indeed, some argue that the growth of grasslands was largely spurred by the actions of hunters and gatherers. Rather than passively respond to their environment, therefore, hunter-gatherers actively shaped it.

Figure 2.7: The use of fire allowed humans to chase game, cook food, and stay warm in cold climates, greatly expanding their security, comfort, and geographic range.

Although most people lived in small, self-sufficient bands, very early trade networks in obsidian, flint, amber, shells, feathers, and other goods indicate that some long distance exchange did occur. Contrary to agricultural societies, in which the vast majority of people spend the bulk of their time laboring, many in hunting and gathering societies had ample free time to do other things, such as gossiping, resting or playing with children. Organized warfare was relatively uncommon, generally small scale, and often highly ritualistic.

Although they were technologically simple compared with contemporary societies, ancient hunting and gathering peoples were not necessarily culturally simple. They possessed great stocks of oral knowledge about how their worlds functioned. Without a complex division of labor, most such cultures were relatively egalitarian, with no classes or private property and leadership confined to a chieftain.

Simple survival required an intimate knowledge of nature, including the ability to identify countless species of plants and animals, understand their behavior, identify and use medicinal plants and herbs, or find water in arid environments. Observations of contemporary hunters and gatherers form much of the basis of knowledge about those of the ancient past. This evidence suggests that religious beliefs were predominantly animistic or nature-worshipping.

Figure 2.8: Paleolithic art, including cave paintings, indicates the capacity for abstract thought and symbolization.

Without writing, people passed down their societies' knowledge in the form of oral folk songs, legends, and myths as told by shamans or storytellers. Because it consisted mainly of stories and music, almost all Paleolithic art has disappeared today. Pigments, jewelry, elaborate arrowheads, spearheads, and fishhooks all point to the existence of a sense of aesthetics. The cave paintings in Lascaux in France and Altamira in Spain, among other places, testify to the use of symbols and abstract thought (Figure 2.8). The widespread dispersal of stone carvings depicting females with large breasts and wide hips, such as the famous Earth Mother of Willendorf (25,000 BC), may indicate that they had a high regard for women.

In hunting and gathering societies, both birth rates and death rates are high. Most people would begin sexual reproduction in their early teens, and average life expectancy rarely exceeded 30 years. Women typically had 10 or more children, but because of high infant mortality rates, a large share died before reaching adulthood (Figure 2.9). Extended families, in which parents, children, aunts and uncles, and cousins lived together, and kinship ties formed the critical bases of local social organization and support networks. Because both birth and death rates were high, population growth rates were generally very low. Some tribes attempted to minimize growth through prolonged lactation or infanticide.

Figure 2.9: Paleolithic societies consisted of family-based clans, with high birth rates, large families, and low life expectancies.

The **Neolithic Revolution** around 10,000 BC produced dense, agricultural societies with complex divisions of labor. The displacement of hunter-gatherer societies by more densely populated agricultural ones occurred in virtually every location where agriculture began. Because agriculture supports more people per unit area than does hunting and gathering, agricultural societies tend to be wealthier and more powerful. The later rise of the nation-state also actively discouraged nomadism, and thus, the lifestyle of these peoples. Today, hunting and gathering is almost extinct. Small pockets can still be found among some groups, including South American tribes in the Amazon forest and parts of Central Africa and Southeast Asia, as well as Namibia (Figure 2.10), the San or !Kung people in Namibia, and some Australian aborigines. But everywhere, hunting and gathering is a dying form of subsistence, as its last practitioners are drawn into urbanized regions and commodified labor markets.

Figure 2.10: These hunters in Namibia offer a contemporary example of the rapidly disappearing Paleolithic lifestyle.

2.3 THE NEOLITHIC REVOLUTION

Beginning approximately 12,000 to 10,000 BC, a widespread series of changes began to occur around the world that initiated the end of the old stone age, or Paleolithic, and the beginning of the new stone age, or Neolithic. By far the most significant of these changes was the invention or discovery of agriculture, one of the most important transformations in the history of the planet. Agriculture greatly changed how people related to nature and one another. It created new types of societies, cultures, and landscapes. Essentially, it allowed people to become food producers, rather than food collectors, and to build small, permanent settlements (Figure 2.11).

Agriculture originated in the domestication of plants and, then, animals. A domesticated plant is deliberately planted, raised, and harvested by humans (Figure 2.12). A domesticated animal depends on humans for food and, often, shelter, and cannot live on its own in the wild. From the human perspective, domesticated plants and animals are superior to their nondomesticated counterparts, because they bear more fruit or provide more milk or meat. Domesticated animals tend to be slower, stupider, and more docile than their wild counterparts. Humans have deliberately engineered these characteristics through selective breeding over long periods.

Figure 2.11: These remains of a Scottish Neolithic village illustrate what housing conditions were like for thousands of years between the end of hunting and gathering and the rise of cities.

Figure 2.12: The domestication of plants meant the beginning of farming, which became the most common occupation across the globe for the next 10,000 years. This recreation of a Neolithic farm illustrates the small buildings, rudimentary technology, and limited lifestyle that revolved around the rise of agriculture.

In all likelihood, agriculture began some 12,000 years ago. From initial centers, agriculture spread to other areas, gradually replacing hunting and gathering. Scholars used to think that agriculture first emerged only in southwestern Asia (present-day Iraq). However, more recent research indicates that it evolved independently in several centers, including Asia (east, southeast, and southwest), Africa (northeast and south of the Sahara), the North American Midwest, Central America, western South America, and southern Europe. Figure 2.13 maps the major centers of domestication and suggests directions of early agricultural diffusion. In each place, people began domesticating plants and animals in response to specific circumstances.

Just as the study of genes is uncovering answers to specific questions about human origins, so it is illuminating many details of the domestication process. Most notably, molecular biology suggests that

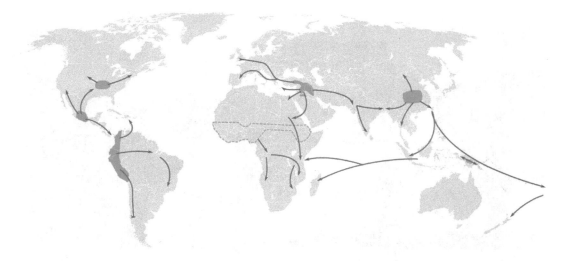

Figure 2.13: Agriculture originated in a few major hearth areas and subsequently spread throughout most of the world, carried by waves of migration over land and sea.

many plant and animal species were domesticated in different places at different times (Table 2.1). Cows, pigs, sheep, goats, yaks, and buffalo were each domesticated at least twice, dogs at least four times, and horses on even more occasions. This means that many different groups of people came up with the idea of domesticating animals. It also suggests that the reason some animal species are not domesticated is not that no one tried, but rather that they proved unsuitable candidates. An important domesticated animal was the horse, which was first domesticated in what is now northern Kazakhstan, where horses were first ridden and used for meat and milk by about 3500 BCE. Horses represent a highly significant form of domestication, as they greatly increased human mobility, and offered a great advantage to tribes that had them over sedentary agricultural peoples who did not. Later, horses were critical to the spread of the Indo-European language family (Chapter 3).

The process of domestication makes use of artificial selection, as opposed to natural selection. Natural selection results in the survival and reproduction of those plants and animals that are best able to cope in a particular environment. Artificial selection involves humans allowing certain plants and animals to survive and breed, because they possess features judged desirable by humans—for example, plants with

TABLE 2.1: MAJOR DOMESTICATED CROPS AND ANIMALS AND THEIR SOURCE AREAS

Fertile Crescent
- Plants: barley, cabbage, dates, figs, grapes, oats, olives, onions, peas, rye, turnips, wheat
- Animals: camels, cattle, dogs, pigs, pigeons

Central Asia
- Plants: almonds, apples, carrots, cherries, flax, hemp, lentils, melons, peas, pears, turnips, walnuts
- Animals: horse, chicken, reindeer, sheep, yak

North China
- Plants: apricots, cabbage, millet, mulberries, peaches, plums, radishes, rice, sorghum, soybeans, tea
- Animals: chickens, dogs, horses, pigs, silkworm

Southeast Asia
- Plants: bamboo, bananas, black pepper, citrus, eggplant, mangoes, sugar cane, yams
- Animals: cat, duck, goose, water buffalo

Ethiopia
- Plants: coffee

Nile River Valley
- Plants: cotton, cucumber, lentils, millet, sesame
- Animals: cats, dogs, donkeys

West Africa
- Plants: watermelon, yam

Central America
- Plants: avocados, beans, cocoa, cotton, pumpkins, red pepper, squash, sunflower, tobacco, tomato
- Animals: dog, turkey

Andes Mountains
- Plants: potatoes, pumpkins, squash, strawberries
- Animals: alpacas, guinea pigs, llamas, vicuñas

Amazon River Basin
- Plants: cassava, peanuts, pineapple, sunflowers, squash, sweet potato
- Animals: dogs, ducks

bigger seeds or animals that are less aggressive. The individual members of plant and animal species that humans favored would reproduce and pass on the desired characteristics, whereas the less desirable traits would be eliminated gradually.

Why human beings created agriculture has been the subject of considerable study. A common interpretation suggests that the Neolithic Revolution occurred at the end of the Pleistocene geologic era, or the Ice Age. At this time, vast continental glaciers that covered much of the Northern Hemisphere retreated, and the ecologies of many places changed radically. In this reading, new crops such as einkorn wheat began to appear in the Middle East as temperatures rose (Figure 2.14), giving rise to new opportunities to find food.

It is notable that agriculture holds only one real advantage over hunting and gathering: It is more efficient and can thus feed more people. Farmers work much harder and longer than hunters and gatherers, but agriculture generates a higher number of calories per unit area. Whereas hunters and gatherers enjoyed a wide array of foods, farmers in early societies were typically reduced to a monotonous diet. Whereas nomadic foragers moved among many places, farmers were stuck in one place all of their lives. Indeed, the shift to agriculture in many ways led to a reduction in the standard of living. Why, then, work long hours in boring, back-breaking labor in the fields? The answer is that once agriculture was adopted, there was no going back: It was the only way to feed the increasingly large populations.

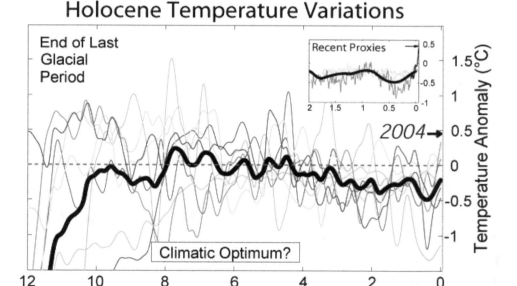

Figure 2.14: The rise of agriculture corresponded with a roughly 10° Centigrade rise in global temperatures, which eradicated the great ice sheets of the Pleistocene and offered opportunities for new crops to emerge in the changed climate.

In addition to agriculture, the Neolithic unleashed numerous other profound changes (Figure 2.15). One of these was the rise of small, settled communities or villages of varying sizes in which people lived year round. Staying put in one place was necessary to tend crops and farm animals, and to store the surplus that agriculture generated. Thus, the Neolithic effectively ended nomadism in vast parts of the world; from the perspective of the Paleolithic, living permanently in one place is highly abnormal. Many small communities became the centers of increasingly complex divisions of labor, with workshops, houses, religious and ceremonial centers, and the like. Only several thousand years later did some of these begin to emerge into true cities.

Within these villages, artistic, scientific, and intellectual life flourished. Astronomy, for example, became important in many early cultures, not only for religious reasons but to yield accurate dates to predict rainy seasons and floods, the rise and fall of water levels on rivers, and the changes in the seasons. Thus, cultures as disparate as the Mayans in Central America, the Chinese, Indians, Cambodians, and others all developed calendars of varying degrees of accuracy. Similarly, the wheel, likely invented in the Middle East in the sixth or seventh millennium BC, revolutionized land transportation, and increased the velocity with which people, goods, and ideas could travel across space, as did the oxcart and sailboat. Agricultural innovations included the plough and the sickle, which raised productivity levels. The invention of cloth during this time allowed goods to be carried and stored more effectively. Evidence of weaving goes back several thousand years, as well, allowing people to make clothes that were stronger, warmer, and

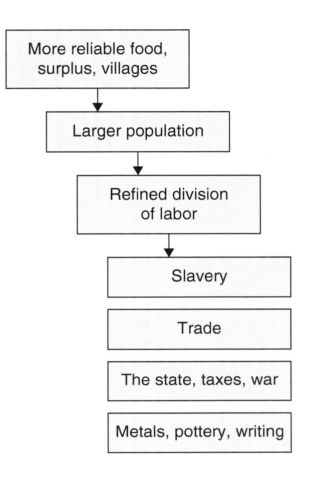

Figure 2.15: Agriculture and the Neolithic Revolution unleashed numerous social and technological changes across the world, including a new division of labor based on slavery and higher levels of inequality.

Figure 2.16: This depiction of slaves in ancient Egypt reflects the new class-based societies that arose on the heels of the Neolithic Revolution.

more comfortable than animal skins.

The Neolithic era also saw the rise of the first class-based societies. Hunter and gatherer groups were typified by high levels of equality, including classlessness. There were no rich people, and wealth was distributed fairly. In contrast, agriculture generated enough calories to allow a small group of people (typically no more than 5% to 10%) to engage in activities other than farming. Agriculture's great advantage over hunting and gathering, after all, was the ability to generate a larger and more reliable stream of food. In the process of creating even a modest surplus or producing more than they consumed, early farmers laid the basis for the rise of small elites that came to dominate all Neolithic societies. For the first time in history, there appeared kings and queens, emperors, generals, priests, soldiers, scholars, and teachers, all of whom reflected the more complex division of labor coming into being (Figure 2.16). Conversely, large numbers of people became slaves. Indeed, slavery was the world's first form of class-based society and remained dominant in parts of the world for millennia (e.g., in Europe until the collapse of the Roman Empire in the fifth century CE). Rather than a simple sexual division of labor, such as that found in nomadic societies, agriculture gave birth to stratified social systems marked by high degrees of inequality.

Central to the rise and perpetuation of class in the Neolithic was the emergence of the state, or government in all its forms. The state allows a small, elite class to manage and control a much larger population of producers: it is the most important means by which the ruling class rules, along with ideologies that naturalize inequality. Early Neolithic societies in many parts of the world developed increasingly elaborate and powerful state systems, including civil servants, religious bureaucracies, and organized militaries. For the first time, warfare as a systematic, organized means of violence among and within societies became a regular feature. With the state, taxes became obligatory, a means of extracting surplus value from producers and transferring it to the elites.

Writing was another innovation of the Neolithic (Figure 2.17). Many forms of writing began to appear around the world shortly after the birth of agriculturally based societies, including among the Egyptians (hieroglyphics), Chinese (pictographs), Sumerians (cuneiform), Mayans (glyphs), Phoenicians (who invented the early alphabet Linear A), and the Indus River valley culture of the early Dravidians. Writing was a profoundly important innovation, freeing information from the constraints of memory and allowing it to be stored and

Figure 2.17: All societies that developed agriculture also developed writing systems, largely so that governments could keep track of tax payments. This example is of Sumerian cuneiform, the oldest known writing system in the world.

Figure 2.18: The rise of organized religion was another aftershock of the Neolithic Revolution, including this example of Egypt's Sobek (left) and Hathor (right); typically religion naturalizes social orders, making them seem god-given and, thus, unchangeable.

transmitted over space and time more effectively. For this reason, historians often date the beginning of history to the rise of written records. Only a small handful of people ever mastered this difficult task, including scribes and some priests. Most people had no need of it, and mass literacy only appeared with the Industrial Revolution many thousands of years later. The most widespread examples of early writing include inventories of goods to be taxed by the state, revealing the close association between class, the state, power, and information. Other applications included commemorations and inscriptions on monuments and tombs, official letters, treaties, declarations, loans and debts, and the production of literary texts, often mythological in nature and designed to celebrate those holding power. Plus, an early example of writing from Mesopotamia was a recipe for beer!

Closely associated with the emergence of early agricultural societies was the development of organized religious systems (Figure 2.18; see Chapter 4). Religion, in the form of various animistic or nature-worshipping beliefs, had long played an important role in hunter and gatherer societies. But with the birth of class societies and the state, religion became much more than a set of stories told by local healers or story tellers. It evolved into an organized system of political and ideological power. Priests played important ceremonial and political roles. They advised the rulers of almost all ancient civilizations, placated gods and made sacrifices, predicted or interpreted signs of danger (e.g., drought, eclipses, or invasion), and spread dominant ideas among the people. Priests were frequently the only literate members of many classical cultures, with access to divine texts. All societies were polytheistic until the invention of monotheism, likely by the ancient Hebrews, in the second millennium BC. Typically, organized religion celebrated a society's current state, making the social order seem god-given and unchangeable. The power of organized religion was also manifested in the landscape, often in the form of large religious temples and monuments. Egyptian and Mayan pyramids and Mesopotamian ziggurats consumed vast amounts of state resources for their construction and maintenance. Religion remained, by far, the dominant ideology in almost all societies until the Enlightenment of the seventeenth and eighteenth centuries. Then, its influence began to wane in the face of growing scientific secularism.

Yet another innovation to emerge from the Neolithic era was the use of metals. Preagricultural societies, as we have seen, relied almost exclusively on stone, wood, or bone for their tools. The extraction and refinement of metallic ores is a complicated process made possible only by the increasingly complex divisions of labor that agriculture facilitated. Metals of different types offered numerous advantages over stone or bone. Metal tools and weapons, for example, were stronger, more flexible, and less brittle. The adoption of metal working is frequently divided into two distinct eras, the Bronze Age and the Iron Age. The Bronze Age, dating roughly between 3000 and 1500 BC, refers to the discovery of bronze, an alloy of copper and tin. Its appearance varied in different parts of the world, and its use probably began first in the Middle East before spreading to Europe, Persia, and Egypt (Figure 2.19), where bronze played a key role in numerous cultures, ranging from the Sumerians and Egyptians to early Greece. Similar Bronze Ages can be dated in China, from which it diffused to Japan, Korea, and India. Bronze Age cultures left behind numerous objects, such as metal containers, tools, and weapons. The Iron Age, lasting between roughly 1500 BC and 500 CE (roughly the end of the Roman Empire), saw iron gradually displace bronze as the preferred metal of choice. The Iron Age was most pronounced in Europe, although similar events occurred in China, India, and Africa. Some Iron Age cultures could make steel, which is iron with large quantities of carbon added to give it strength and flexibility. However steel remained rare and expensive until the Industrial Revolution of the late eighteenth and nineteenth centuries (Chapter 5).

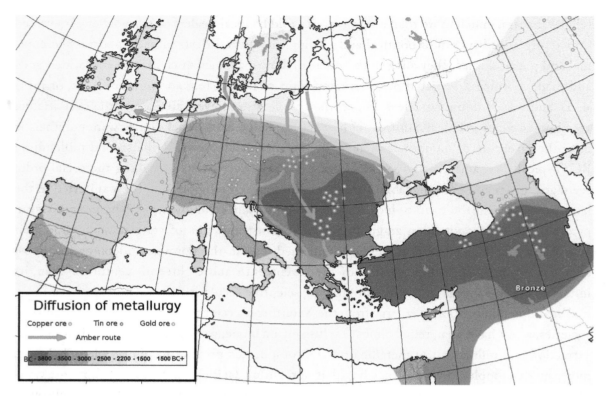

Figure 2.19: The Bronze and Iron Ages swept across Neolithic Europe, improving tools and weapons and, thus, productivity and social relations.

2.4 PREINDUSTRIAL AGRICULTURE

Since the Neolithic Revolution, various societies have fed themselves through an assortment of preindustrial agricultural systems, many of which are still used in several parts of the world. Thus, although preindustrial agriculture was important to the rise of civilizations, it also remains an important part of the world's agricultural landscapes today.

Preindustrial or nonindustrial agricultural systems differ from industrialized ones in a variety of respects. Most importantly, preindustrial systems do not use the inanimate sources of energy that are vital to industrialized agricultural systems (i.e., fossil fuels), and therefore, consume much less energy per unit of output (are less energy intensive). Rather, work in preindustrial farming systems is accomplished entirely through human or animal labor power, including tasks, such as tilling the soil with a plough, planting seeds, weeding, and harvesting. Thus, these types of farming are much more labor intensive. In societies fed predominantly through preindustrial agriculture, the vast majority of people work as farmers or peasants. Second, because preindustrial societies are often not fully capitalist, food grown in this way is mostly grown for local consumption (subsistence), rather than for sale on a market.

Preindustrial agricultural systems played an enormous role in the slave-based and feudal social systems that existed for millennia in most of the world. Roman *latifundia*—large estates worked by slaves—for example, formed the backbone of agricultural production during the empire. The expansion of medieval agriculture into the dense soils of northern Europe was made possible by the introduction of the heavy plow and the three-field system, in which one-third of farmland was periodically left unplanted to improve its productivity (Chapter 5). The manorial system that formed the social and economic basis of feudal Europe involved peasants and serfs who leased land from large landowners, paying rent with a fraction of their output. Variations of peasant-based production continue to be important in many contexts.

Today, various forms of preindustrial agricultural systems are still found throughout the world, with large variations in the types of crops grown, the methods used to plant and harvest them, and their productivity, labor relations, and relative vulnerabilities to drought or other hazards.

NOMADIC HERDING

Although not technically a form of agriculture, many observers classify **nomadic herding** in this category, although it involves only the domestication of animals, not crops (Figures 2.20 and 2.21). Typically, nomadic herders measure their wealth in terms of livestock (generally cattle, goats, or reindeer). They follow their herds in annual migratory cycles, such as **transhumance**, the movement between summer pastures in higher elevations and winter pastures in lower ones. Nomadic herding has been slowly vanishing throughout the world over the past two centuries, but contemporary examples include the Masai of East Africa, Mongols in Mongolia and northern China, the Tuareg of northern Africa, and the Lapps of northern Finland.

SHIFTING CULTIVATION

The best known example of preindustrial agriculture is **slash-and-burn**, also known as **swidden** or **shifting cultivation** (Figure 2.22). This form is found only in tropical areas, such as parts of Central America, the Amazon rain forest, West and Central Africa, and Southeast Asia.

Figure 2.20: Nomadic herding involves the domestication of animals but not plants. It is still practiced in a few areas in the world, such as northern Scandinavia, Oman, East Africa, and Tibet, as illustrated in this photo.

Roughly 50 million people continue to be fed this way in these regions. Due to heavy rainfall and the subsequent leaching of nutrients, tropical soils are generally quite poor, and most nutrients are stored in the living tissues of vegetation. The first step in slash-and-burn, therefore, is to cut down existing trees and bushes in a given plot of land and burn them. This process releases nutrients into the soil through the ash. Crops are then planted for several years. However, because the rate of nutrient extraction exceeds the rate of replenishment, the site can only be used for a brief period—generally 2 to 6 years—before the farmers must move on to a new site. Abandoned sites may gradually recover if left unplanted for long enough. Should rapid population growth occur and periods between plantings become reduced, the soil may permanently decline in fertility. This form of farming was widely practiced in the Mayan kingdoms prior to the Spanish conquest, and declining soil fertility may have played a role in the collapse of the Mayan states.

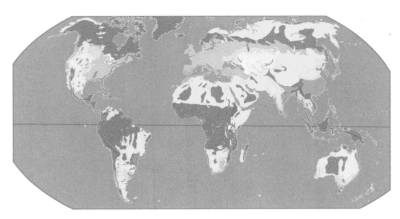

Figure 2.21: Nomadic herding is usually found in areas too cold or too dry for agriculture, as indicated in yellow on this map.

ASIAN RICE PADDY CULTIVATION

A third form of preindustrial agriculture, Asian rice paddy cultivation, is widely practiced throughout a region stretching from Japan, Korea, and southern China throughout Southeast Asia into eastern India. Rice is the staple crop for billions of people in Asia (Figure 2.23), and its cultivation in the current form goes back millennia. Young rice plants require standing pools of water. To create spaces for this to occur, East Asian societies

Figure 2.22: Slash-and-burn agriculture, also called swidden or shifting cultivation, is a preindustrial form of farming well suited to the tropics. Cutting and burning the vegetation releases the nutrients stored there, temporarily enhancing soil fertility. It is still practiced in parts of Central and South America, Africa, and Southeast Asia.

54 HUMAN GEOGRAPHY: A SERIOUS INTRODUCTION

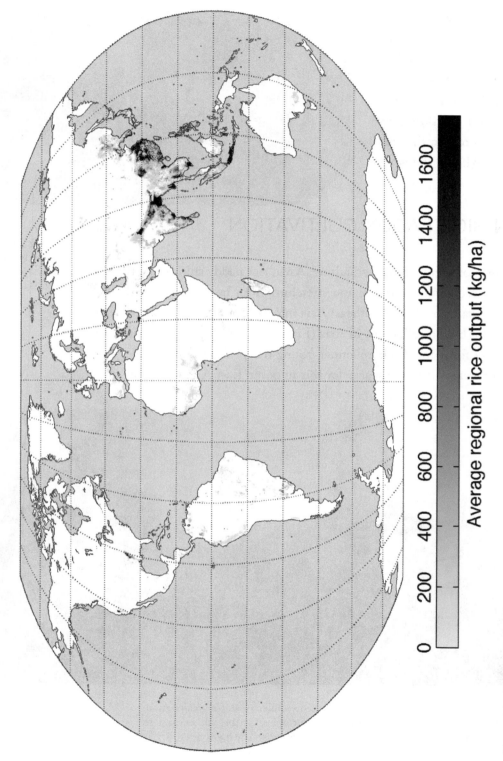

Figure 2.23: Rice is the world's most commonly grown crop and dominates Asian diets.

carved countless terraces over thousands of years out of hillsides along contour lines of equal elevation to control the flow of water with vast networks of dikes and small levees (Figures 2.24). Furrows are often dug using water buffalos, a common beast of burden in Southeast Asia. Often, small fish are grown in these pools of water as a source of protein for farmers. The planting of rice is exceedingly laborious and has significant seasonal variations in the amount of time required. The supply of water may rely on monsoon rainfalls; if the monsoon fails, people can starve.

Preindustrial agricultural systems have functioned effectively for thousands of years, and continue to do so in many parts of the developing world. Famed agriculturalist and demographer Esther Boserup showed that historically, rising populations in such places often stimulated productivity growth. In most places, preindustrial systems are marginalized or threatened by the expansion of globalized, capitalist, industrialized farming systems, and are dying out. These threats include imports of subsidized grains from Europe or North America. However, preindustrial systems' advantages include a diversity of crops and freedom from dependence on pesticides and petroleum. Thus, it may be helpful to view these not as remnant forms of an earlier way of life, but as historical adaptations to particular social and environmental contexts: as nonindustrial rather than preindustrial.

Figure 2.24: Rice is the staple crop of Asia and the most widely grown crop in the world. Societies in East, South, and Southeast Asia developed elaborate rice terraces to plant and grow rice, which requires much water in its younger stages.

2.5 THE URBAN DAWN AND THE RISE OF CIVILIZATIONS

As various preindustrial agricultural systems became widespread over large parts of the world and, with them, complex, class-based divisions of labor, the **social surplus** of such societies grew accordingly, and with it, the ability to sustain a growing number of nonagricultural workers. (*Social surplus* refers to the crops and materials that these farmers produced beyond the amount they required for their own survival.) The interactions among agriculture, the surplus, class, and the state were all fundamental to a phenomenon closely associated with the Neolithic Revolution: the rise of cities. Small villages had existed in many regions for thousands of years and remained the norm for the majority of the world's people until very recently. But the growth of large, densely populated societies that could feed themselves regularly and predictably gave rise to the world's first true cities, an event that dramatically changed the economic and cultural landscapes of the Bronze and Iron Age worlds. True cities require an agricultural surplus, for city dwellers rely on the wealth generated by farmers in the fields. But cities came into their own as distinct centers of power, ideology, culture, and innovation, in turn changing the peoples and places that surround them.

It was only with the agricultural or Neolithic revolution that what we might now term civilization appeared: a particular type of culture that includes a relatively sophisticated economy, political system, and social structure. The first civilizations in the world inevitably appeared in areas where agriculture also evolved, frequently along the banks of rivers. For example, in Mesopotamia—the area between the Tigris and Euphrates Rivers, which today is part of southern Iraq—irrigated agriculture, water management, and land reclamation were probably being practiced as early as 5,000 years ago. The fact that this "cradle of civilization," as it is often known, introduced regulations on water use, which indicates the complexity of its social organization, and the city of Babylon became a great administrative, cultural, and economic center. Similar river-based cultures developed in Egypt, India, and China.

But what specific changes gave rise to the beginnings of civilization? There are many possible responses to this question. Some suggested explanations for the development of civilizations merit more detailed comment.

However they came about, early civilizations represented major cultural advances and brought many changes. Agriculture and urbanization have obvious physical dimensions, such as cities, but civilization also involves profound nonmaterial changes: the rise of powerful elites, labor that serves the elite, the disappearance of the egalitarian family-based group, the rise of armies and the technology of war, the growth of bureaucratic systems, opportunities for patronage for artists and scholars, and private land ownership. Religion was often the basis of early power.

The city-civilization link is an ancient one. The two words share the Latin root *civitas*. Civilizations create cities, but cities mold civilizations. The earliest cities probably date from about 3500 BCE and developed out of large agricultural villages. It is no accident that the emergence of cities coincided with major cultural advances, such as the invention of writing, pottery, and the wheel. The complex divisions of labor that accompanied and gave rise to cities necessitated new forms of production, transportation, and communication.

Scholars debate the origins of urbanization and **urbanism**—the urban way of life. Most cities probably originated in at least one of four ways. First, cities were initially established in agricultural regions, and city life did not become possible until agriculture freed some group members from the need to produce food. Thus, in some areas, the first cities reflected the production of an agricultural surplus, possibly as a result of irrigation schemes. Catal Huyuk in Turkey and Jericho in Israel/Palestine reflect this process. Other early cities of this type were located in the Tigris–Euphrates region (present-day Iraq), the Nile Valley (Egypt), the Indus Valley (Pakistan), the Huang He Valley (China), Mexico, and Peru, and were associated with the rise of agriculture in these regions. These cities were the residential areas for those not directly involved in agriculture. Populations generally ranged from 2,000 to 20,000, although Ur on the Euphrates and Thebes on the Nile may have numbered as many as 200,000. The city of Caral, located about 200 kilometers north of Lima, Peru, has been dated as early as 2627 BCE—about 1,000 years before any other city in the Americas. The inhabitants of Caral appear to have supported themselves by practicing irrigation, growing squash, beans, and cotton but not corn.

It is important to understand early urban growth in the context of the unique geographic and historical circumstances in which cities began. Cities began in a limited number of places in the world, all of which had previously established productive agricultural systems (Figure 2.25; Table 2.2). The world's major centers of early urbanization are discussed below. In each case, the emergence and growth of these centers depended on their location within their respective physical environments and wider networks of trade.

MESOPOTAMIA

Lying between the Tigris and Euphrates Rivers (Figure 2.26), Mesopotamia (Greek for "land between the rivers") is widely considered the world's first center of urban growth. This region is included as part of the Fertile Crescent, the band of territory that stretches toward the west to include contemporary Syria, Lebanon, and Israel. Many different cultures and civilizations occupied this area over time. With the domestication of barley and wheat, the first Mesopotamian culture to arise was the Sumerians, around roughly 3500 BC. They built cities such as Ur, Uruk, Eridu, and Lagash, all noted in the Old Testament of the Bible, which were likely the first cities in the world. The wheel appeared here around the same time. The ziggurats, or stepped pyramids that the Sumerians constructed, demonstrated the importance of organized religion in their culture (Figure 2.27). The Sumerians also invented an early system of writing, or cuneiform, much of which consisted of symbols pressed into clay. They were the first to divide the year into 12 months, the month into 4 weeks, the day into 24 hours, and the hour into minutes. Every

58 HUMAN GEOGRAPHY: A SERIOUS INTRODUCTION

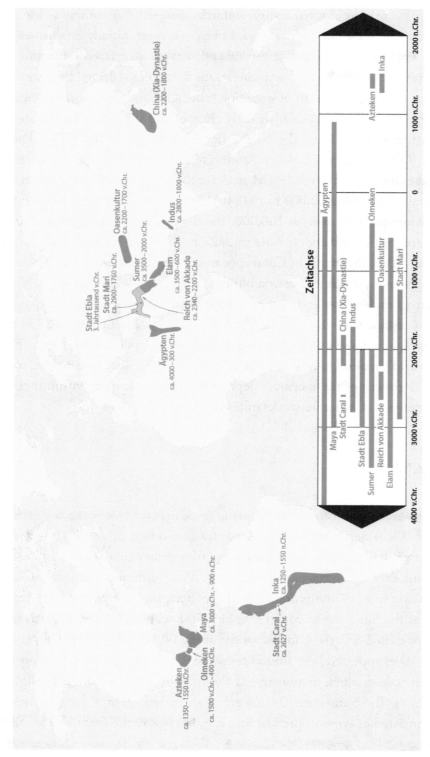

Figure 2.25: Major locations of the world's first cities. Not surprisingly, this map closely resembles that of agricultural domestication. The surplus resources generated by agriculture enabled the creation of urban centers, where a small, nonfarming elite could live off of the labor of farmers.

TABLE 2.2: CHRONOLOGY OF SOME MAJOR EARLY CIVILIZATIONS

Egypt	3000–332 BCE
Minoan	3000–1450 BCE
China	3000 BCE–present
Andean	3000–1800 BCE
Indus	2500–1500 BCE
Mesopotamia	2350–700 BCE
Mycenaean	1580–1120 BCE
Olmec	1500–400 BCE
Greece	1100–150 BCE
Rome	750 BCE–375 CE
Monte Alban	200 BCE–800 CE
Moche and Nazca	200 BCE–700 CE
Teotihuacan	100–700 CE
Maya	300–1440 CE
Toltec	900–1150 CE
Chimo and Inca	1100–1535 CE
Aztec	1200–1521 CE
Benin	1250–1700 CE

time you look at a clock or watch you are indebted to them. The Sumerians were eventually displaced by other early cultures, such as the Babylonians (roughly 1900 BC), who constructed the famed city of Babylon, the world's largest metropolis at the time. Under their famed King Hammurabi (approximately 1750 BC), the Babylonians codified and standardized their laws, inscribing them on public monuments for all to see. Ultimately, by about 1500 BC the Babylonians were conquered by a series of other cultures, such as the Assyrians, Hittites, and Persians.

EGYPT

Stretched along the banks of the Nile River, Egyptian civilization has always depended on the river (Figure 2.28). The classical Greek geographer and historian Herodotus noted, not without reason, that "Egypt is the gift of the Nile." The Nile regularly flooded its valley, refreshing the fertility of the soils there, an event that made knowledge of the

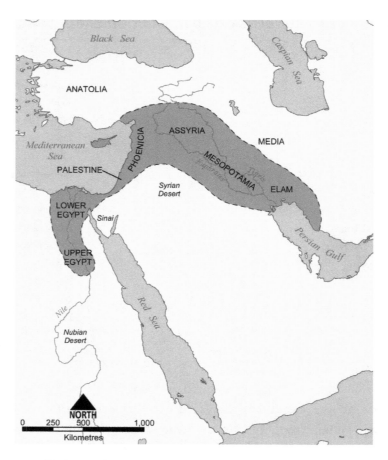

Figure 2.26: The Fertile Crescent in the present-day Middle East includes Mesopotamia, now modern Iraq, and the Levant lands of the Eastern Mediterranean Sea, or Syria, Lebanon, and Israel. This region was likely the site of the world's first agricultural domestication and urban centers.

seasons, rainfall, and the river's rise and fall crucial to this culture. Along its banks toiled countless peasants, who grew wheat, barley, and flax and irrigated their fields. These farmers generated the agricultural surplus that sustained the region's urban centers (Figure 2.29).

Around roughly 3100 BC, the Old Kingdom of Egypt emerged. The legendary King Menes supposedly unified the upper (southern) and lower (northern) Egyptian kingdoms into a coherent entity. During this period, the pyramids were constructed (Figure 2.30) as tombs for the pharaohs. Also at this time, Egyptian writing, or hieroglyphs (Greek for "sacred symbols"), began to be used. Scribes wrote hieroglyphs on papyrus and carved them into stone walls. Old Kingdom

Figure 2.27: Ziggurats were the pyramids of Sumeria and testify to the significant role of organized religion in many early urban centers.

architecture is covered with hieroglyphic inscriptions, which include myths and religious tales, celebrations of kings and battles, and royal edicts. After the collapse of ancient Egyptian culture, hieroglyphs were indecipherable until the turn of the nineteenth century, when the French linguist Louis Champollion translated them. He based his work on the Rosetta Stone, which was discovered in 1799 CE and contained similar inscriptions in three different languages. Soon thereafter, modern Egyptology was born.

Like its counterparts elsewhere in the world, ancient Egyptian culture painted detailed portraits of its many gods and goddesses. Ra, the sun god, was the most important of these, traveling across the sky by day, dying in the west at night, only to be reborn in the east the next morning. Other deities included Osiris, Isis, Horus, Sobek, and the sky goddess Nut.

At the peak of ancient Egyptian culture was the pharaoh, a king who assumed divine properties (making Egypt's human-made social structure seem divinely mandated). Pharaohs founded and led the dynasties that structured ancient Egyptian history. Their burial tombs,

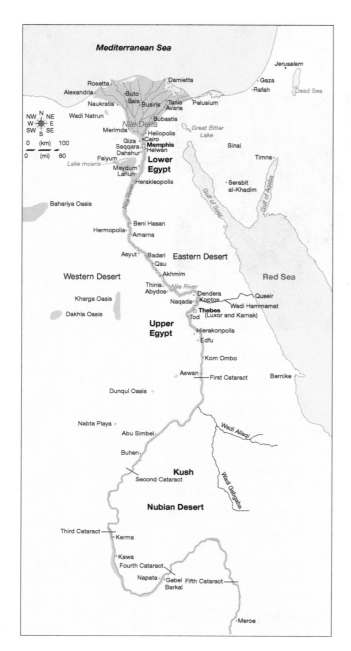

Figure 2.28: Ancient Egyptian civilization emerged on the banks of the Nile River and remained tightly clustered there.

the pyramids, concentrated vast quantities of state resources into specific sites. Designed to protect the pharaoh on his voyage through eternity (as mummification did his body), the pyramids ranged in size and height, depending on the wealth of their builders. Roughly 60 were built during the height of the Old Kingdom. Estimates of the labor required for their construction range between 10,000 and 20,000 men working for 20 years. Most are concentrated in northern Egypt, particularly on the outskirts of modern Cairo. One of these sites, the Giza complex, includes the Great Pyramid, the Chefren Pyramid, and the Sphinx.

The New Kingdom, which eventually displaced the Old Kingdom, witnessed the shift of power and pyramid construction to the south of the country, including Luxor. The most famous of the New Kingdom pharaohs, Ramses II, led enormous military campaigns south to subdue the Kushitic kingdoms and north into what is now Palestine and Lebanon. A great builder, Ramses also constructed the enormous statues of pharaohs at Abu Simbel (Figure 2.32), originally located on the banks of the Nile but subsequently moved by the United Nations in the 1970s to protect it from flooding.

Ancient Egyptian society eventually collapsed by the third century BC, when it was overrun by a series of foreign powers. Alexander the Great made the region part of his empire, establishing the city of Alexandria at the mouth of the Nile as his capital. Alexandria became

CHAPTER TWO: EARLY HUMAN CULTURES 63

Figure 2.29: Peasants, slaves, and farmers formed the backbone of ancient Egypt, generating the foods that fed everyone else.

Figure 2.30: The pyramids, all constructed during the Old Kingdom, were monumental tombs for pharaohs and exemplified the ability of such societies to form divisions of labor that freed large numbers of men from farming to construct them.

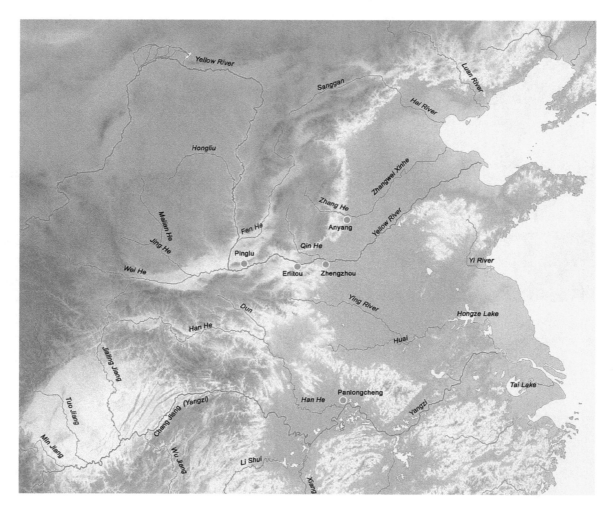

Figure 2.31: The origins of Chinese civilization were concentrated on the banks of the Yellow River (Huang He), one of two major rivers that flow through China. Chinese culture began with the cultivation of wheat and barley and adopted rice as it came to dominate the warmer, moister regions of southern China and the Yangtze River Valley.

an important port city and center of learning for centuries thereafter. The Romans conquered Egypt in 30 BC, when Augustus Caesar made it his personal domain and the breadbasket of the empire. Later, in the seventh century, the Arabs overran the country, founding the city of Cairo and giving it its current Islamic composition.

Figure 2.32: The Great Wall of China was built to protect the country, and the vital Silk Road trading link, from attacks by Mongol and Turkic tribes to the north. It was rebuilt several times in its history; this version is from the Ming Dynasty in the seventeenth century.

CHINA

A third great center of early civilization was China, the world's oldest continuously inhabited culture. China had been populated for millennia during the Palaeolithic era, but ancient Chinese culture began to take shape around 3500 BC in the Huang He River Valley in northern China (Figure 2.31). Wheat and barley were the staple crops cultivated here; only later in southern China did rice cultivation become widespread, giving rise to the enormous number of paddies that cover the landscapes of the southern part of the country.

Bronze Age China was defined politically by the Shang dynasty, China's earliest (1500 BC), which expanded to cover the central part of the nation. Some of the country's earliest urban centers, such as Anyang, arose during this era. Early Chinese history was often marked by squabbling kingdoms. In 221 BC, the famous emperor Shih Huang Ti unified the country, founding the Chin dynasty (from which the word "China" is derived). It was around this time that Confucianism became institutionalized as state ideology (Chapter 4), naturalizing the social order and prescribing strict roles for every member to follow. The Chin also began building the Great Wall to protect themselves from raids by the nomadic peoples to the north (Figure 2.32); the wall was rebuilt many times, most recently by the Ming dynasty in the fourteenth century. One of ancient China's most memorable sites is the former capital of Xian, where artisans built a terra cotta army of individually designed soldiers to accompany the emperors in the afterlife. Other early dynasties of note included the Han, roughly contemporaneous with the Roman Empire, which expanded Chinese control westward along the important Silk Road routes.

Early China was famous as a source of innovations. Indeed, China was, on average, much larger and wealthier than Europe for most of its history. Not until the European conquests that began in the eighteenth century did China fall behind. For example, the Chinese invented papermaking and bloc printing. The Arabs, capturing Chinese soldiers in a war in the eighth century CE, learned of these innovations and eventually transmitted them to Europe. Other Chinese inventions included the field pump (to draw water up from rivers to irrigate crops), the crossbow, the iron plough, the umbrella, the compass, the wheelbarrow, movable type, paper money, the stirrup, the windmill, and gunpowder (created to frighten Mongol horses with firecrackers during their attacks). Most of these items found their way westward along the Silk Road. They gradually percolated into medieval Europe and helped to reshape those societies.

THE INDUS RIVER VALLEY

Along the banks of the Indus River in present-day Pakistan (Figure 2.33), a fourth major early center of civilization arose around 2600 BC, only to collapse suddenly around 1700 BC. The people who populated this region were likely the Dravidians, an ethnic group speaking a distinct family of languages (Chapter 3). Today Dravidians live only in southern India, considerably far to the south of the Indus.

The Dravidians created a number of flourishing urban centers along the Indus, of which the two most famous were Harappa and Mohenjo-Daro (Figure 2.34). Each likely contained tens of thousands of people, fed by vast networks of wheat fields nearby. Because the writing of this culture has yet to be deciphered, much remains unknown about them. Clearly they had extensive trading ties with India, Central Asia, and the Middle Eastern world, exchanging goods by caravan and by sea routes. They seem to have had an elaborate calendar and knowledge of advanced mathematics, including prime numbers.

The collapse and disappearance of the Indus River Valley culture around 1700 BC has been the subject of considerable speculation. Some hold that it fell due to prolonged drought and

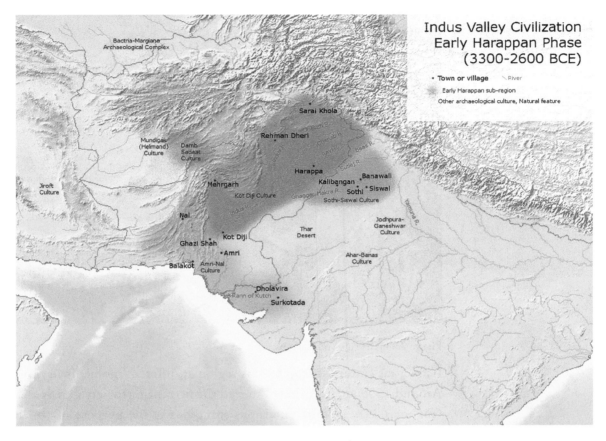

Figure 2.33: Along the banks of the Indus River in what is now Pakistan, an urban civilization of competing city-states arose and flourished. Harappan culture, as it is sometimes called, developed widespread trade relations and advanced mathematics.

ensuing famine. But the most persistent explanation is that the Indus River Dravidians fell victim to invaders from the northwest, the Indo-Europeans. This group, which had domesticated the horse, went on to conquer the northern half of the Indian subcontinent. They drove the Dravidians to their present home in the southern part of the peninsula (Chapter 3).

MESOAMERICA

Mesoamerica, the isthmus that includes Mexico and Central America, was also an important early center of urbanism and civilization.

Figure 2.34: The ruins of Harappa.

Numerous cultures occupied this region at different moments in time. Around 1500 BC, the Olmecs started the first major society in this area (Figure 2.35), building enormous stone heads, the purpose of which has never been discovered. Between 200 and 700 CE, the city of Teotihuacan, located near what is now Mexico City, established control over a vast region of southern Mexico (Figure 2.36); in the city's borders, its residents built giant pyramids, known as the Pyramid of the Sun and Pyramid of the Moon.

The best known and longest lasting of the Mesoamerican cultures were the Maya, who inhabited a series of city-states in the Yucatan Peninsula and what is today Belize and Guatemala (Figure 2.37). Dozens of Mayan cities dominated the coastal lowlands and the highlands of the interior, including Calakmul, Caracol, Tikal, Chichen Itza, and others. Mayan societies sustained themselves through widespread use of slash-and-burn agriculture or shifting cultivation. They grew their staple crop, maize (corn), in fields irrigated with a vast network of canals. It was through this process that much of the surplus food was generated to sustain the Mayan elite, the kings, priests, and military figures who dominated these cultures. The Maya built a series

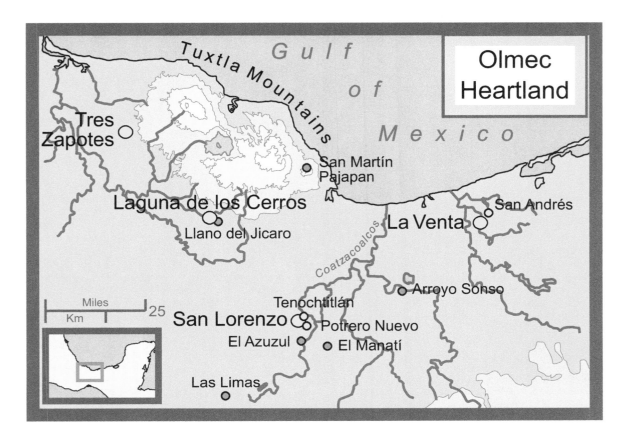

Figure 2.35: The Olmecs were the first of many MesoAmerican civilizations, and are famous for the giant stone heads they left behind.

of pyramids as ceremonial temples (Figure 2.38). Many were larger than the Egyptian pyramids. Mayan culture also exhibited a profoundly intellectual side, with a writing system that combined phonetic and pictographic symbols (Figure 2.39), a famously accurate calendar structured around multiple cycles of creation and destruction, detailed observations of the movement of the planets (especially Venus), and knowledge of mathematics. The Mayans could also be violent, and the city-states engaged in frequent warfare with one another.

The collapse of Mayan society around 900 CE has been hotly debated. Some maintain that it was due to prolonged droughts that diminished the food supply. Others hold that overcultivation of the soils through swidden agriculture contributed to loss of vegetation cover, increased soil erosion, and declining crop fertility. Yet other

Figure 2.36: Teotihuacan was a single mammoth city that dominated much of southern Mexico for centuries. Its ruins today are located near contemporary Mexico City.

explanations point to foreign invaders. It is clear that, by 100 CE, the civilization was at an end, although southern Mexico and Guatemala continue to have large Mayan populations today.

The successors of the Mayans were the Aztecs, who conquered a large swath of southern Mexico (Figure 2.40). They established an important but short-lived empire in the fourteenth century. This civilization vanished during the Spanish colonial onslaught of the sixteenth century. Unlike the Maya, who formed independent city-states, the Aztecs had a single capital city, Tenotchtitlan, at present-day Mexico City. According to Aztec mythology, the city was founded on a site where an eagle was observed eating a snake. The motif appears on the Mexican flag today. From their capital, the Aztecs established military authority over numerous surrounding tribes, many of which resented their control. They also supervised extensive trade routes in the region

Figure 2.37: Like the classical Greeks in Europe, the Maya did not form a single empire, but inhabited a group of competing city-states.

that exchanged goods, such as jade, obsidian, seashells, parrot feathers, and precious metals.

Many observers view the Aztecs as having a particularly warlike culture. Often they fought for the purpose of capturing slaves and prisoners for sacrifice. Thus, the empire wielded a large army, including an elite guard (the Mexica, from which the name "Mexico" was derived). Prisoners were often sacrificed in large numbers during important religious occasions, their blood used to quench the thirst of the sun god, Huitzilopochtl. Typically, priests cut the heart out of the

Figure 2.38: The Mayan pyramids were every bit as large and complex as those of ancient Egypt and served important ceremonial roles.

living victim and threw the body down the pyramid steps to the masses below (Figure 2.41).

In the 1520s, the Spanish conquistador Hernando Cortes led a small group of soldiers into Tenotchtitlan and overthrew the Aztec emperor, Moctezuma. How did such a tiny group of foreigners manage such a feat? In *Guns, Germs, and Steel*, Jared Diamond argues that the Spanish had numerous technological advantages, including horses, iron swords and shields, guns, and attack dogs. Others note that tribes conquered by the Aztecs assisted the Spanish. The Europeans also brought smallpox with them. This disease spread rapidly throughout the Americas, as the indigenous peoples had no biological resistance to this foreign virus. Disease, alone, effectively depopulated much of Mexico and paved the way for the Spanish conquest.

CHAPTER TWO: EARLY HUMAN CULTURES 73

Figure 2.39: Like all urban societies with complex divisions of labor, the Mayans developed a writing system to commemorate important events, keep government documents, and monitor taxes.

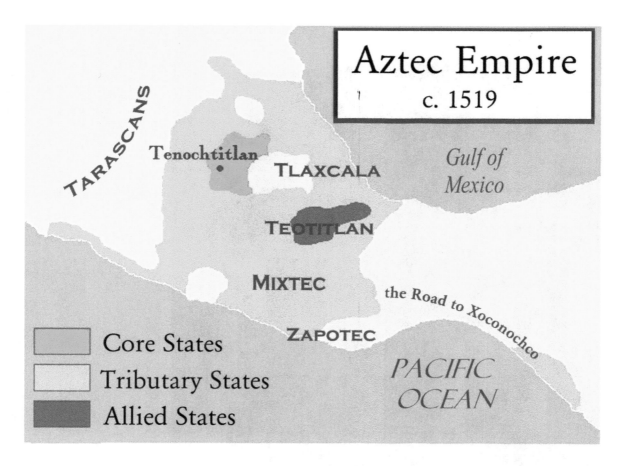

Figure 2.40: Unlike the Mayans, who were concentrated on the Yucatan, the Aztecs formed an empire that occupied the valleys of central Mexico near what is now Mexico City.

SOUTH AMERICA

Between the Andes Mountains and the western shore of South America, a variety of cultures arose, gave birth to civilizations, and eventually, fell. Most of these cultures relied heavily on the potato, which they domesticated and grew, cultivating 700 different types adapted to different elevations. Chavin culture, for example, arose around 1700 BC in coastal Peru. The Chavins supplemented potatoes with seafood, notably anchovies. The Nazca people drew elaborate chalk designs of surprisingly large size and accuracy on the desert landscapes. The Moche people built large temples and practiced human sacrifice.

Figure 2.41: This image of human sacrifice, taken from a rare surviving codex, indicates that it was an important part of Aztec culture, as human blood was needed to satisfy the sun god. Upon commemorating new pyramids, tens of thousands of people were sacrificed.

But, it was the Inca who created the largest empire of South America, stretching along the coast for almost 2,000 miles (Figure 2.42) from the jungles of Colombia to the Atacama Desert. In the Quechua language that they spoke, the empire was called Tawantinsuyu ("land of the four quarters"). Like the Aztecs in Mexico, the Inca conquered numerous neighboring peoples in an empire that lasted only about two centuries. From their capital city of Cuzco, high in the Andes, the Inca administered an empire that was connected by a vast network of roads, along which traveled messengers on foot, llama and alpaca caravans carrying foodstuffs, slaves, and soldiers. The Inca had no wheel or system of writing, but they did develop a counting system using *quipu*, or woven cloth with a systematic pattern of knots. They worshipped the sun god, Inti, and believed their leader, the Sapa Inca, was a demi-god.

Figure 2.42: The Inca Empire stretched for almost 2,000 miles along western South America.

The Inca met a fate essentially identical to that of the Aztecs. In 1526, the Spanish conquistador Francisco Pizarro led a small group of men and captured the last chief of the Incas, Atahualpa. After holding him for ransom, they executed him. Deprived of its leader, the empire began to crumble rapidly. The Spanish accelerated this process with their advantage in military technology. Smallpox also took a horrendous toll and exterminated millions of Incas. The Spanish enslaved many survivors, working two million of them to death in the silver mines of the Andes. One such was the city of Potosi, the largest silver-producing location on the planet.

In 1911, the last Incan city, Machu Picchu, was discovered high in the Andes mountains (Figure 2.43). Far from the area of Spanish control, it escaped the destruction that other Incan cities met, and remains an important monument to this day to the power and sophistication of this civilization.

NORTH AMERICA

Many earlier views of native cultures in North America incorrectly portrayed the continent as populated only by hunters and gatherers. In fact, long before Christopher Columbus, several centers of urbanization existed in what is now the United States. The largest of these was

Figure 2.43: Machu Picchu was a famous Incan city located high in the Andes. Discovered only in 1911, it testifies to the wealth and sophistication of Incan civilization. It is one of the major tourist attractions in the world today.

the Mississippian mound-building culture. It was named after the groups of tribes that built thousands of burial mounds of varying sizes over a large area extending from Minnesota and Ohio to northern Florida (Figure 2.44). Many of these peoples relied on corn as their staple crop, which had spread northward after being domesticated in Mexico. The largest pre-Columbian North American urban center was Cahokia. It flourished between 900 and 1200 CE near what is now the city of St. Louis. At its peak, Cahokia may have had 12,000 residents. Its remnants today are home to some of the largest mounds built by that culture.

Far to the southwest, in the Sonoran Desert of Arizona and New Mexico, a series of small urban centers populated by the Anasazi people arose between 450 and 1350 CE (Figure 2.45). In an unforgivingly dry and hot climate far from major rivers (unlike most early urban cultures), the Anasazi constructed a series of impressive villages (Figure 2.46). Nestled at the base of enormous cliffs, these settlements included multistory buildings and sweat lodges used for ceremonial purposes. Elaborate pottery and weaving testify to their skills. Likely because of prolonged drought, or perhaps invasion from Mexico, the Anasazi people vanished in the fourteenth century. Their descendants are the Hopi and Navajo today.

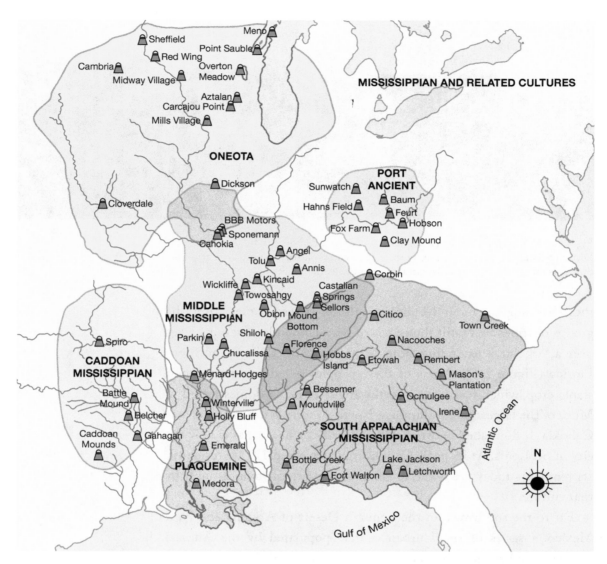

Figure 2.44: Cahokia was the largest of several thousand pre-Columbian settlements along the banks of the Mississippi and Ohio Rivers.

SUB-SAHARAN AFRICA

Although Egypt is often seen as the most elaborate early African society, several parts of Africa south of the Sahara also established flourishing kingdoms, as well (Figure 2.47). In western Africa, for example, along the Niger River, the Mali, Ghana, and Songhai kingdoms played important roles in organizing vast numbers of people, growing a variety of staple crops (e.g., yams), and building urban centers.

Many of these kingdoms became wealthy by taxing the trading caravans that crossed the Sahara desert using camels, connecting the Arabs to the north with the black-skinned peoples of the coastal areas (Figure 2.48). From the north, the Arabs provided salt. From the south, Africans provided gold, ivory, textiles, and slaves. Along these routes, Islam also entered western Africa, where it plays an important cultural role today. The most famous of the western African cities was Timbuktu, located at the great bend of the Niger River. Timbuktu became important not only for its commercial success, but as an intellectual center as well, with numerous Islamic universities (Figure 2.49). When the Europeans set up maritime trade routes in the

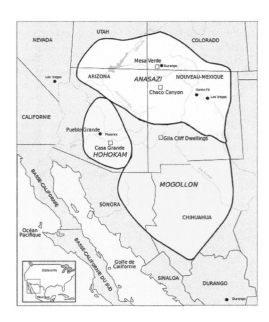

Figure 2.45: The Anasazi peoples created a series of pueblos or villages in the harsh deserts of the southwestern United States.

Figure 2.46: The Pueblo peoples of the southwestern United States built and lived in villages in the face of enormous rock cliffs. Their buildings had several stories and included sweat lodges for ceremonies.

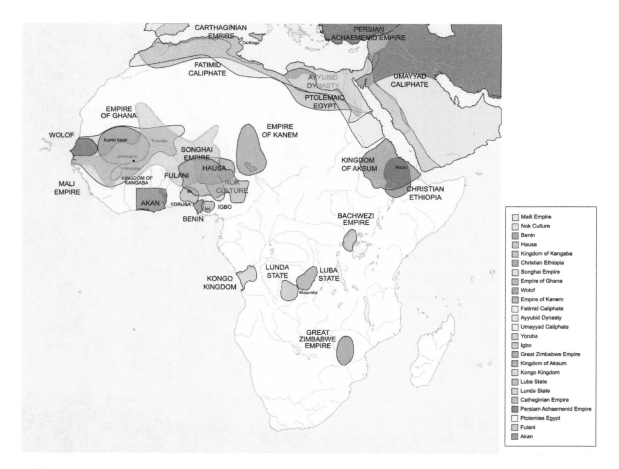

Figure 2.47: Sub-Saharan Africa was the site of a number of early civilizations, the best known of which are those in West Africa, Ethiopia, and Zimbabwe.

sixteenth century, they effectively put an end to the trans-Saharan caravans. As a result, the great cities of the interior declined in importance and wealth.

In eastern Africa, a similar series of kingdoms flourished in what is now Ethiopia (Figure 2.50). The Kingdom of Axum, for example, dominated the area from about 100 to 1000 CE. This part of Africa had long-standing ties across the Red Sea with the Arabian Peninsula, particularly Yemen. From Arabia, both Christianity, and later, Islam, arrived in this part of the continent. By the eleventh century, as Islam shifted the region's wealth and power north, Axum entered into a long period of terminal decline.

Figure 2.48: The trans-Sahara trade routes connected West African civilizations with the Arabs to the north, generating great wealth for cities along the way, such as Timbuktu.

Figure 2.49: On the banks of the Niger River, Timbuktu emerged as the largest and most famous of precolonial sub-Saharan African cities, a center of trade and Islamic scholarship.

In southern Africa, the Kingdom of Zimbabwe developed a series of urban centers in its own right between 1100 and 1500 CE (Figure 2.51). With exports of slaves, ivory, gold, and cattle hides, the Zimbabwean kings connected this part of the country with Arab traders along the coast. Their goods also tied them loosely to distant centers of wealth across the Indian Ocean. This flourishing international trade network lasted for centuries until the European arrival in the sixteenth century. The modern country of Zimbabwe takes its name from this empire.

EUROPE

Although it was not one of the first places in the world to undergo urbanization, southern Europe nonetheless also emerged as the site of a series of important and influential civilizations. Indeed, it is

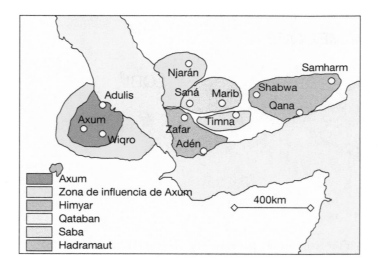

Figure 2.50: The Kingdom of Axum, in what is now Ethiopia, was a Christian society with extensive relations across the Red Sea. It eventually fell to Islam, introduced from Yemen.

Figure 2.51: Remains of the Zimbabwe kingdom, which at one time held great sway over southern Africa.

impossible to understand the geography of Europe today without a preliminary knowledge of this era.

In Bronze Age Greece, some of the earliest signs of metal working and other artifacts of urban culture were found on the Cycladian Islands and also on the Isle of Crete, home to the Minoan culture (Figure 2.52). The Minoans were a maritime-oriented culture that relied heavily on the seas for sustenance. This civilization flourished by means of its extensive trading ties throughout the eastern Mediterranean Sea, notably with the Phoenicians and Egyptians. Maritime trade routes can be deciphered from patterns of shipwrecks, some of which still carry the amphora, or clay vessels used to hold goods such as wine or olive oil. Although nineteenth-century historians sought to minimize the influence of Middle Eastern cultures on early Greece to enhance the sense of a unique set of European origins, more recent scholarship points to important influences from its neighbors to the south, including pottery and the introduction of the Phoenician alphabet, or Linear A. On Crete, the Minoan imperial center, Knossos, was renowned for its elaborate palaces and festivals that included

Figure 2.52: Bronze Age Greece included several societies, such as those on the Cycladian Islands, Crete, and Mycenae.

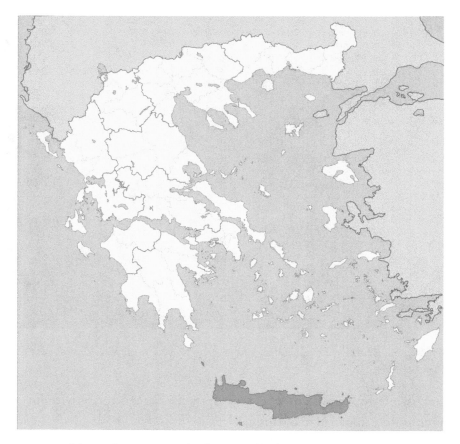

young athletes dancing on the horns of bulls. Minoan culture came to a sudden, violent end in 1635 BC when an enormous volcanic explosion on the island of Thera likely sent a huge tidal wave over the island, effectively exterminating its population. As Chapter 1 noted, human geographies do not exist in isolation from the natural environment, including natural hazards. This event serves as a poignant reminder of that theme.

On the Greek mainland, Mycenaean culture connected the series of city-states that later developed into classical Greece. Located high in the hills, Mycenae was a military fortress that controlled a vast part of the Peloponnesian Mountains and coastal areas. The influx of Doric and Ionian Greeks around 1100 BC, part of the broader Indo-European conquest of Europe (Chapter 3), witnessed rival city-states emerge. One was Troy, located near the narrow strait between the Aegean and Black Seas. Troy, of course, was destroyed at the end of the Trojan

War, an event portrayed by the blind poet Homer in his famous *Iliad*. Famed German archaeologist Heinrich Schliemann discovered the ruins of both Mycenae and Troy in the late nineteenth century.

Classical or Hellenic Greece, which flourished between 800 and 400 BC, witnessed a series of city-states rise in the coastal lowlands not only of Greece, but throughout much of the northern Mediterranean. Greek colonies arose in present-day Spain, southern France, southern Italy and Sicily, Turkey, and the shores of the Black Sea (Figure 2.53). The city-state, or *polis*, formed the primary unit of Greek society. (The word *polis* appears today in names, such as Minneapolis and Indianapolis.) Like the Mayans, the Greeks inhabited independent city-states that traded and often feuded with one another. The largest of the city-states was Athens. At its peak in the fifth century BC, Athens became a major center of learning, science, philosophy, art, and scholarship (Figure 2.54). Wars among the city-states weakened the Greeks. The largest such conflict was the Peloponnesian War between Athens and Sparta, which greatly weakened both. In addition, the

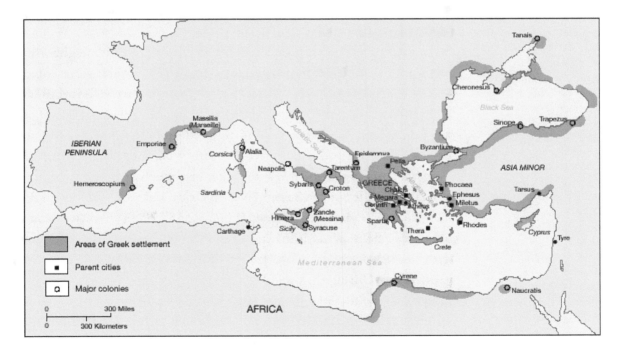

Figure 2.53: Classical Greek city-states were scattered across the Mediterranean and Black Seas.

Figure 2.54: In ancient Athens, the largest of the Greek city-states, the Acropolis consisted of a series of temples that overlooked the city.

Greeks faced frequent invasions and threats of conquest by the Persian empire to the east.

In the fifth century BC, Greece fell to its more powerful neighbor to the west, Rome. The Roman Republic, and later Empire, represented one of world's largest social formations before the Industrial Revolution. Thus, it was the most advanced society that relied on slaves for most of what it produced. The republic and empire lasted from roughly 700 BC to 476 CE. At its peak in the second century CE, the empire stretched from Scotland in the northwest to encompass much of western Europe, all of the Mediterranean Sea and North Africa, and parts of the Middle East, such as Palestine and Iraq (Figure 2.55). The system was connected by extensive maritime trade routes, as well as land-based routes that crossed the large network of well-built roads that the Romans constructed. During the period of the Pax Romana, or "Roman peace," (roughly 27 BC to 180 CE), the region experienced rising prosperity for several centuries. Throughout the empire, the Romans built roads, bridges, theatres, temples, arches, aqueducts, and monuments of varying sizes.

Figure 2.55: The Roman Empire represented the peak of the type of ancient society based on slave labor. It dominated the Mediterranean realm and western Europe for almost a millennium. Rome may have had a population of one million people, and it consumed vast resources drawn from throughout the empire.

Under the Romans, cities grew in much of southern and western Europe in a process that diffused from the southeast to the northwest, leaving a huge impact on the subsequent human geography of Europe. Examples of Roman cities, some established on Celtic sites, include Lyon and Paris in France; Londinium (London) in Britain; Belgrade in Serbia; Vienna in Austria; Trieste in Italy; and Toledo in Spain.

But by far, the largest of these was Rome itself, which, at its height, may have had upward of one million people. Enormous aqueducts channeled large volumes of water to the city from surrounding hills (Figure 2.56). In Rome, some of the largest and most famous monuments and temples were to be found, such as the Forum, the Coliseum (Figure 2.57), and the Pantheon.

Figure 2.56: Great builders, the Romans constructed aqueducts, such as this famous one in Seville, Spain, to provide their cities with water.

Figure 2.57: The Coliseum was the largest building the Romans ever made, and it served as a center for popular amusement, including staged naval battles and gladiatorial combat.

In the fifth century CE, the Roman Empire collapsed. Many factors contributed to its fall. Famed eighteenth-century historian Edward Gibbon blamed the growth of Christianity, which became the state religion in the third and fourth centuries. Shortages of precious metals led to gradual inflation and long-term debt. Overspending on the military sapped investments in other spheres. Several civil wars wracked the empire and destabilized it. Slave rebellions, such as that of Spartacus, presented a constant threat. Finally, attacks by Germanic tribes to the north, and by the Asian Huns, succeeded in dismantling the empire in the late fifth century. After Rome fell, the average city size in Europe declined for the next millennium.

■ KEY TERMS

Domestication: the process of changing plants and animals, so that they cannot reproduce without human intervention, making them suitable for human use

Hunting and gathering: the lifestyle or mode of production characterized by following the migratory cycles of animals and collecting but not producing food supplies from the environment

Neolithic Revolution: the process that began roughly around 10,000 BC typified by domestication, agriculture, and the rise of settled communities

Nomadic herding: a lifestyle that involves driving herds of herbivores along seasonal routes

Nomadism: a way of life in which people continually migrate from place to place, with no long-term communities in one location

Paleolithic: the "Old Stone Age," or period before the discovery or invention of agriculture, that encompassed the vast bulk of time people have lived on Earth

Shifting cultivation: also known as slash-and-burn or swidden, this type of preindustrial agriculture involves burning vegetation to release the nutrients, then growing crops on the fields for temporary periods

Slash-and-burn: another name for shifting cultivation or swidden

Social surplus: a society's output that exceed consumption, allowing a small minority to live off of the labors of others

Swidden: another name for shifting cultivation or slash-and-burn

Transhumance: annual migratory cycles in which livestock are switched between low and high altitudes during the winters and summers, respectively

Urbanism: the culture and way of life that accompanies and is embedded in cities

■ STUDY QUESTIONS

1. Where did humans evolve? Why did this species enjoy so much success?

2. Define the term Paleolithic and the lifestyle it encom-passed.

3. Where in the world is the Paleolithic lifestyle still to be found today?

4. What advantage did agriculture have over hunting and gathering?

5. Define domestication. Name five places in the world where domestication occurred.

6. What are five consequences associated with the Neolithic Revolution?

7. What are the relations between a surplus and the rise of class-based divisions of labor?

8. What is transhumance?

9. Why do farmers practicing swidden burn the vegetation?

10. Why do rice farmers in Asia create terraced hillsides?

11. Why is shifting cultivation well suited to tropical environments?

12. How and why are Asian rice paddies constructed?

13. What role did river valleys play in the rise of cities?

14. What did an agricultural surplus have to do with the rise of cities?

15. Where is the Fertile Crescent?

16. On a world map, show six places where cities originated.

BIBLIOGRAPHY

Anthony, D. (2007). The Horse, the Wheel, and Language: How Bronze Age Riders From the Eurasian Steppes Shaped the Modern World. *Princeton, NJ: Princeton University Press.*

Bauer, S. (2007). The History of the Ancient World: From the Earliest Accounts to the Fall of Rome. *New York: Norton.*

Bellwood, P. (2004). First Farmers: The Origins of Agricultural Societies. *Hoboken, NJ: Wiley-Blackwell.*

Bellwood, P. (2014). First Migrants: Ancient Migration in Global Perspective. *Hoboken, NJ: Wiley-Blackwell.*

Cavalli-Sforza, L. (2000). Genes, Peoples, and Languages *(M. Seielstad, Trans.). New York: North Point Press.*

Harari, Y. (2015). Sapiens: A Brief History of Humankind. *New York: Harper Collins.*

Kelly, R. (2013). The Lifeways of Hunter-Gatherers: The Foraging Spectrum. *New York: Cambridge University Press.*

Lascelles, C. (2014). A Short History of the World. *London: Crux Publishing.*

McCarter, S. (2007). Neolithic. *London: Routledge.*

Roberts, J., & Westad, O. (2013). The History of the World. *New York: Oxford University Press.*

Shipman, P. (2013). The Invaders: How Humans and Their Dogs Drove Neanderthals to Extinction. *Cambridge, MA: Belknap Press.*

Spinden, H. (1999). Ancient Civilizations of Mexico and Central America. *Mineola, NY: Dover Publications.*

Trigger, B. (2007). Understanding Early Civilizations: A Comparative Study. *New York: Cambridge University Press.*

Zohary, D., Hopf, M., & Weiss, E. (2013). Domestication of Plants in the Old World: The Origin and Spread of Domesticated Plants in Southwest Asia, Europe, and the Mediterranean Basin. *New York: Oxford University Press.*

… CHAPTER TWO: EARLY HUMAN CULTURES 93

IMAGE CREDITS

- Figure 2.1a: Copyright © Greg Hume (CC BY-SA 3.0) at https://commons.wikimedia.org/wiki/Hylobatidae#/media/File:MuellersGibbon.jpg.
- Figure 2.1b: Copyright © Zyance (CC BY-SA 2.5) at https://commons.wikimedia.org/wiki/Pongo#/media/File:Zoo_z01.jpg.
- Figure 2.1c: Copyright © Raul654 (CC BY-SA 3.0) at https://commons.wikimedia.org/wiki/Gorilla#/media/File:Gorilla_gorilla_gorilla8.jpg.
- Figure 2.1d: Clker-Free-Vector-Images, "People," https://pixabay.com/fr/famille-l-homme-femme-gar%C3%A7on-312018/. Copyright in the Public Domain.
- Figure 2.1e: Copyright © Aaron Logan (CC by 2.5) at https://commons.wikimedia.org/wiki/Pan_(genus)#/media/File:Lightmatter_chimp.jpg.
- Figure 2.2: eatwell.in, https://www.flickr.com/photos/eatwellin/4837363686/.
- Figure 2.3: Altaileopard, http://commons.wikimedia.org/wiki/File:Spreading_homo_sapiens.jpg. Copyright in the Public Domain.
- Figure 2.4: Copyright © Dark Tichondrias (CC BY-SA 3.0) at http://commons.wikimedia.org/wiki/File:Unlabeled_Renatto_Luschan_Skin_color_map.png.
- Figure 2.5: Copyright © Nilenbert (CC BY-SA 3.0) at http://commons.wikimedia.org/wiki/File:Range_of_Neanderthals.png.
- Figure 2.6: Copyright © Depositphotos/Tepic.
- Figure 2.7: Copyright © Hans Splinter (CC BY-ND 2.0) at https://www.flickr.com/photos/archeon/4469498464/.
- Figure 2.8: Copyright © Peter80 (CC BY-SA 3.0) at http://commons.wikimedia.org/wiki/File:Lascaux_04.jpg.
- Figure 2.9: Copyright © Musetress (CC by 3.0) at http://commons.wikimedia.org/wiki/File:HK_Museum_of_History_TST_Prehistoric_HK_03.JPG.
- Figure 2.10: Copyright © Mopane Game Safaris (CC BY-SA 4.0) at https://commons.wikimedia.org/wiki/File:Bosquimanos-Grassland_Bushmen_Lodge,_Botswana_08.jpg.
- Figure 2.11: Copyright © Bob Jones (CC BY-SA 2.0) at http://commons.wikimedia.org/wiki/File:Skara_Brae_neolithic_village_-_geograph.org.uk_-_1340360.jpg.
- Figure 2.12: Copyright © Depositphotos/ya-mayka.
- Figure 2.13: Copyright © Joey Roe (CC BY-SA 3.0) at http://commons.wikimedia.org/wiki/File:Centres_of_origin_and_spread_of_agriculture.svg.
- Figure 2.14: Copyright © Robert A. Rohde (CC BY-SA 3.0) at https://commons.wikimedia.org/wiki/File:Holocene_Temperature_Variations.png.
- Figure 2.16: Copyright © Jo Schmaltz (CC BY-SA 2.0) at http://commons.wikimedia.org/wiki/File:Abu-Simbel_temple3.jpg.
- Figure 2.17: "Cuneiform," https://commons.wikimedia.org/wiki/File:Cuneiform_tablet-_a-sheer_gi-ta,_balag_to_Innin-Ishtar_MET_DP109172.jpg. Copyright in the Public Domain.
- Figure 2.18: Copyright © Fanny Schertzer (CC BY-SA 4.0) at http://commons.wikimedia.org/wiki/File:Sobek_and_Hathor_-_Kom_Ombo_Temple.jpg.

- Figure 2.19: Osado, http://commons.wikimedia.org/wiki/File:Diffusion_m%C3%A9tallurgie_es.png. Copyright in the Public Domain.
- Figure 2.20: Copyright © McKay Savage (CC by 2.0) at https://commons.wikimedia.org/wiki/Category:Nomads#/media/File:India_-_Ladakh_-_Trekking_-_077_-_sending_out_the_herds_in_the_monring_(3896556490).jpg.
- Figure 2.21: Copyright © Miyuki Meinaka (CC BY-SA 3.0) at http://commons.wikimedia.org/wiki/File:Agricultural_Map_by_Whittlesey,_D.S.png.
- Figure 2.22: Copyright © Dirk van der Made (CC BY-SA 3.0) at https://commons.wikimedia.org/wiki/File:DirkvdM_santa_fe_scorched.jpg.
- Figure 2.23: Copyright © AndrewMT (CC BY-SA 3.0) at http://commons.wikimedia.org/wiki/File:RiceYield.png.
- Figure 2.24: Lynn Betts, https://commons.wikimedia.org/wiki/File:TerracesBuffers.JPG. Copyright in the Public Domain.
- Figure 2.25: Copyright © Maximilian Dörrbecker (CC BY-SA 2.5) at http://commons.wikimedia.org/wiki/File:Karte_Hochkulturen.png.
- Figure 2.26: Copyright © NormanEinstein (CC BY-SA 3.0) at https://commons.wikimedia.org/wiki/File:Fertile_Crescent_map.png.
- Figure 2.27: Copyright © Kaufingdude (CC BY-SA 3.0) at https://commons.wikimedia.org/wiki/File:Zig_close.JPG.
- Figure 2.28: Copyright © Jeff Dahl (CC BY-SA 2.5) at http://commons.wikimedia.org/wiki/File:Ancient_Egypt_map-en.svg.
- Figure 2.29: http://commons.wikimedia.org/wiki/File:Maler_der_Grabkammer_des_Sennudem_001.jpg. Copyright in the Public Domain.
- Figure 2.30: Copyright © Ricardo Liberato (CC BY-SA 2.0) at http://commons.wikimedia.org/wiki/File:All_Gizah_Pyramids.jpg.
- Figure 2.31: Copyright © Ismoon (CC BY-SA 4.0) at http://commons.wikimedia.org/wiki/File:Northeast_early_bronze_age_sites_map.png.
- Figure 2.32: Copyright © Depositphotos/Sumners.
- Figure 2.33: Copyright © Avantiputra7 (CC BY-SA 3.0) at http://commons.wikimedia.org/wiki/File:Indus_Valley_Civilization,_Early_Phase_(3300-2600_BCE).png.
- Figure 2.34: Copyright © Hassan Nasir (CC BY-SA 3.0) at https://commons.wikimedia.org/wiki/File:Harappa_Ruins_-_III.jpg.
- Figure 2.35: Copyright © Madman2001 (CC BY-SA 3.0) at https://commons.wikimedia.org/wiki/File:Olmec_Heartland_Overview_v2.svg.
- Figure 2.36: Copyright © Deror_avi (CC BY-SA 4.0) at https://commons.wikimedia.org/wiki/Category:Pir%C3%A1mide_de_la_Luna#/media/File:Moon_Pyramid_IMG_7304.JPG.
- Figure 2.37: Copyright © Depositphotos/Sateda.
- Figure 2.38: Copyright © Rose Vekony (CC BY-SA 3.0) at https://commons.wikimedia.org/wiki/File:Castillo-tzompantli.jpg.
- Figure 2.39: Copyright © Einsamer Schütze (CC BY-SA 3.0) at https://commons.wikimedia.org/wiki/File:British_Museum_Mesoamerica_002.jpg.

CHAPTER TWO: EARLY HUMAN CULTURES   95

- Figure 2.40: Copyright © Madman2001 (CC BY-SA 3.0) at https://commons.wikimedia.org/wiki/File:Olmec_Heartland_Overview_v2.svg.
- Figure 2.41: Codex Magliabechiano, http://commons.wikimedia.org/wiki/File:Codex_Magliabechiano_(141_cropped).jpg. Copyright in the Public Domain.
- Figure 2.42: Pruxo, http://commons.wikimedia.org/wiki/File:Expansion_Imperio_Inca4.jpg. Copyright in the Public Domain.
- Figure 2.43: Copyright © Martin St-Amant (CC by 3.0) at https://commons.wikimedia.org/wiki/File:Panorama_du_Macchu_Picchu_et_des_environs_2.jpg.
- Figure 2.44: Copyright © Herb Roe (CC BY-SA 3.0) at http://en.wikipedia.org/wiki/File:Mississippian_cultures_HRoe_2010.jpg.
- Figure 2.45: Copyright © Historicair (CC BY-SA 3.0) at http://commons.wikimedia.org/wiki/File:Anasazi.svg.
- Figure 2.46: Copyright © Tobi 87 (CC BY-SA 4.0) at https://commons.wikimedia.org/wiki/File:Cliff_Palace-Colorado-Mesa_Verde_NP.jpg.
- Figure 2.47: Copyright © Jeff Israel (CC BY-SA 3.0) at http://commons.wikimedia.org/wiki/File:African-civilizations-map-pre-colonial.svg.
- Figure 2.48: Aa77zz, http://commons.wikimedia.org/wiki/File:Trans-Saharan_routes_early-pt.svg. Copyright in the Public Domain.
- Figure 2.49: Copyright © KaTeznik (CC BY-SA 2.0 FR) at https://commons.wikimedia.org/wiki/File:Djingareiber_cour.jpg.
- Figure 2.50: Copyright © Rowanwindwhistler (CC BY-SA 4.0) at http://commons.wikimedia.org/wiki/File:AxumYElSurDeArabiaHaciaEl230.svg.
- Figure 2.51: Copyright © Simonchihanga (CC BY-SA 4.0) at https://commons.wikimedia.org/wiki/File:Great-Zimbabwe-still-standing_strong.jpg.
- Figure 2.52: Copyright © Fulvio314 (CC BY-SA 3.0) at http://commons.wikimedia.org/wiki/File:Greece_(ancient)_Crete.svg.
- Figure 2.53: Regaliorum, "Greek Colonization Archaic Period," https://commons.wikimedia.org/wiki/File:Greek_Colonization_Archaic_Period.png. Copyright in the Public Domain.
- Figure 2.54: Copyright © Depositphotos/Hurricanhank.
- Figure 2.55: Copyright © FJ-de (CC BY-SA 3.0) at https://commons.wikimedia.org/wiki/Category:Maps_of_the_Roman_Empire#/media/File:Portail_rome_antique_fr.gif.
- Figure 2.56: Copyright © Manuel González Olaechea y Franco (CC BY-SA 3.0) at http://commons.wikimedia.org/wiki/File:AcueductoSegovia04.JPG.
- Figure 2.57: Copyright © Andreas Tille (CC BY-SA 4.0) at https://commons.wikimedia.org/wiki/File:ColosseumAtEvening.jpg.

Chapter Three

THE GEOGRAPHY OF LANGUAGES

MAIN POINTS

1. Languages are not simply modes of communication, but symbolic systems for producing and sharing meaning.

2. There are six major families of languages in the world, of which the Indo-European is, by far, the largest, encompassing about one-half of the world's population, largely the result of colonialism. Other major language families are Afro-Asiatic, Niger-Congo, Ural-Altaic, Sino-Tibetan, and Malayo-Polynesian. The world's most commonly spoken language, including those who speak it as a second language, is English, although the most commonly spoken at home is Mandarin Chinese.

3. Conflicts within countries over which languages are used in government and taught in schools can be severe, even threatening to destroy a state.

4. The number of languages spoken across the world, currently about 6,500, is declining rapidly in the face of national school systems, globalization of the media, and the extinction of many small local cultures, typically tribes in remote areas.

Have you studied a foreign language? If not, you should. If you have, you can appreciate how difficult it can be to master a different grammar and vocabulary. Moreover, studying a foreign language offers a window into a different way of thinking, of representing the world. Language, one of the central pillars of any culture, is profoundly geographical in nature. Understanding the distribution of the world's languages is key to appreciating the complexity of the world's cultures and how they interact. Language is one of the major dimensions of any culture, with profound impacts on how people think and act. The evolution of languages over time and space serves as a mirror for other social processes, such as migration, invasions, and colonialism. It also helps us understand contemporary political events, such as struggles over the use of a particular language (such as French in Canada). The geography of the world's languages emerged gradually over a long period of time. In many ways, it can help us understand how the world was structured before European colonialism began in the sixteenth century and how the world's cultures were reshaped by colonialism.

This chapter delves into the geographies of language. First, it summarizes the role of language as a symbolic system that does much more than communicate; it structures the way in which conscious thought is organized. Next, it charts the distribution of the largest language families, or groups of related tongues. These often emerged out of the Neolithic Revolution to define the linguistic landscapes of vast parts of the world, only to be reshaped by the tsunami of European colonialism. Then, it focuses on linguistic conflicts. The chapter concludes by turning to language death, the contemporary disappearance of so many of the world's tongues in the face of globalization. As Chapter 1 noted, humans are conscious beings, and in sketching the contours of language, this chapter hopes to explain how systems of meaning and communication are unevenly distributed across the world's surface.

3.1 WHAT IS LANGUAGE?

Language is not identical to speech, which is the verbal form that language takes. A language may exist without speech, such as in the printed text or the way in which you think. Language may be understood in several ways: (1) as a means of organizing thought; (2) a way of communicating, that is, producing and sharing meaning; and (3) as a vehicle for understanding one's sensory impressions of the world, that is, of converting sensations into perceptions. In a sense, language is the architecture of consciousness, one of the defining qualities of being human beings. Because language structures how we think, it plays an enormous role in shaping perceptions and behavior. For native speakers, of course, speaking a language is effortless and takes little or no conscious thought, although infants take several years to do so. Precisely because it is so deeply embedded in our brains, learning a foreign language is a difficult and time-consuming process; learning to *think* in another language, rather than just speak it, is very challenging. For this reason, language as a structured symbolic system has figured prominently in philosophy for decades. In this light, language is not simply a medium for communicating but an obstacle as well, a force that enters into the making of social and spatial reality. Language affects individuals' sense of their identity and their membership in a larger group that speaks that language. Language is, thus, simultaneously a psychological, social, and cultural phenomenon; it exists within the individual human mind but is also a fundamental part of how groups of people are organized and communicate with one another.

Geographers are interested in the role of language in the representation of space: Language is how we bring space into consciousness. Language is a window into the structure of human consciousness and spatial perception, for it shapes how human beings give meaning to space and their sense of place. Feminists have emphasized the gendered nature of language and how it reinforces or challenges gender roles (Chapter 13). Language is always intimately associated with power relations and how people are socialized. Postmodernists hold that language always oversimplifies the world, for reality is more complex than any language can admit. A broad consensus emerging from these various perspectives is that all worldviews are inherently and necessarily partial, contingent, and situated in context. Such lines of thought have enormous implications for the ways in which truth is conceived (e.g., as a mirror of the world or as part of a situated worldview) and how explanation is justified. Language is, thus, a powerful and complex phenomenon, for humans do not simply use language, they are, in turn, shaped by it.

Languages, by definition, are mutually intelligible to their speakers. However, many languages exhibit local and regional variations called **dialects**, which are, at least in theory, understandable to everyone speaking a common language. Dialects, which are spoken by native speakers of a language, should not be confused with **accents**, which refer to the ways in which nonnative speakers of a language pronounce its words. Almost all widely spoken languages have dialects that vary among and within countries. For example, English is spoken in both the United States and Britain, but Americans drive trucks and live in apartments whereas the British drive lorries and live in flats. In the U.S., English spoken in the South is quite different from that heard in New England. Spanish spoken in Argentina is quite different from that found in Cuba or Mexico. French in Quebec varies considerably from that spoken in Paris, which, in turn, is different from that in Marseilles. Portuguese spoken in Brazil is not the same as that spoken in Portugal. Dialects often have class connotations associated with them, and the strongest dialects are often spoken by the least well-educated segments of a population.

Finally, **creoles** are languages formed from mixtures of different tongues but are spoken by native speakers. For example, in the former French colony of Haiti, the creole consists of a mixture of African, French, Spanish, and Arawak Indian words. In Suriname, once a Dutch colony in South America, the creole was invented by children from many different cultural heritages, and mixes Portuguese, African, Dutch, Spanish, English, and several indigenous tribal languages.

Languages are never static but exist in a state of constant flux, changing over time. Languages change due to the influence of various social forces (e.g., invasions), the adoption of slang words, new technologies (e.g., printing, the Internet), changing cultural and social roles, and the borrowing of words and ideas from other cultures. Dialects, in some respects, represent new languages coming into being, such as when various dialects of Latin in the fifth century CE evolved into Romance languages, such as Spanish, French, Italian, and Romanian.

3.2 GEOGRAPHY OF THE WORLD'S MAJOR LANGUAGE FAMILIES

Today, there are roughly 6,500 languages spoken in the world, although that number is rapidly decreasing due to reasons we shall explore shortly. Linguists and cultural geographers typically speak of roughly eight major language families, based on historical relationships and similarities, as well as several smaller ones. Language families, which have been studied extensively by linguists, anthropologists, historians, and others, are organized around several common properties, including shared root words that give rise to related vocabularies, rules of grammar (e.g., conjugation of verbs), the use of gender, representations of time (e.g., past and future tenses), and the various orders in which subjects, verbs, and objects appear. In addition, several languages, unrelated to any other, are called **isolates**. These language families and isolates appear across different parts of the planet's surface (Figure 3.1). It should be noted that not all linguists agree on the relations among and within these families, and the position of languages within them has changed over time.

Language is not the same thing as speech and can be written, as well as used orally. The world uses a variety of writing systems in this regard (Figure 3.2), including several alphabets (e.g., Roman, Hebrew, Arabic) as well as pictographic systems (e.g., Chinese).

INDO-EUROPEAN

By far the largest and most widespread of the major language families, the Indo-European family was first identified by the eighteenth-century linguist William Jones, a British judge

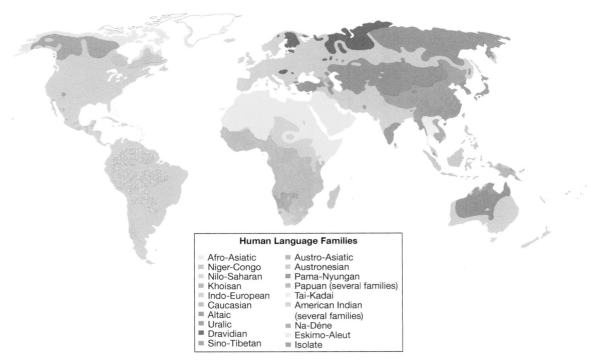

Figure 3.1: Most of the world's tongues fall into a few major language families, of which the Indo-European is the largest. The precolonial distribution of languages was markedly changed by the onslaught of colonialism starting in the sixteenth century.

stationed in India who was the first to identify commonalities among tongues spoken in South Asia, Iran, and Europe. The historical rise of the Indo-European language family was enormously important in structuring the world's linguistic and cultural landscapes. Starting with the migrations of the **Aryans** around 2000 BC, perhaps as a result of their domestication of the horse, Indo-Europeans moved outward from their homeland, which may have been either in Ukraine or near the Caucasus Mountains. (Indeed, names such as "Ireland" and "Iran" are derived from "Aryan.") Some people call them "Caucasians" because of their possible homeland there, although the term has acquired a racial, rather than linguistic, meaning. The expansion of the Indo-Europeans has been the subject of great controversy: Some historians and linguists insist they were a fierce, nomadic, predatory people who quickly conquered all who lay before them, whereas others maintain that their growth and diffusion occurred slowly along with the expansion of their farming sites. Both processes were probably at work.

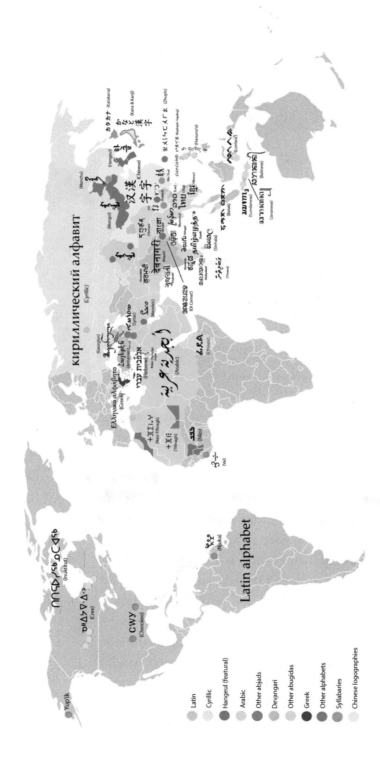

Figure 3.2: The world uses a variety of writing systems. Whereas the Latin alphabet predominates in most of Europe, the Americas, much of Africa, Australia, and parts of Asia, Cyrillic is used in Russia, Ukraine, and Mongolia. The Arabic alphabet is used not only in Arabic-speaking countries but also in other Muslim countries, such as Iran, Afghanistan, and Pakistan. The Chinese writing system, which is pictographic but not an alphabet, diffused to neighboring countries, such as Japan, Korea, and Vietnam, but was often displaced by indigenous writing systems (e.g., Hangul in Korea) or the Latin alphabet (as in Vietnam).

One group of Indo-Europeans moved east into northern India (Figure 3.3) and may have been responsible for the destruction of the Indus River Valley civilization (Chapter 2). (Southern India remains populated by the Dravidians, an entirely different linguistic group.) The invaders spoke and wrote texts in **Sanskrit**, the language of the ancient Hindu sacred writings (Chapter 4). Although this context seems far removed from Britain and, therefore, English, Sanskrit and English are distantly related. For example, the Sanskrit word *punj*, meaning "land of five rivers" (and the basis of the area known as the Punjab) is distantly related to the English word "punch" (originally the "drink of five fingers"). The Sanskrit word "deva" is cognate with (related to) the English word "divine." Sanskrit played a role in northern India similar to that played by Latin in southern Europe, that is, as the written basis for several diverse spoken tongues that diverged over time. Sanskrit was also adopted as a writing system by some non–Indo-European cultures, including Tibet, Myanmar (Burma), Thailand, Cambodia, and Laos (Figure 3.2).

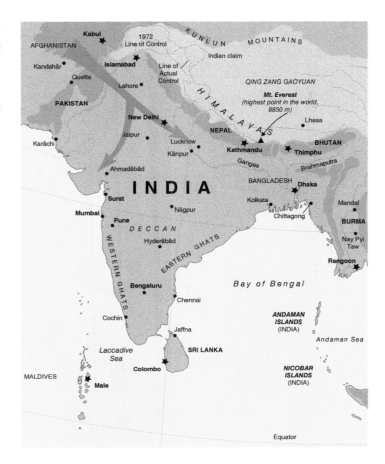

Figure 3.3: The early Indo-Europeans, sometimes called Aryans, spread their languages both to South Asia and to Europe.

Sanskrit became the basis of the Indic branch of languages, such as Hindi (the most commonly spoken language of India and almost its national tongue), Punjabi, Bengali, Bihari, Urdu, Gujarati, Oriya, Marathi, and Nepali (Figure 3.4). Today, in addition to English, many people still read and write Sanskrit-based languages in South Asia. Because South Asia has a very large population, this branch of the Indo-European family includes hundreds of millions of speakers. For example, more than 400 million people speak Hindi (with many dialects), and

104 HUMAN GEOGRAPHY: A SERIOUS INTRODUCTION

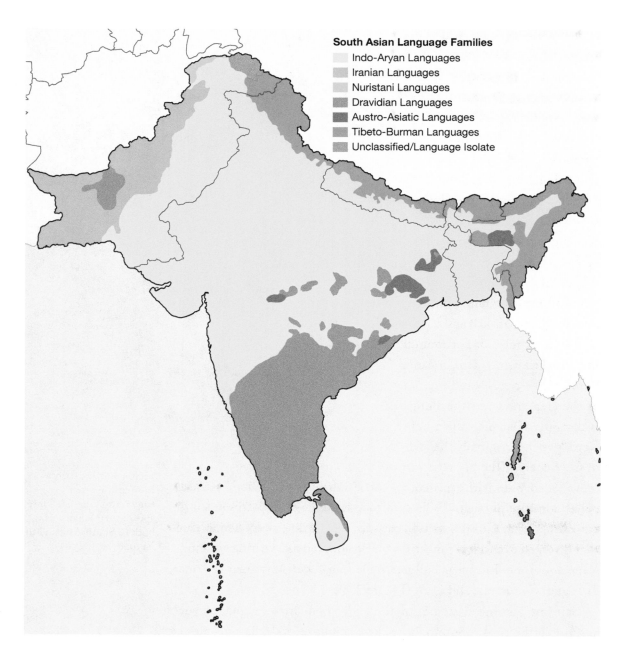

Figure 3.4: As a result of the Aryan conquests, northern South Asia speaks a variety of Indo-European languages, such as Hindi, Bengali, Gujarati, and Marathi.

240 million speak Bengali, including those in the Indian state of Bengal and the country of Bangladesh (literally "land of the Bengal speakers").

Other Indo-Europeans spread from their Caucuses homeland to settle in the Middle East. Early examples include the Hittites, who occupied the Anatolian peninsula, and the Hyksos, who invaded the Old Kingdom of Egypt (Chapter 2). Indo-Europeans in this region gave rise to the Iranic family of languages including Farsi (formerly Persian), Kurdish, Armenian, and Pashto. Modern Iranians speak the Indo-European language, Farsi, not Arabic. Farsi has much more in common linguistically with English than it does with Arabic, and is also spoken in parts of Afghanistan. The Kurds are an Indo-European people who converted to Islam (Chapter 4) and today are scattered among several countries in the Middle East (Iraq, Turkey, Iran, and Syria), where they form a nation without a state (Chapter 9). Balochi is an Iranic language spoken in Balochistan in southern Pakistan.

The other major branch of early Indo-Europeans moved into Europe, where their language diverged into several groups (Figure 3.5). It may surprise some people to learn that languages spoken in places as distant as Iceland and Bangladesh are related, but the historical and linguistic evidence is irrefutable. For example, the ancient Indo-European word "bare," meaning "to carry," was incorporated in different ways in several European languages, including the Anglo-Saxon "bare," "burden," "birth," and "bring." In Latin-based languages, it became "fer," such as "ferry," "fertile," "transfer," and "prefer." In Greek, it became "pher," such as "pheromone," "paraphernalia," and "amphora."

European language families include the Latin-based **Romance languages** (Italian, Spanish, Portuguese, Galician/Gallegan, Catalan, French, Romansch, and Romanian), which arose during the disintegration of the Roman Empire in the fifth century CE (shown in blue in Figure 3.5). Roman government officials traveling through the empire in its waning days complained about the varieties of Latin being spoken in the places they encountered; they were witnessing the birth of different languages out of the mother tongue. The Catholic Church also adopted Latin and played a major role in its diffusion throughout much of Europe; later, Latin terms became widely used in the sciences, as well. The Roman alphabet was adopted not only by speakers of Romance languages, but also by Germanic cultures as well. Given the powerful role of western European colonial powers, the Roman alphabet has become the most common script in the world today. Although they have existed as separate languages for centuries, the Romance languages remain closely related to one another in grammatical structure and, often, in vocabulary. Each has numerous local dialects. For example, there are substantial differences between the varieties of Italian spoken in Sicily and in the north of the country. French was long divided between the northern part, in which the word *oui* meant "yes," and the southern part that used the word *oc* (giving rise to the regional term Occitan). Catalan, spoken in eastern Spain,

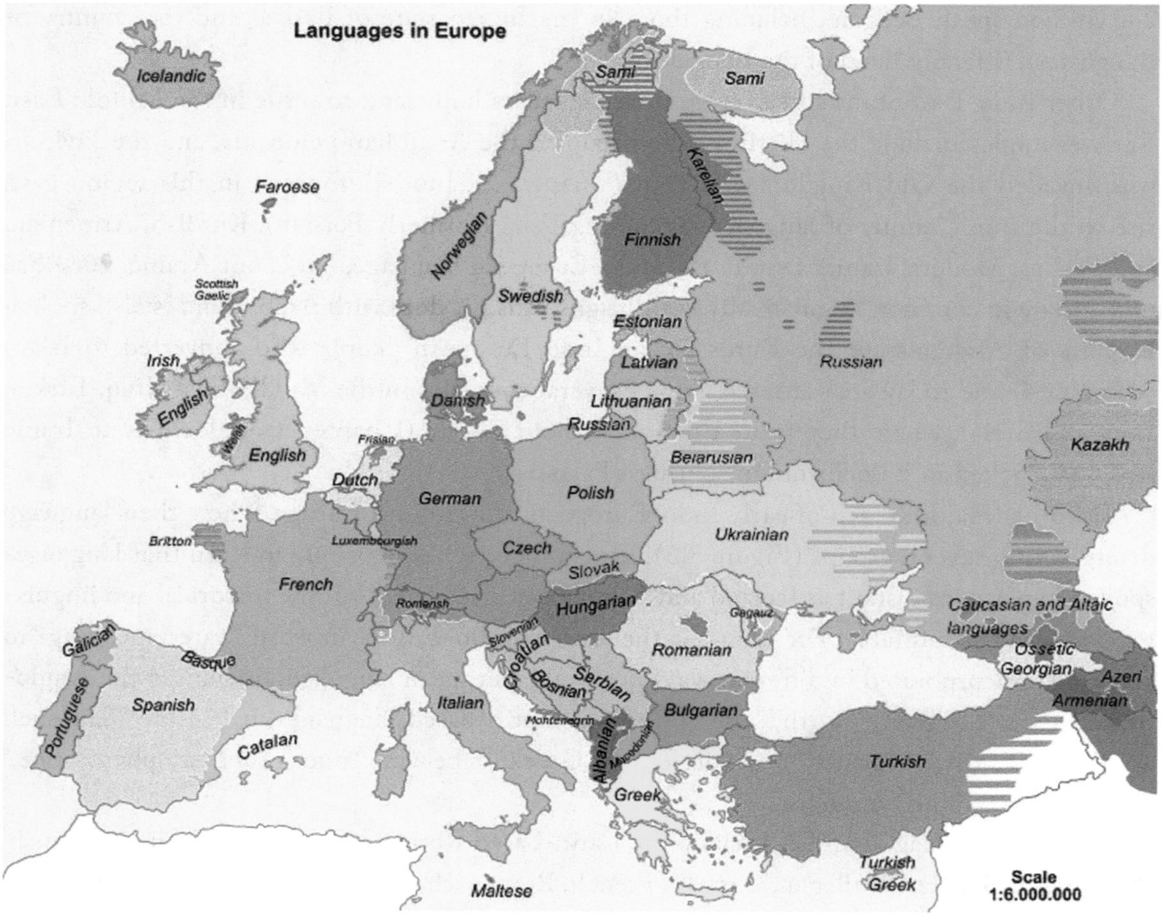

Figure 3.5: With a few exceptions (such as Basque, Finnish, and Hungarian), Europe is dominated by Indo-European languages. The major groups include the Romance (Latin-based), Germanic, Slavic, Celtic, and Baltic families, as well as Greek and Albanian, each of which forms its own.

resembles both Spanish and French. In northwestern Spain, Galician (or Gallego) was shaped by the Celtic people who lived there at the time of the Roman conquest. Romanian emerged from the Roman province of Dacia; although the Romans occupied it for only 150 years, it remains closer to the original Latin than does modern Italian.

Farther north, the Germanic languages include German, Dutch, the Scandinavian tongues of Swedish, Danish, Norwegian, and Icelandic (but not Finnish), and English (shown in light colors in Figure 3.5). There are several dialects of German, which is also spoken in Austria and most of

Switzerland. German has numerous dialects, not only among various German-speaking countries, but within Germany itself, such as the differences between upper German, spoken in the southern part of the country, and lower German spoken in the north.

Celtic is a language family that was once widespread throughout western Europe, but was greatly reduced in size and number by the expansion of the Romans. Today this family of languages is confined to the "Celtic fringe" of the British Isles and France (Figure 3.6), and includes Scottish and Irish Gaelic, Welsh, Breton in the Brittany peninsula in Western France, and extinct tongues such as Cornish. Celtic also influenced the Galician dialect in northwestern Spain. Given the hegemony of English throughout the British Isles (and of French in France), this language group is in danger of disappearing. In Wales, only 20% of the population speaks Welsh, and in Scotland, only 1% speaks Gaelic.

Other Indo-European language groups in Europe include Greek and Albanian, which form separate, independent categories. Early Doric and Ionian Greeks settled on both sides of the Aegean Sea (Chapter 2). The Greeks, of course, developed their own alphabet distinct from the Romans, one that incorporated Phoenician elements and became known as Cyrillic (after St. Cyril). It was later adopted by Russians, Serbs, and Ukrainians, when they also adopted Orthodox Christianity (Chapter 4), and is one of the most commonly used writing systems in the world today (see Figure 3.2).

In eastern Europe and Russia, the Slavic family of languages includes a northern branch (Polish, Russian, Belorussian, Ukrainian, Czech, Slovak, and Slovenian) and a southern branch (Serbian, Croatian, Macedonian, and Bulgarian). Finally, the Baltic group includes Lithuanian and Latvian (but not Estonian) (shown in green in Figure 3.5).

Figure 3.6: The Celtic languages are another branch of Indo-European languages that are on the verge of extinction; they include Gaelic in Scotland, and Welsh in Ireland, and Breton in France.

A BRIEF SKETCH OF ENGLISH

A brief historical geography of the English language illustrates the numerous forces that have shaped it over time and space. English is fundamentally a Germanic language in origin and began with the invasion of the former Roman province of Britannia in the fifth century CE by Germanic tribes, such as the Angles, Saxons, Jutes, and Frisians (Figure 3.7). Indeed, Frisian, spoken on the islands of the

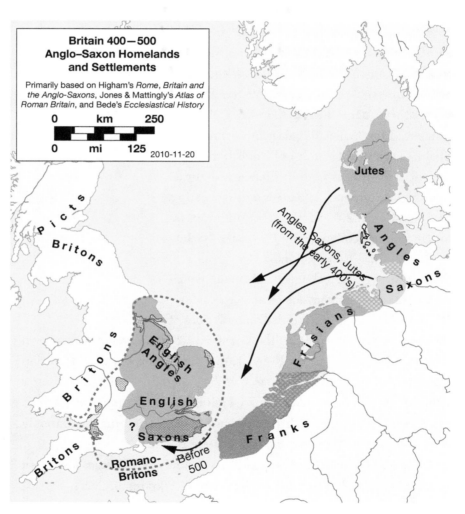

Figure 3.7: English began in the sixth century CE when the Angles, Saxons, Jutes, and others left Germany and Denmark and invaded what is now England.

same name in the Netherlands, remains very similar to English in vocabulary to this day. The Anglo-Saxons, as they came to be called, displaced the indigenous Celtic peoples of southern Britain, and the area became known as England ("land of the Angles"). Early English originally retained much of its Germanic grammar (e.g., using vowel changes, such as "steal" to "stole" or the letter "d" to indicate past tense) but gradually lost it over time. Some Latin words crept in with the introduction of Christianity.

In the eighth and ninth centuries, the Vikings conquered northern England and added a second layer of (northern) Germanic vocabulary (Table 3.1); for example, the Norse words "anger," "ill," and "skill" were added to the English counterparts "wrath," "sick," and "craft," respectively. Because it has long borrowed words from other languages, there has never been a "pure" form of English.

Starting in 1066 with the Norman conquest of England, French became the language of power, commerce, and the nobility, eventually contributing up to 40% of the words in the English vocabulary. Sometimes the new words became synonyms for existing ones (Table 3.2). The French words "close," "odor," "annual," and "power," for example, paralleled the English words "shut," "smell," "yearly," and "might," respectively. The Normans were wealthier and more powerful than the English, who tended their livestock for them; thus, Norman words for meats became the norm ("beef," "veal," "mutton," "pork," "poultry," and "venison"), whereas English words for the

TABLE 3.1: EXAMPLES OF NORSE WORDS THAT ENTERED ENGLISH AND THEIR SAXON COUNTERPARTS

NORSE	SAXON
Anger	wrath
Raise	rear
Ill	sick
Bask	bathe
Skill	craft
Skin	hide
Dike	ditch
Skirt	shirt
Scatter	shatter
Skip	shift

TABLE 3.2: EXAMPLES OF FRENCH WORDS THAT ENTERED ENGLISH AND THEIR SAXON COUNTERPARTS

FRENCH	SAXON
Close	shut
Reply	answer
Odor	smell
Annual	yearly
Demand	ask
Chamber	room
Desire	wish
Power	might
Ire	wrath
Pork	pig
Beef	cow

TABLE 3.3: EXAMPLES OF OLD ENGLISH, NORSE, NORMAN FRENCH, AND LATIN WORDS IN ENGLISH

Old English
the, you, I, we, that, it, is, yes, was, this, but, on, he, have, year, good, come, only, think, work, man, woman, son, fish, bird, house, blood, foot, tooth, nose, eye, eat drink, come, see, know, sleep, star, sun, moon, rain, cloud, fire, night, mother, father

Norse
they, them, there, get, take, skin, shirt, skirt, sky, egg, kid, anger, crawl, weak, loan

Norman French
habit, local, merit, exercise, routine, familiar, study, parliament, government, administer, crown, royal, subject, peasant, noble, baron, people, judge, attorney, evidence, excuse, acquit, plead, verdict, punish, prison, economy, finance, profit, revenue, property, estate, heir, military, battle, offense, defense, army, peace, religion, saint, prayer, faith, devout, salvation, clergy, baptism, button, blanket, cushion, chair, paper, dress, boot, jewel, dance, tennis, racquet, bacon, sugar, sausage, bisque, fruit, salad

Latin
kettle, wall, chalk, mile, street, wine, pea, abbot, alter, candle, angel, disciple, hymn, pope, priest, psalm, school

---

animals from which the meat was derived were retained ("cow," "calf," "sheep," "pig," "chicken," and "deer," respectively). The dialect of French that the Normans introduced, which became known as Anglo-Norman, differed from that spoken in the rest of France, particularly in Paris. Why didn't England become a French-speaking country? The answer is likely that the Norman nobility only made up about 2% of the population, and over time, processes such as intermarriage gradually diluted the French political and cultural influence.

English evolved steadily throughout the medieval period (which is very difficult for modern speakers to understand) and the Renaissance (e.g., Shakespeare). During the Renaissance, interest in classical languages led to a large influx of Latin words, particularly formal ones used in science and scholarly activities, but also in music, literature, food, and education. Many Greek words entered as well, (e.g., "psychiatry," "asteroid," "geography," "chronology," "geology," and "bibliography") and became common in medicine and the sciences. Much of the odd spelling of English words, which is notoriously difficult for nonnative speakers to learn, is due to William Caxton, who introduced the printing press to England in 1472 and standardized the spellings used at the time. Indeed, as with many other languages, English changed far more rapidly before the invention of the printing press than it did afterward; printing greatly slows down linguistic evolution (Table 3.3).

Throughout the sixteenth through the twentieth centuries, English expanded worldwide with the growth of British, then American, global economic and military dominance (Figure 3.8). In doing so, it acquired a number of loan words from other languages (Table 3.4). English became dominant in Canada, the United States, Australia, New Zealand, and South Africa, and is also widely spoken in former British colonies, such as India, Pakistan, Singapore, Malaysia, and Hong Kong, as well as other parts of Africa. For these reasons, English has the largest vocabulary of any language in the world. However, its unique historical trajectory has meant that English lacks any closely related languages in the way that the Romance languages, for example, have certain similarities to one another. There are many dialects of English in Britain, Canada, and the United States; in Britain, dialects were long associated with class status, whereas, in the United States, it is easy to differentiate dialects, such as those spoken in the South, New England, or New York and New Jersey.

Figure 3.8: Countries where English is either predominant or commonly spoken today. The dominance of English is due to centuries of British colonialism and the contemporary economic and cultural might of the United States.

TABLE 3.4: LOAN WORDS ENGLISH HAS ACQUIRED FROM OTHER LANGUAGES

Africa: okra, gumbo, impala, banana, bogus, boogie, jazz, yam, bozo, chimpanzee, impala, safari, zombie

Arabic: guitar, charlatan, alfalfa, algebra, algorithm, giraffe, magazine, zero, mattress, saffron, coffee

Australian aborigines: boomerang

Chinese: ginseng, tofu, lo mein, kung fu, tea, wok, gung ho, coolie, lychee, silk, bok choy, kumquat

Dutch: smuggle, aardvark, pickle, bumpkin, landscape, cookie, cruise, brandy

Finnish: sauna

Greek: labyrinth, odyssey, titanic, mentor, nemesis

Hawaiian: ukulele, aloha, wiki

Hindi/Urdu: pajama, pundit, orange, bandanna, bottle, cheetah, chutney, guru, jungle, punch, khaki, shawl, sorbet, shampoo, typhoon, veranda

Hungarian: paprika, goulash, couch

Indonesian/Malay: amok, ghetto, cockatoo, bamboo, batik, gecko, gong, ketchup, paddy, ratan

Inuit: kayak

Italian: pasta, pesto, pimiento, broccoli, balcony, sonnet

Japanese: tycoon, tsunami, sushi, haiku, geisha, bonsai, futon, kudzu, ninja, judo, karate, sudoku, karaoke, soy, tempura, ramen

Korean: taekwondo, kimchi

Mongol: horde

Native American: powwow, teepee, moccasin, chocolate, moose, squash, skunk, hurricane, potato, tobacco, jaguar, piranha, cannibal, coyote, llama, papaya

Persian: bazaar, beige, bronze, carcass, divan, gizzard, jasmine, kabob, lemon, mummy, rank, sandal, scarlet, shah, tiger, tulip, turban

Portuguese: monsoon, molasses, flamingo, coconut, albino, caramel, cashew, cobra, savvy, zebra

Quechua: pampas

Russian: tsar, mammoth, steppe, gulag, pogrom, taiga, vodka

Spanish: rodeo, plaza, cilantro, chili, adobe, alligator, armadillo, burrito, canyon, galleon, macho, machete, patio, piñata, poncho, sombrero, stampede, taco, tuna, vigilante, tango

Tagalog: boondocks, yo-yo, cooties

Tamil: pariah, curry, anaconda, avatar, mango, cash, peacock, sugar, teak

Turkish: macramé, Balkan, pasha, shaman, shish kebab

Yiddish: schmooze, schlock, schlong, kvetch, hutzpah, schmaltz

Welsh: flannel, bard

With the expansion of European colonialism, Indo-European languages spread throughout much of the world. South Africa was originally colonized by the Dutch, who were called Boers, as is their language, Afrikaans, a dialect of Dutch. Spanish became widespread throughout Latin America, with many local dialects (Figure 3.9), as did Portuguese in Brazil and, to a lesser extent, parts of Africa, such as Angola, Mozambique,

and Guinea Bissau (Figure 3.10). Islands of Portuguese are also found in Asia, as they were the first Europeans there (Chapter 5), including Timor and Macau. French was implanted in Quebec and some Caribbean islands.

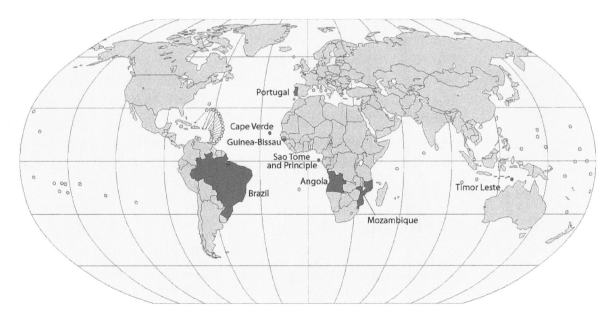

Figure 3.10: Countries where Portuguese is predominant or commonly spoken today. The Portuguese were the last Europeans to end their colonies in Africa.

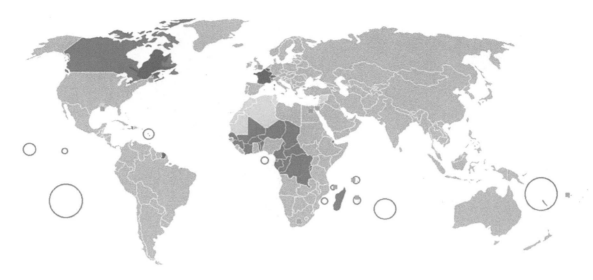

Figure 3.11: Countries where French is predominant or commonly spoken today, including Canada (particularly Quebec) and former colonies in Africa.

Many other inhabitants of the French empire, such as in parts of West Africa, Algeria, and Indochina, also learned French (Figure 3.11).

Today, almost half of the world's people speak an Indo-European tongue. Table 3.5 illustrates the relative sizes of major European languages, most, but not all, of which are Indo-European. English, in particular, has more speakers than any other tongue (when second-language speakers are included). English is unquestionably the world's dominant language in commerce, trade, scholarly publications, airlines, international finance, and tourism, a **lingua franca** used as a medium of communication by speakers of different languages even if it may not be spoken at home. Never in world history has one language dominated the planet's linguistic landscape as much as English does today.

AFRO-ASIATIC

A second major language family, Afro-Asiatic, extends across the Middle East and North Africa (Figure 3.12). This group includes most of the extinct or nearly extinct languages of the ancient Middle East, such as Canaanite, Phoenician, Assyrian, and Aramaic (the language

TABLE 3.5: MAJOR EUROPEAN LANGUAGES SPOKEN WORLDWIDE

	NO. OF SPEAKERS (MILLIONS)	PERCENTAGE OF WORLD POPULATION
Spanish	470	5.8
English	400	5.5
Portuguese	220	3.1
Russian	170	2.4
German	95	1.4
French	80	1.1
Italian	64	0.9
Polish	45	0.6
Ukrainian	40	0.5
Romanian	24	0.4
Dutch	23	0.3
Hungarian	13	0.2
Greek	13	0.2
Czech	11	0.1
Swedish	9	0.1

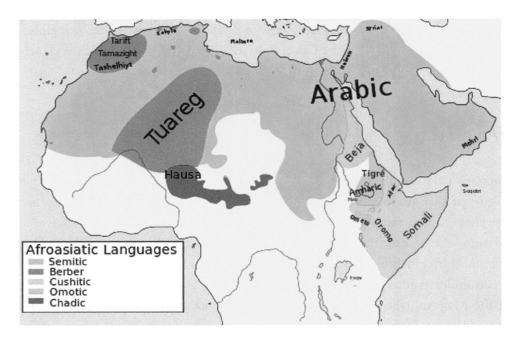

Figure 3.12: The Afro-Asiatic language family encompasses the Semitic languages of Hebrew and Arabic, as well as a broader group of languages that includes Berber and Cushitic. This family is spoken across most of the Middle East, North Africa, and the Horn of Africa (Ethiopia and Somalia).

of Jesus, and of which pockets survive today). The dominant branch of Afro-Asiatic is **Semitic**, whose surviving members today include Arabic and Hebrew, as well as Maltese, spoken on the Mediterranean island of Malta. Hebrew was the early Middle Eastern language spoken by the Jews, who were crushed by the Romans and subsequently spread throughout Europe (Chapter 4). Hebrew nearly became extinct during this period, but was revised in the late nineteenth century as part of the rise of Zionism, the Jewish nationalism that led to the establishment of Israel in 1948. Today, Hebrew is spoken by roughly eight million people, mostly in Israel and in Jewish communities throughout the world.

Arabic is spoken by more than 290 million people stretching from the Indian Ocean to the Atlantic, and has many dialects that are often so different from one another that it is difficult for people from, say, Morocco and Oman to communicate with one another. Often Arabs speak both "Standard Arabic," the language of business, commerce, and the media, as well as local colloquial variants. Arabic and Hebrew, which are linguistically related, both have unique alphabets and are read from right to left. Arabic is the language in which the Muslim holy book, the Koran (Chapter 4), is written, and therefore, Arab script was adopted by many non-Arab peoples, such as Iranians and Urdu-speaking Pakistanis.

Other branches of this family include Berber, widespread in Northwest Africa (a people who predate the Arab influx there and form an important minority in northwest Africa). In eastern Africa, Ethiopians speak the Semitic language of Amharic, and have their own alphabet. Somali, Afar, and Oromo are part of the Cushitic branch of Afro-Asiatic, which is closely related to and overlaps geographically with Semitic.

URAL-ALTAIC

Ural-Altaic comprises a third major language family (some linguists see these as two distinct groups). The origins of this group, probably near the Altai Mountains of Mongolia, are lost in prehistory. Speakers of this family may be descendants of several waves of migration that generated populations across Eurasia who speak loosely related tongues (Figure 3.13). One branch, called Finno-Ugric, includes Finnish and Estonian in northern Europe, which are so close that they can understand one another's radio stations. A second branch is Hungarian, the language of the Magyars who settled in eastern Europe in the eighth century (medieval Europeans misnamed them for the Huns, a distantly related group that invaded Europe five centuries earlier; hence the name "Hungary").

A third branch of this family is the Turkic languages, which emanated from the Turkish migrations into Central Asia and Anatolia in the ninth and tenth centuries. The Turks

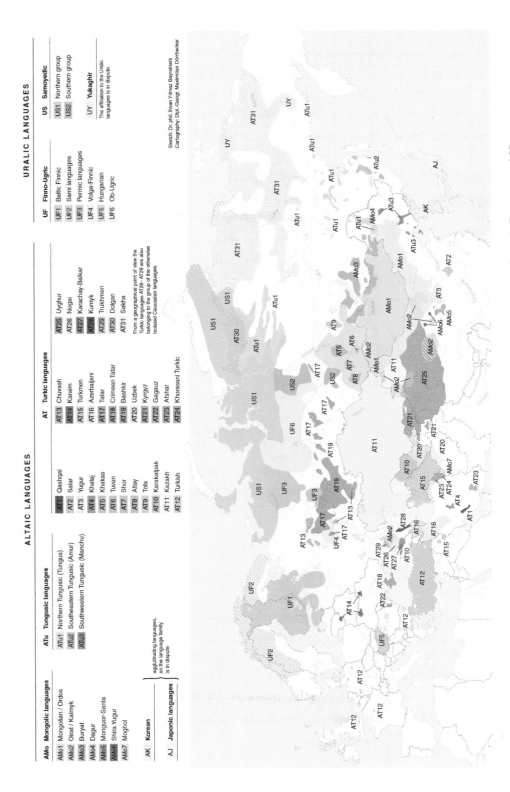

Figure 3.13: The Ural–Altaic language family stretches across Eurasia. It includes diverse tongues, such as Finnish, Estonian, and Hungarian in eastern Europe; Turkish and the Turkic languages of Central Asia; Mongolian; and indigenous Siberian, such as Tungus and Samoyed. Korean and Japanese possibly form part of this family, although linguists continue to debate this question.

converted to Islam (Chapter 4) but their language is completely different from Arabic. In addition to Turkish, other existing Turkic languages include Azerbaijani (or Azeri), Kazakh, Uzbek, Turkmen, Kirgiz in Central Asia, and Uighur in western China.

Yet another Altaic branch is Mongol and Tatar, still spoken by the numerous tribes that inhabit the Asian steppes. Manchu, formerly spoken in Manchuria, is now nearly extinct, as are the indigenous tongues of Siberia such as Samoyed and Tungic, which were marginalized by the expansion of the tsarist Russian empire.

Finally, although it is controversial, many linguists assign Japanese and Korean to this family as well. Although Japan and Korea both historically adopted elements of Chinese writing (made possible only through a pictographic script), the Japanese and Korean languages are completely different from Chinese. For example, in Chinese, the pitch and intonation ("tone") of a word affect its meaning, but this is not the case in Japanese and Korean. Japanese and Korean share about 15% of their vocabulary. Korea adopted its own, unique Hangul alphabet in the fifteenth century, which has 24 characters, whereas Japan retained but modified the original Chinese system (Kanji) and added two smaller pictographic writing systems, Hiragana and Katakana, much later.

NIGER-CONGO

In Africa the area south of the Sahara desert offers a very complex mosaic of tongues from several language families (Figure 3.14). In addition to Afro-Asiatic languages in the north (i.e., Arabic and Berber) and in the northeast (i.e., Amharic and Somali), it has smaller families, such as Nilo-Saharan (including, for example, Nubian, Maasai, and Dinka). In Southwest Africa, there are the few surviving click languages of the Khoisan family (e.g., !Kung, in which the exclamation mark stands for a click of the tongue). Khoisan languages, at one time, were widespread over much of southern Africa but, like hunters and gatherers everywhere, were displaced by the influx of an agricultural society, the Bantus, who originated in the central part of the African continent and moved south.

Today, the bulk of the many languages spoken throughout the African continent fall under the Bantu, or Niger-Congo, language family, which includes thousands of tongues. Arising from the migrations of agriculturalists from Central Africa around the time of Christ, this family includes languages as diverse as Mande in western Africa; Kikuyu in Kenya; and Tswana, Nbele, and Zulu in southern Africa. Along the eastern part of the continent, Swahili has long formed a lingua franca, including words of several different languages, including Bantu tongues and Arabic. Swahili is widely spoken in countries such as Tanzania, where it is the national language.

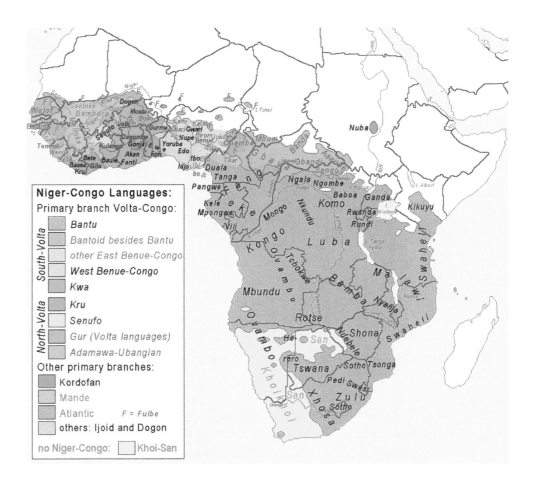

Figure 3.14: Although Afro-Asiatic languages, such as Arabic and Berber, dominate Africa north of the Saharan desert and the Horn region in the east, Sub-Saharan Africa has a complex mix of tongues. The Niger-Congo or Bantu family covers most of the continent. The Khoisan family, today confined to Southwest Africa, such as Namibia, used to be spoken over a much broader area. These hunter-gatherers were pushed into the desert by the expanding Bantu agriculturalists. In Madagascar, an Asian language, Malagasy, is spoken.

SINO-TIBETAN

In eastern Asia, the Sino-Tibetan language family is the most commonly spoken (Figure 3.15). The core of this family is Chinese, which embraces a variety of languages that are not mutually intelligible; although these are sometimes called "dialects," which implies their

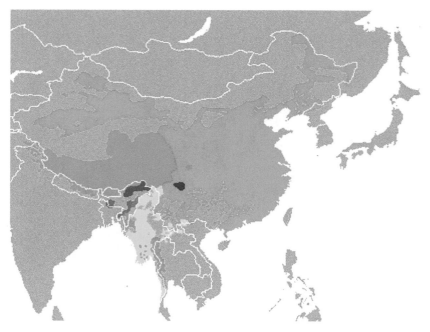

Figure 3.15: The Sino-Tibetan language family includes multiple tongues in China, Tibet, and Myanmar.

speakers can understand one another, Chinese is not in fact, one language but several different ones (Table 3.6).

Communication among several different varieties of Chinese is made possible by a common **pictographic writing system**, in which meanings are linked to symbols (Figure 3.16). In contrast, an alphabet's symbols stand for sounds. (Some Chinese characters contain

TABLE 3.6: NUMBER OF SPEAKERS OF CHINESE LANGUAGES (MILLIONS)

Mandarin	935
Wu (Shanghainese)	80
Cantonese	59
Jin	48
Min Nan	47
Xiang	38
Hakka	31
Gan	22
Min	11
Min Dong	9

phonetic components.) Thus, speakers of different varieties of Chinese can communicate by writing if not orally. This fact contributed to the relatively high levels of literacy found historically in China. Moreover, the Chinese writing system was adopted by several non-Chinese neighboring cultures, such as Korea, Japan, and Vietnam.

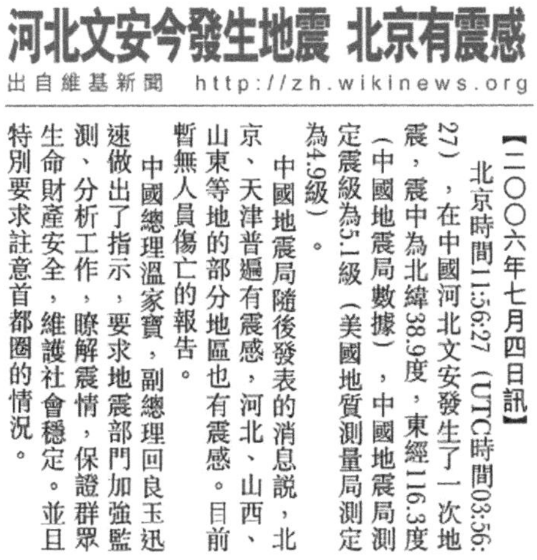
河北文安今發生地震 北京有震感

出自維基新聞 http://zh.wikinews.org

【二〇〇六年七月四日訊】北京時間11:56:27（UTC時間03:56:27），在中國河北文安發生了一次地震，震中為北緯38.9度，東經116.3度（中國地震局數據），中國地震局測定震級為5.1級（美國地質測量局測定為4.9級）。

中國地震局隨後發表的消息說，北京、天津普遍有震感，河北、山西、山東等地的部分地區也有震感。目前暫無人員傷亡的報告。

中國總理溫家寶，副總理回良玉迅速做出了指示，要求地震部門加強監測、分析工作，瞭解震情，保證群眾生命財產安全，維護社會穩定。並且特別要求註意首都圈的情況。

Figure 3.16: In the pictographic writing system adopted by the Chinese (as well as several neighboring countries), there is no alphabet; symbols are linked to the meaning, rather than the sound, of words.

The most common form of Chinese is Mandarin, the dominant language of northern and southwestern China and the most commonly spoken first language at home in the world. Given China's great size (Chapter 8), almost one billion people speak Mandarin. The Beijing-based dialect of Mandarin is the official national tongue of China and is spoken by 70% of the country, often forming a lingua franca for other groups. Many dialects of Mandarin exist, such as Szechwanese, Hebei, and Wu or Shanghainese (Figure 3.17). Mandarin is also spoken by most people on the island of Taiwan, as well as Singapore. In southeastern China, the dominant language is Cantonese ("Canton" is the Anglicized name for the province Guangdong). Most of the Chinese laborers shipped to the United States in the nineteenth century spoke Cantonese, as did the overseas Chinese in Southeast Asia (Chapter 5). Other southeastern variations of Chinese include Hakka and Min/Fujienese or Taiwanese, which is spoken on both sides of the Taiwan Strait. Given the dominance of Mandarin, however, speakers of Cantonese or other languages must learn it to succeed in Chinese government or business circles.

Common to the Sino-Tibetan group (and the Nilo-Saharan languages of Africa) is the use of tones, in which pitch is part of the meaning of the word. Thus, Mandarin has four tones and Cantonese has roughly seven. In nontonal languages, it is possible to speak in a monotone and still be understood; in tonal languages this is simply impossible. (To Western ears, this makes these languages difficult to learn and gives them a certain musical quality). Nontonal languages use tones para-linguistically, that is, to supplement the meaning of a

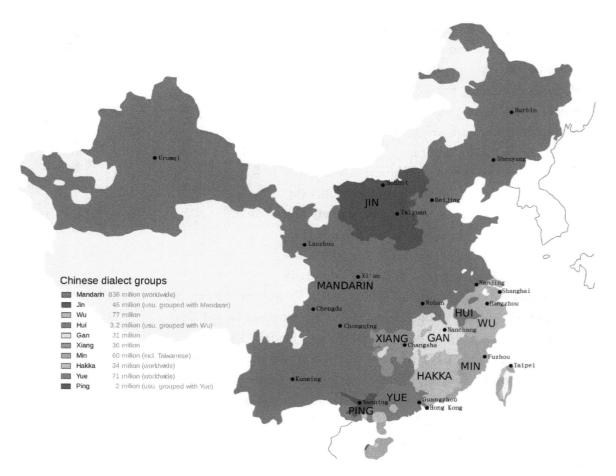

Figure 3.17: "Chinese" is not one language, but a group of related ones. Mandarin, the most common, is the national language of China. Spoken mostly in Northern China, it is widely used in government and commerce. In southern China, other forms include Cantonese, Fukienese, Wu, and Hakka. A common writing system unites them, which is possible only because it is pictographic and does not use an alphabet.

word, such as when a speaker raises his or her pitch to ask a question. In tonal languages, the pitch helps to define a word.

The Sino-Tibetan group also includes Tibetan and Burmese, because of Tibetan migrations down the Irrawaddy River into Burma (now Myanmar) in the ninth century. Because of the influence of nearby India, Tibet and Burma adopted the Sanskrit, not Chinese, system of writing.

AUSTRONESIAN OR MALAYO-POLYNESIAN

A sixth major family is Austronesian, also known as Malayo-Polynesian, a diverse group that extends across much of Southeast Asia into the Pacific islands of Polynesia and Micronesia (Figure 3.18) and includes Hawaiian and New Zealand's Maori. Originating among indigenous, pre-Chinese tribes in Taiwan (who today make up 1% of that island's population), it expanded starting 6,000 years ago, as waves of colonizers settled the islands toward the south and rode canoes across thousands of miles of ocean. Prior to European colonialism and the spread of Indo-European languages, this family was the most widely spread group in the world.

Today the largest members of this group are the Malay languages of Malaysia and Indonesia (also called Bahasa, each with countless dialects, such as Javanese, Sundanese, Madurese, Acehnese, Balinese, and Bornean). Collectively, Malay is the eighth-most commonly spoken language in the world. It also includes the numerous tongues of

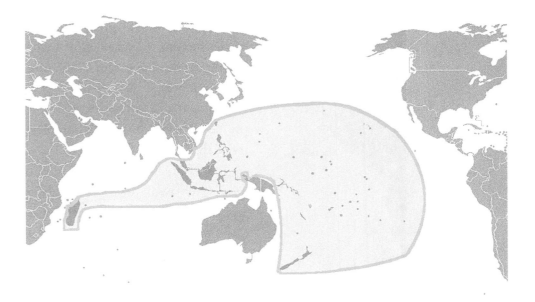

Figure 3.18: As a result of ancient seafarers who crossed both the Indian and Pacific Oceans, the Malayo-Polynesian language family includes tongues as diverse as Malagasy, Malay/Indonesian, Tagalog, Maori, and Hawaiian.

the Philippines, of which Tagalog is the best known (because it was spoken around Manila when the Spanish colonists arrived). Around 700 CE, Indonesian sailors crossed the Indian Ocean and settled in Madagascar, so the indigenous Malagasy language is also part of this family, although the island also was influenced by the Bantu tongues on the African mainland.

OTHER LANGUAGE FAMILIES

Several other smaller families are worth noting. Southern India is home to a sizeable population that does not speak Indo-European languages, but Dravidian tongues, such as Tamil, Telugu, Kannada, and Malayalam (Figure 3.19). Recall the Dravidians are likely the descendants of the people who inhabited the cities of Harappa and Mohenjo-Daro on the Indus River (Chapter 2) and were driven south by the Aryan invasions. Today, given India's great population size, more than 200 million people speak these languages. Table 3.7 gives estimates of the number of speakers of major South Asian languages, including both members of the Indo-European and Dravidian families, testifying to the bewildering linguistic complexity of this part of the world.

The Indochinese peninsula is home to two distinct language groups, Austro-Asiatic (Vietnamese, Cambodian) (Figure 3.20) and Thai-Kradai (or Kadai) (Thai, Lao) (Figure 3.21). Both families may have originated in southern China and became distinct groups in their own right over several thousand years. Like Sino-Tibetan languages,

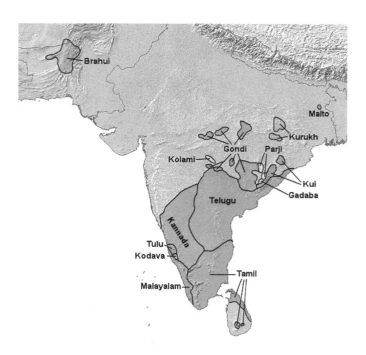

Figure 3.19: The Dravidian languages of southern India, including Tamil, Telugu, Malayalam, are not Indo-European and are spoken by hundreds of millions of people.

TABLE 3.7: NUMBER OF SPEAKERS OF SOUTH ASIAN LANGUAGES

LANGUAGE	SPEAKERS (MILLIONS)
Hindi	422
Bengali	202
Punjabi	96
Telugu	74
Marathi	72
Tamil	70
Urdu	66
Gujarati	49
Pashto	39
Kannada	38
Malayalam	38
Oriya	33
Bhojpuri	29
Maithili	27
Sindhi	26
Awadhi	22
Saraiki	17
Nepali	17
Sinhalese	16
Chittagonian	16
Assamese	16
Marwari	14
Deccan	11
Konkani	7

these are tonal, that is, the pitch is part of the meaning of the word. These cultures were deeply affected by India and China, as some (e.g., Burma, Thailand, Cambodia) adopted Sanskrit writing, whereas Vietnam used Chinese characters for centuries before the arrival of French colonialists in the nineteenth century.

In Australia, the Andaman Islands, Fiji, and New Guinea, descendants of ancient migrants who spread to the region in prehistory include various aboriginal tribes who, today, make up 1% of the world's population (Figure 3.22). They speak 20% of the world's languages in an enormously diverse group called Indo-Pacific, although some linguists dispute the existence of this group. These languages are spoken

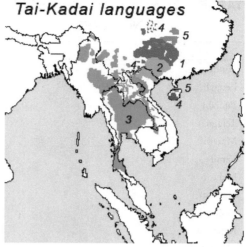

Figure 3.20: Indochina is a very linguistically complex region and includes two families, one of which is Austro-Asiatic (spoken in Vietnam and Cambodia). To add to this complexity, Sanskrit was introduced from India into what is now Thailand, Laos, and Cambodia; the Chinese script was introduced to Vietnam; and following European colonialism, the Roman alphabet is also found widely in the area.

Figure 3.21: In addition to Austro-Asiatic, spoken in Vietnam and Cambodia, the other major language family of Indo-China is Thai-Kradai (or Kadai), which is spoken in Thailand and Laos.

Figure 3.22: The Melanesian languages, which are ancient, are spoken by the dark-skinned aboriginal peoples of Australia, New Guinea, Fiji, and the Andaman Islands, who differ from the lighter-skinned Polynesians. Although the total number of speakers is small, this group is so diverse that it includes one-fifth of the world's languages.

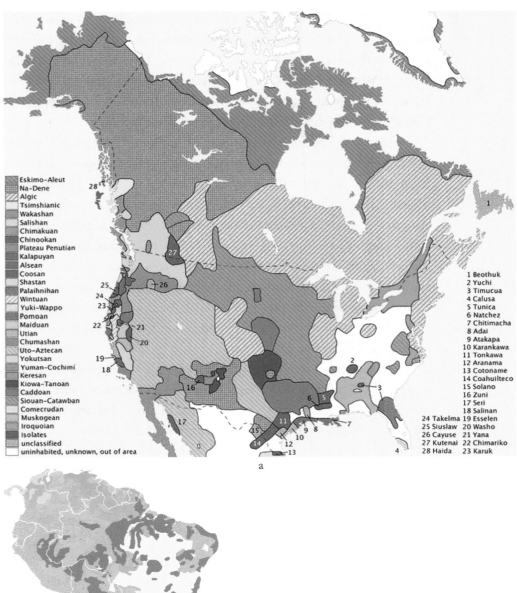

Figure 3.23ab: The New World had an enormous array of different languages before the onset of European colonialism, which annihilated most of them. Most such tongues today are spoken by very small groups of speakers.

by a group called Melanesians, dark-skinned peoples who have inhabited the region for thousands of years and are culturally very different from the Polynesians who settled the Pacific Islands and are of Asian descent. Because it has so many different tongues, each spoken by a tribe in a different valley, New Guinea is a linguist's paradise, with a tiny share of the world's people but one-fifth of its languages.

In the New World, a huge range of indigenous languages existed prior to the mass extermination unleashed by the European conquest: More than a dozen families existed in North America (e.g., Iroquoian, Siouan, Salishan, Athabascan, Mayan) (Figure 3.23a) and in South America (Andean, Chibchan, Macro-Carib) (Figure 3.23b). Although some tongues, such as Mayan, which has numerous variations and dialects, continue to have millions of speakers, many smaller ones are spoken by only a handful of speakers and are in serious danger of disappearing within a few generations. In South America, roughly 10 million people in Peru and Ecuador speak Quechua, the language of the Incas (Chapter 2).

Finally, as mentioned, several isolates (unique languages unrelated to any other) continue to survive. Examples include Basque, spoken in northern Spain and southwestern France (Figure 3.24), and Kartvelian (or Caucasian) tongues, such as Georgian, in the extremely complex linguistic topography of the Caucasus Mountains (Figure 3.25).

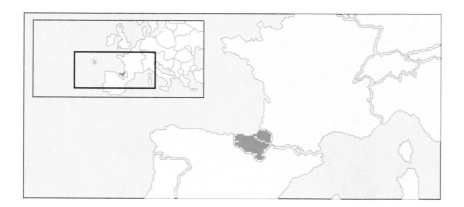

Figure 3.24: Basque, spoken in northern Spain and southern France, is what linguists call an isolate (i.e., a languages unrelated to any other and not belonging to any known family). It is likely the last surviving prehistoric language in Europe from the period before the Indo-European invasions.

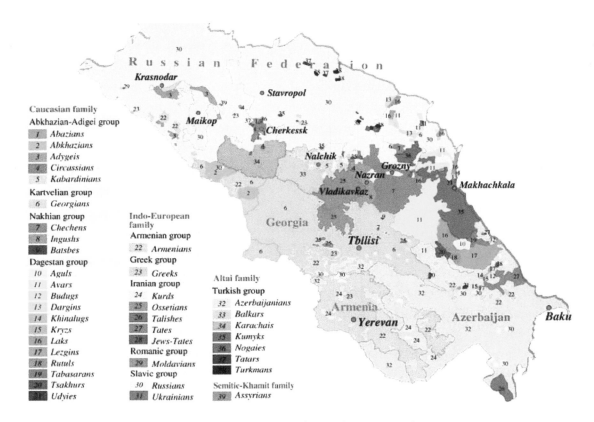

Figure 3.25: The Caucuses are, linguistically, an enormously complex region, and include Kartvelian languages, such as Georgian, as well as Indo-European languages, such as Russian and Armenian, and Turkic languages, such as Azerbaijani.

3.3 LINGUISTIC CONFLICT

Languages are closely tied to a culture's identity and are often a source of pride for nationalists (Chapter 9). For this reason, forces that limit the teaching or learning of languages can provoke enormous resentment and hostility. Moreover, linguistic differences among and within countries are often tied to variations in wealth and power. Linguistic minorities, therefore, often feel that the cultural majority, or at least those with power, fail to take their language seriously and struggle arduously to keep their tongue visible. In short, languages are often a source of political and ethnic conflict.

Many more dialects existed prior to the rise of the nation-state than do today. However, national institutions, including public schools and the media, typically elevated the dialect spoken in the national capital at the expense of other dialects. For example, the Roman version of Italian became dominant in Italy; Parisian French grew at the expense of dialects in the south; Castilian Spanish was imposed on the ethnic minority regions of the country's north. The privileged dialect (usually spoken by elite groups) became the national language, annihilating local differences in vocabulary and pronunciation but integrating diverse groups into a single nation. Starting in the sixteenth century, the wider availability of printed materials supported the spread of the national language. These newly established national languages connected speakers of (for example) huge local varieties of "Englishes," "Spanishes," and "Germans."

Most countries have multiple languages spoken within their borders, and large numbers of people in the world are bilingual or even trilingual. The political status of languages varies widely depending on government policies and the degree to which majority cultures are tolerant of different forms of speech. In countries such as Switzerland, different linguistic groups live together peacefully with no single group seeking to form an independent country. However, in some countries at various times, teaching, or even speaking, a language was against the law. Thus, in Spain under the dictatorship of Francisco Franco, minority languages such as Basque, Catalan, and Galician were essentially forbidden in the face of the dominant Castilian dialect. (after his death in 1975, they flourished). Resurgence of the Basque language, completely unrelated to any other, became a major goal of the Basque nationalist movement, which, at times, resorted to violence. Similarly, in France, speakers of Bretton have long resented French domination of their culture. In such cases, teaching languages in the schools became a non-negotiable demand. Some Scottish nationalists feel similarly about teaching Gaelic. Demands for linguistic visibility are an integral part of secessionist and independence movements.

At other times, linguistic conflicts threaten to tear a country apart. The most famous case is Belgium, where the Walloon speakers of the south, who speak a dialect of French, have long jostled with the Flemish-speaking population of the north, who speak a dialect of Dutch (Figure 3.26). This linguistic difference is accompanied by differences in incomes, unemployment, and access to government power, with the Flemish population enjoying higher incomes and lower unemployment rates than their French-speaking brethren to the south. In Canada, the French-speaking population has frequently rebelled against the Anglophone majority, with some demanding an independent Quebec. The compromise was to adopt two official languages in that country.

Other examples of linguistic conflict abound. In India, people in the south have demonstrated against the widespread use of Hindi, a northern Indian language. In Indonesia, people living on the outer islands resent the Javanese domination of the government and economy. In

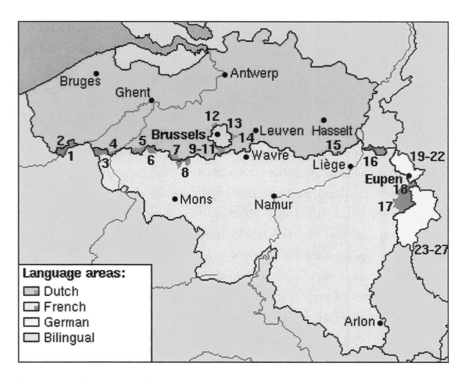

Figure 3.26: Belgium is a famous example of countries where linguistic divisions have been a major source of political conflict.

the United States, an "English only" movement is opposed to the growing Spanish-speaking population.

3.4 LANGUAGE DEATH

Today, roughly 6,500 languages remain in the world. Most, however, have very few speakers and are not written. The total number was much larger in the distant past and has been steadily declining for centuries. As we have seen, the rise of the nation-state often led to a homogenization of cultures and dialects. The movement of people from diverse rural environments to the cities, or rural-to-urban migration, helps to homogenize languages. National school systems often use only one preferred language. In the Americas, disease, genocide, government-run boarding schools in the United States, and cultural assimilation annihilated large numbers of Native American languages. Colonialism and globalization also contributed to the decline, particularly with the dominance of English, across the planet. Many languages today on the brink of vanishing have only a small handful of mainly elderly speakers, such as Manchu in northeastern China.

Today, 96% of the planet speaks one of the top 20 languages (Figure 3.27), and many observers predict that 50% of all languages will disappear within the next century (one every two weeks). The bulk of those disappearing are small tribal tongues spoken by a handful of people in remote areas; as the young learn the language of their nation-state and the elderly pass away the language disappears. This decline represents a crisis in cultural diversity that deprives humanity of the varied ways of viewing the world inherent in having different languages. Languages are not simply a means of communicating; they inform the ways their speakers understand the world, contributing different senses of reality, such as how they depict the passage of time. Languages are windows into the thoughts and worldviews of entire groups of people, and language death is the slow sacrifice of varied ways of interpreting the world, to the detriment of all humanity.

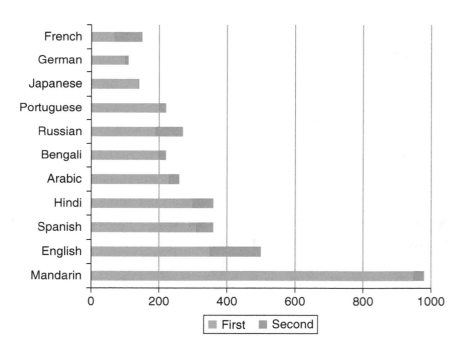

Figure 3.27: Chinese, particularly Mandarin, is the world's most commonly spoken language, followed by English and Spanish (which are essentially tied), then Hindi and Arabic. About 96% of the planet speaks one of the 20 most common languages.

■ KEY TERMS

Accent: the influence of the linguistic background on the speech of a nonnative speaker

Aryans: the early Indo-Europeans who invaded the Indian subcontinent about 1700 BC

Creole: a language spoken by native speakers that is formed by mixing earlier parent languages

Dialect: a regional variation of a language

Isolates: languages with no linguistic relatives, such as Basque

Lingua franca: a language used for purposes of exchange among different cultural groups who may not use it at home, including Swahili in East Africa and, often, English

Pictographic writing: a writing system without an alphabet, in which symbols stand for meanings rather than sounds

Romance languages: languages derived from the Romans (i.e., Latin-based), including Italian, French, Spanish, Catalan, Portuguese, Romanian, and Romansch

Sanskrit: the verbal and written language of the early Indo-Europeans, which became the basis of many Hindu sacred texts

Semitic: pertaining to the Afro-Asiatic language family, including Hebrew and Arabic

■ STUDY QUESTIONS

1. Define language. How is it different from speech?

2. What is a dialect? Give two examples of English dialects.

3. Where in the world are the six largest language families spoken?

4. Who were the Aryans?

5. Summarize the Anglo-Saxon, Viking, and Norman influences on the evolution of English.

6. Which languages are the Romance languages, and what does the name mean?

7. Why is the Celtic family of languages dying?

8. What are the major Slavic languages?

9. What is a lingua franca? Give two examples.

10. How did the Khoisan (!San) become marginalized in southwestern Africa?

11. How can it be claimed that languages as diverse as Finnish, Hungarian, Turkish, and Japanese are related?

12. What is a tonal language?

13. How does a pictographic writing system differ from one using an alphabet?

14. Is Malagasy linguistically related to Hawaiian? If so, how is it related?

15. What are two examples of linguistic conflict?

16. Why is the number of languages worldwide rapidly declining?

BIBLIOGRAPHY

Austin, P. (Ed.). (2008). One Thousand Languages: A Linguistic World Tour. *Berkeley: University of California Press.*

Beine, H., & Nurse, D. (2008). A Linguistic Geography of Africa. *Cambridge: Cambridge University Press.*

Crystal, D. (2014). Language Death. *Cambridge: Canto Classics.*

Desbiens C, & Ruddick, S. (2006). Speaking of geography: Language, power, and the spaces of Anglo-Saxon "hegemony." Environment and Planning D: Society and Space, 24(1), 1–8.

Desforges L., & Jones, R. (2010). Geographies of languages/languages of geography. Social & Cultural Geography, 2(3), 261–264.

Gamkrelidze, T., & Ivanov, V. (1990). The early history of Indo-European languages. Scientific American *(March), 110–116.*

Dryer, Matthew S., & Haspelmath, M. (Eds.). (2005). The World Atlas of Language Structures. *New York: Oxford University Press.*

Laponce, J. A. (1987). Languages and Their Territories. *Toronto: University of Toronto Press.*

Radding, L., & Western, J. (2010). What's in a name? Linguistics, geography and toponyms. Geographical Review, 100, *394–412.*

Renfrew, C. (1989). The origins of Indo-European languages. Scientific American *(October), 106–114.*

Williams, C. (Ed.). (1988). Language in Geographic Context. *Philadelphia: Multilingual Matters.*

Williams, C. (2004). The geography of language. In P. Trudgill (Ed.), Sociolinguistics—An International Handbook of the Science of Language and Society *(pp. 130–145). Berlin: Walter de Gruyter.*

Wright, R. (1991). Quest for the mother tongue. Atlantic Monthly *(April), 39–68.*

IMAGE CREDITS

- Figure 3.1: Mike A Mitchel jr, http://commons.wikimedia.org/wiki/File:Image-Human_Language_Families_(wikicolors).png. Copyright in the Public Domain.
- Figure 3.2: Copyright © Nickshanks (CC BY-SA 3.0) at https://commons.wikimedia.org/wiki/File:World_alphabets_%26_writing_systems.svg.

- Figure 3.3: Adapted from https://commons.wikimedia.org/wiki/File:Indian_subcontinent_CIA.png and http://go.hrw.com/ndNSAPI.nd/gohrw_rls1/pKeywordResults?ST9%20India%20Migration.
- Figure 3.4: Copyright © Kitkatcrazy (CC BY-SA 3.0) at http://commons.wikimedia.org/wiki/File:South_Asian_Language_Families.jpg.
- Figure 3.5: Maicco, http://commons.wikimedia.org/wiki/File:Rectified_Languages_of_Europe_Map.png, ~1. Copyright in the Public Domain.
- Figure 3.6: Copyright © Celtic_Nations (CC BY-SA 2.5) at http://commons.wikimedia.org/wiki/File:Map_of_Celtic_Nations_(alternate).svg.
- Figure 3.7: Copyright © Notuncurious (CC BY-SA 3.0) at http://commons.wikimedia.org/wiki/File:Britain.Anglo.Saxon.homelands.settlements.400.500.jpg.
- Figure 3.8: Copyright © Sulez raz (CC BY-SA 4.0) at https://commons.wikimedia.org/wiki/File:Countries_with_English_as_Official_Language.png.
- Figure 3.9: Spitfire19, http://commons.wikimedia.org/wiki/File:Map-Hispano.png. Copyright in the Public Domain.
- Figure 3.10: Copyright © Jocostinha2011 (CC BY-SA 3.0) at http://commons.wikimedia.org/wiki/File:Portuguese_Speaking_Countries-_Program_ePORTUGUESe_WHO.jpg.
- Figure 3.11: aaker, http://commons.wikimedia.org/wiki/File:New-Map-Francophone_World.PNG. Copyright in the Public Domain.
- Figure 3.12: Copyright © Miskwito (CC BY-SA 3.0) at http://commons.wikimedia.org/wiki/File:Afroasiatic_languages-en.svg.
- Figure 3.13: Copyright © Maximilian Dörrbecker (Chumwa) (CC by 2.5) at http://commons.wikimedia.org/wiki/File:Linguistic_map_of_the_Altaic,_Turkic_and_Uralic_languages_(en).png.
- Figure 3.14: Ulamm, http://commons.wikimedia.org/wiki/File:Niger-Congo_map.png. Copyright in the Public Domain.
- Figure 3.15: Copyright © JorisvS (CC BY-SA 4.0) at http://commons.wikimedia.org/wiki/File:Sino-tibetan_languages_-_branches.png.
- Figure 3.16: Copyright © Shizhao (CC BY-SA 2.5) at http://commons.wikimedia.org/wiki/File:Hebei_Wen%27an_jin_fasheng_dizhen_-_Beijing_you_zhengan.png.
- Figure 3.17: Copyright © Wyunhe (CC BY-SA 3.0) at http://commons.wikimedia.org/wiki/File:Map_of_sinitic_dialect_-_English_version.svg.
- Figure 3.18: Copyright © Christophe Cagé (CC BY-SA 3.0) at http://commons.wikimedia.org/wiki/File:Langues-autronesiennes.png.
- Figure 3.19: Copyright © Dravidianhero (CC BY-SA 3.0) at http://commons.wikimedia.org/wiki/File:Classification_of_Dravidian_languages.png.
- Figure 3.20: Rursus, http://commons.wikimedia.org/wiki/File:Se_asia_lang_map.png. Reprinted with permission.
- Figure 3.21: Amble, http://commons.wikimedia.org/wiki/File:Taikadai-en.svg. Copyright in the Public Domain.
- Figure 3.22: Kahuroa, http://commons.wikimedia.org/wiki/File:Melanesian_Cultural_Area.png. Copyright in the Public Domain.

- Figure 3.23a: Copyright © Ishwar (CC by 2.0) at http://commons.wikimedia.org/wiki/File:Langs_N.Amer.png.
- Figure 3.23b: Davius, http://commons.wikimedia.org/wiki/File:South_America_04.png. Copyright in the Public Domain.
- Figure 3.24: Copyright © Homeruniverse (CC BY-SA 3.0) at http://commons.wikimedia.org/wiki/File:Basque_Country_location_map.png.
- Figure 3.25: Archaeogenetics, "Map Caucasus Languages," https://www.flickr.com/photos/archaeogenetics/2717784687.
- Figure 3.26: Copyright © Gpvos (CC BY-SA 3.0) at http://commons.wikimedia.org/wiki/File:Faciliteitengemeenten_en.png.

Chapter Four

THE GEOGRAPHY OF RELIGIONS

MAIN POINTS

1. Religions are complex systems of cosmology, morality, and metaphysics that powerfully shape people's worldviews and behavior. The three largest religions in the world include Christianity, Islam, and Hinduism. Other major faiths include Confucianism, Daoism, Buddhism, and Shintoism.

2. The geography of religious beliefs also includes those who do not adhere to them. Generally, with the exception of the United States, wealthier countries tend to be more secular, whereas poorer ones tend to be more devout.

For many people, religion is one of the central pillars of everyday life, offering a source of comfort and moral values and a community of fellow believers. Few forces in human history have exerted so powerful an effect on culture. Religions have bred wars and simultaneously inspired great acts of compassion and charity. Some of the world's most beautiful architecture is deeply religious in origin, including mosques, temples, and cathedrals.

Similar to language, religion helps people understand and interpret the world around them. Essentially, a religion consists of a set of beliefs and associated activities that facilitate appreciation of our place in the world. Typically, religions include an account of how the universe began (and sometimes, how it will end), elements of the supernatural and an afterlife, and an explanation of suffering and death. In many instances, religious beliefs generate sets of moral and ethical rules for how people should live (although one does not need to be religious to be ethical). Generally, a religion's core set of beliefs find expression in many forms, including texts, rituals, everyday behavior, diet, symbols, and, of course, landscapes, such as sacred sites and buildings.

4.1 MAJOR WORLD RELIGIONS

A handful of very large religions dominate the world's religious landscapes (Table 4.1). **Ethnic religions** are those generally confined to a specific region and the people who inhabit it, often a given ethnic group (e.g., Hinduism, Judaism), and typically do not aggressively proselytize. **Universalizing religions**, in contrast, seek to establish themselves over as large a region as possible via conversion (e.g.,

TABLE 4.1: WORLD'S LARGEST RELIGIONS, 2015

RELIGION	FOLLOWERS (MILLIONS)	PERCENTAGE OF WORLD POP.
Christianity	2,300	31.2
Islam	1,800	24.1
Hinduism	1,100	15.1
Buddhism	500	6.9
Judaism	14	0.2

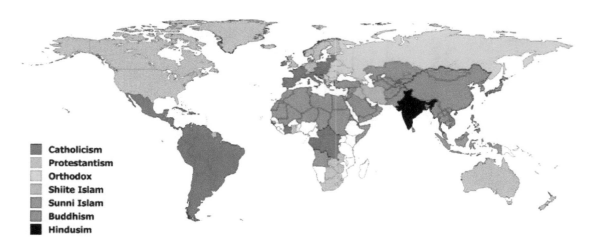

Figure 4.1: The world's major faiths are unevenly distributed. Christianity dominates Europe and Russia, the Americas, and Australia and New Zealand. Islam is found in a huge belt across Northern Africa, the Middle East, Central Asia, and parts of Southeast Asia. Hinduism is primarily confined to India. In East Asia, Buddhism serves as a unifying force.

Christianity, Islam). The world's four largest religions—Christianity, Islam, Hinduism, and Buddhism—emerged from particular hearth areas and spread over vast areas. Today, about a third of the planet is Christian, roughly a quarter is Muslim, 15% are Hindu, 14% have no formal religious beliefs, and the rest are distributed among several smaller faiths. Each religion has a unique geography (Figure 4.1) that results from historical processes that unfolded over long periods of time.

JUDAISM

The first monotheistic religion (worshipping one single god), Judaism originated about 2500 BCE in the Near East, which makes it the oldest Western religion. The Jewish tradition eventually gave rise to Christianity and Islam, making it influential far beyond the number of its adherents. The most sacred text of Judaism is the Torah, the first five books of the Bible: Genesis, Exodus, Leviticus, Numbers, and Deuteronomy. Also important is the Talmud, a set of writings about

Figure 4.2: The ancient Jewish kingdoms of Israel and Judah form the basis for contemporary Zionist claims to Israel as the homeland of the Jews.

Jewish ethics, laws, customs, and history. The ancient Hebrews were one of a series of Semitic peoples (see Chapter 3) in the Middle East who engaged in prolonged periods of wandering and migration, much of which is recounted in the Old Testament of the Bible.

Abraham, for example, led the Hebrews into Egypt around 2000 BC, and Moses supposedly led them out about 800 years later (although some archaeologists dispute this story because of a lack of empirical evidence). In Palestine, the Hebrews established the kingdoms of Israel and Judah (Figure 4.2). These fell victim to a series of foreign conquerors, including the Assyrians, Chaldeans, Greeks, and Romans. Nonetheless, the existence of these early Hebrew kingdoms formed the basis for the Zionist claim to Palestine in the twentieth century and today. As the capital of ancient Israel and the site of the Holy Temple, Jerusalem remains a holy city to Jews, as it is today to Christians and Muslims.

Following the Roman destruction of Jerusalem in 70 CE, the Jews were driven out of their homeland and eventually dispersed throughout Europe and the Middle East. The entire body of Jews living outside Israel, in Europe and elsewhere, is known as the Diaspora. Later, the term diaspora came to refer to any group of people living outside of their homeland. In Europe, for more than 1,500 years, Jews were the only non-Christian minority and often suffered terrible repression. Jews were typically not allowed to own farmland, and thus, were more urbanized than most Christians, often concentrated in skilled craft trades. In cities, they were forced to live in ghettos, which originally referred to the Jewish quarter of a city. They were also frequently the victims of violent anti-Semitism, blamed whenever there was a famine or plague, and often massacred in genocidal pogroms (Figure 4.3). In the late medieval period, many were expelled from western Europe in a wave of Catholic fundamentalism, leading them to be concentrated in Germany, eastern Europe, and Russia when the twentieth century began.

In the late nineteenth century, a unique form of Jewish nationalism, called **Zionism**, began. Zionists advocated, among other things, a revival of the Hebrew language, which had almost dropped into extinction during the long years of the Diaspora. They also began to agitate for an independent Jewish homeland because of the persecution they endured elsewhere. After some dispute, they laid claim to Palestine. This territory became a British colony after the collapse of the Ottoman Empire at the end of World War I (Chapter 5). Not all Jews were or are Zionists; some disagree on political grounds, and some very religious Jews hold that a Jewish state is contrary to divine will. Nonetheless, Zionism became the dominant political face of Judaism in the twentieth century.

Figure 4.3: Anti-Semitic pogroms witnessed countless Jews killed at the hands of Christians.

The Nazi atrocities against the Jews, including the murder of six million as part of Hitler's attempted genocide (Figure 4.4), provided the ultimate evidence for Zionists that they would never be safe or free without a state of their own. Sympathetic to their plight, much of the world agreed with them. After a brief struggle to force Great Britain to give up its Palestine colony, Jewish nationalists founded the state of Israel in 1948. By virtue of Israel's "law of return," all Jews are allowed to become citizens whenever they wish.

Judaism contains significant internal divisions reflecting theological and ideological differences. These differences include those between Sephardic Judaism, concentrated in Spain and in Arab countries, and

Figure 4.4: The Nazis' "final solution" lay in the extermination of six million Jews in concentration camps, which later helped fuel the Zionist search for a Jewish homeland.

Ashkenazi Judaism, located elsewhere in Europe. Ashkenazi Judaism experienced a number of divisions, especially with the eighteenth-century rise of Hasidic Judaism in western Ukraine and the emergence of ultra-Orthodox Judaism initially in Hungary. With the establishment of Israel, Ashkenazi Jews initially played a key role in the early socialist Zionism that dominated the country's politics. Later, waves of Sephardic Jews moved to Israel from Arab countries. They formed the basis for a more conservative wing of Israeli politics, such as the Likud political party and smaller religious ones.

There are, today, perhaps only 20 million Jews in the world, a small number in comparison with the other major world faiths. In addition to Israel, there are significant pockets of Jews in Europe and the United States (Figure 4.5). Jews make up about 2% of the population of the United States, notably in cities such as New York, Miami, and Los Angeles.

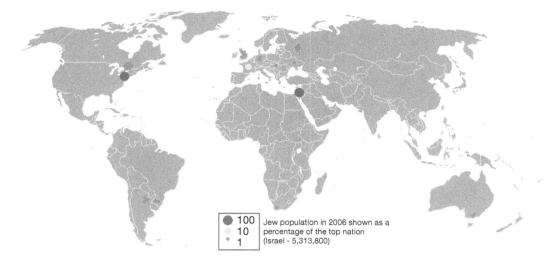

Figure 4.5: Judaism is the world's smallest of the major religions. Most Jews are concentrated in Israel and North America, with other populations in Europe and Russia.

CHRISTIANITY

Christianity began as an offshoot of Judaism when disciples of Jesus of Nazareth, a Jew, accepted that he was the Messiah. (The title "Christ," a Greek term, was added later, and reflects Christianity's mixture of Middle Eastern and Greek origins.) Because of Alexander the Great's influence several centuries earlier, Greek was widely spoken in the Middle East, which is why the New Testament was originally written in Greek, whereas the Old Testament was first written in Hebrew. Wandering throughout the Roman colony of Palestine, Jesus preached that his followers owed their allegiance only to God, not to earthly powers, a stance that ultimately led to his crucifixion in 30 CE. Gradually, Christianity became a universalizing faith separate from Judaism. The religion spread slowly during Jesus's lifetime, but following his death, missionaries carried it, initially, to areas around Jerusalem and, then, through the Mediterranean to Cyprus, Turkey, Greece, and Rome. The first four books of the New Testament, Matthew, Mark, Luke, and John, were written during the late first century CE, roughly three generations after Jesus died.

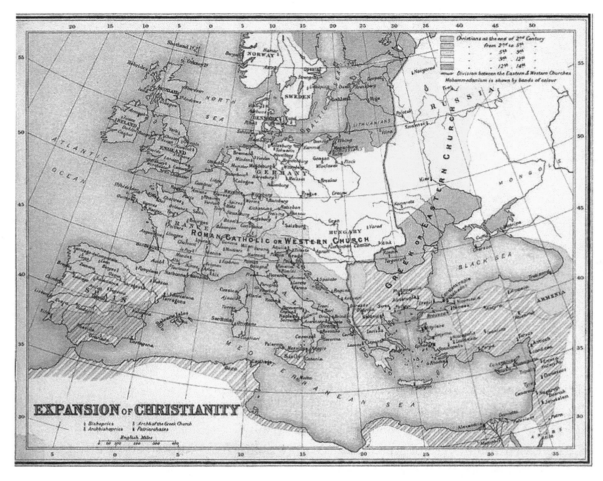

Figure 4.6: Christianity achieved much of its power by diffusing throughout the late Roman Empire, eventually becoming the state religion in the fourth century CE.

In Rome, where Christians were initially persecuted savagely, Christianity gradually took hold among the elites until it became the state religion in the fourth century CE (Figure 4.9). Some have argued that the collapse of the Roman Empire was attributable to the spread of Christianity; Edward Gibbons's famous book *The Decline and Fall of the Roman Empire* blames the religion for the collapse of the virtues that built the empire. Others have held that the religion would never have succeeded had it not captured the Roman state. With the collapse of the empire, Christianity spread slowly into northern Europe

(Figure 4.6), carried by missionaries to northern Europe. The last parts of England became Christianized under Alfred the Great during the struggle against the Vikings in the ninth century. St. Patrick is held to have brought Christianity to Ireland. Eventually, Christianity reached Scandinavia and the Baltic countries, which it did not succeed in converting until 1300 CE.

In medieval Europe, Christianity reigned supreme for more than a millennium from approximately the fifth to the seventeenth centuries (see Chapter 5). Indeed, until the Enlightenment of the seventeenth and eighteenth centuries brought about a secular Western culture, Europe was essentially synonymous with Christendom. The pope, located in Rome, inherited many of the powers and the status of the old Roman emperors, and it was in Rome during the Renaissance that the Vatican became the center of the Roman Catholic faith. The pope was the foremost political authority in Europe from roughly the fifth to the seventeenth centuries and owned vast amounts of land, controlled armies, raised taxes, and could excommunicate kings or those who offered resistance. Beneath him was a vast network of bishops and archbishops that controlled territories in their own right, forming a religious hierarchy that paralleled the secular hierarchy of kings. At the local level, priests played an important role in spreading the faith, often acting as judges, teachers, physicians, and administrators. The church's cathedrals testified to its enormous power and wealth (Figure 4.7).

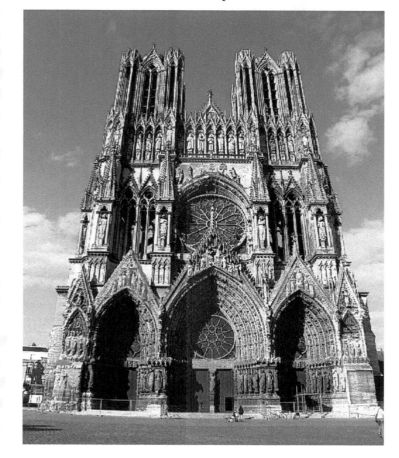

Figure 4.7: The majesty of cathedrals reflected the power of the Catholic Church in Europe between the sixth and sixteenth centuries.

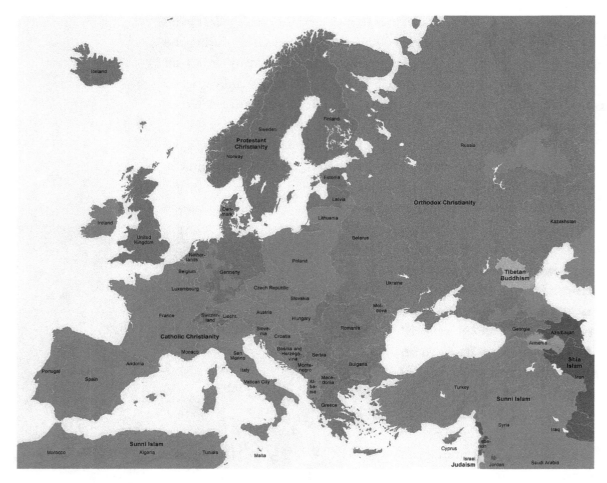

Figure 4.8: Over time, Christianity split into three major groups, each concentrated in a different part of Europe. Catholicism remains dominant in the Iberian Peninsula, France, Ireland, Italy, and a swath of countries in eastern Europe. Northwestern Europe is mostly Protestant. The Orthodox Church is found in southeastern Europe, Ukraine, and Russia, as well as parts of the Middle East.

Gradually, however, tensions emerged between two parts of Christian Europe: the Latin-speaking western region and a Greek-speaking eastern one. This division had characterized the Roman Empire, as well. Essentially, Christians in the east rejected claims by bishops of Rome to have authority over the entire Church. As a result of these differences, a major east–west division between the Roman Catholic (western) and Eastern Orthodox forms of Christianity occurred in 1054 CE, and the line dividing these two remains the most basic religious boundary in contemporary Europe (Figure 4.8). The two versions of Christianity competed for converts throughout the Slavic areas of Europe. The Orthodox branch became, and remains, dominant in Greece, Serbia, Bulgaria, Russia, and parts of Ukraine. There are Orthodox minorities in Lebanon (e.g., Maronites) and Egypt (Coptic Christians, who form about 12% of the population there).

The second great divide in Christianity occurred with the Protestant Reformation of the 1500s, which produced a third major version of Christianity. Started by Martin Luther, who protested corruption in the Catholic Church, the Reformation was an effort to reform Roman Catholic dogma, teachings, and practices and an attempt by some northern European leaders to challenge the wealth of the Catholic Church. Protestants established a number of independent churches, notably the Lutheran in Germany and Scandinavia, Zwinglian and Calvinist in Switzerland, and Anglican in England. In Scotland, the Presbyterians took hold, and in the Netherlands, Protestantism became part of the Dutch struggle for independence from Catholic Spain. As a result, most of northwestern Europe is Protestant today, whereas southern Europe (e.g., Spain, Italy, and France) remains predominantly Catholic. Thus, European Christianity has three major branches, each of which is located in a different part of the continent (Figure 4.8).

Both Catholic and Protestant versions of Christianity spread to other parts of the world over the course of European colonialism and movement overseas (Chapter 5; Figure 4.1). Catholicism, for example, was carried by the Spanish and Portuguese to Latin America and the Philippines, where it remains the dominant faith, and by the French to Quebec and parts of Africa. Today, the vast bulk of Catholics live in the developing world. The British exported various Protestant faiths to their colonies in North America, South Africa, and Australia and New Zealand.

In the United States, these denominations included Anglicans (called Episcopalian in the United States after the Revolutionary War), Presbyterians, Methodist, Baptist, and Lutherans. Today, the United States is roughly 55% Protestant (of various faiths) and 30% Catholic (particularly in areas of Irish, Italian, Polish, or Latin American immigration) (Figure 4.9).

ISLAM

The world's second-largest religion, Islam, also arose in the Middle East. Its emergence in the seventh century CE makes it the youngest of the world's major faiths. It was founded by Muhammad, who was born in Mecca in 570 CE. In 610, he received the word of God from the angel Gabriel at the Ka'aba, the huge black volcanic stone in the Muslim holy city of Mecca.

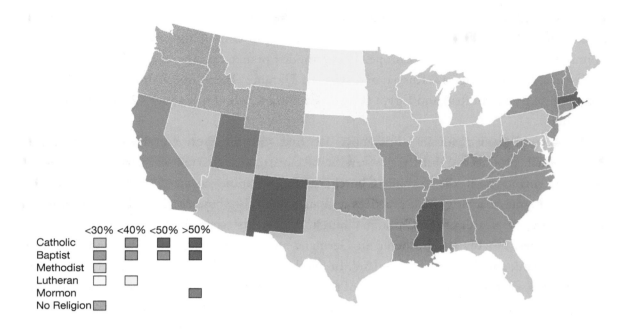

Figure 4.9: The U.S. population is about 55% Protestant and 30% Catholic, with the rest including practitioners of a variety of faiths (e.g., Jews, Muslims, Hindus, Buddhists), as well as nonbelievers. Different forms of Protestant and Catholic are distributed unevenly around the country, reflecting, in large part, the historical geography of immigration to the country.

Although Muhammad was illiterate, he told his followers the angel's message, recorded in the Muslim holy book, the **Quran** (or **Koran**). "**Muslim**" simply means a follower of Islam, regardless of his or her culture or linguistic heritage. Persecuted for his teachings and banished to the City of Yathrib (later renamed Medina), Muhammad eventually returned to Mecca in triumph. He died in 632 CE.

Although many Westerners are not aware of it, Islam shares much of the same biblical narrative of both Christianity and Judaism. All three faiths worship one God, trace their lineage to the prophet Abraham, and originated in the Middle East. Islam includes many beliefs and variations, but all Muslims subscribe to the "Five Pillars of Islam." These are the following:

1. Belief that there is no God but Allah, and Muhammad is the messenger of God. (Allah is Arabic for god).

2. Prayer five times per day facing the holy city of Mecca (Figure 4.10). Prayer times, regularly spaced from sunrise to sunset, were traditionally announced by mullahs calling the faithful from minarets. In addition to daily prayer times, Friday is the Muslim day of prayer, much as Saturday is for Jews and Sunday for Christians.
3. Fasting during the month of **Ramadan**. During Ramadan, the ninth month of the Muslim (lunar) calendar, the faithful do not eat, drink, smoke, or have sex during the hours of daylight (exceptions are made for the seriously ill and small children). Each night, the fast is broken with a feast.
4. The giving of alms, or charity. Islam has a long-standing egalitarian impulse manifested in care for the poor and incapacitated. Giving of alms also indicates that the material world means less than the spiritual one.

Figure 4.10: Devout Muslims answer the call to prayer five times per day, facing Mecca.

5. Pilgrimage, or **haj**, to Mecca once during one's lifetime, if one can afford to do so. Every year, millions of Muslims from around the world visit the Ka'aba in Mecca to reaffirm their faith (Figure 4.11). For many pilgrims from poorer countries, the haj may be the one and only time they embark on an airplane. Saudi Arabia has a special status in Islamic culture as the guardian of the holy cities of Mecca and Medina.

Beyond the Five Pillars, Islam has a rich tradition of law and scholarship. Islamic religious law, or **sharia**, is practiced unevenly among and within Muslim countries. In some Muslim states, the law is secular (for example, Turkey), whereas in others, no separation of church and state exists (for example, Saudi Arabia). The *hadith*, sayings attributed to Muhammad, form another guide to proper behavior. They include, for example, "The ink of the scholar is worth more than the blood of the martyr," and "An hour of study is worth more than a year of prayer." Islam's rich history of learning is exemplified by the

Figure 4.11: Muslims on the haj swirl around the Ka'aba in Mecca, the epicenter of Islam.

status of Arabic universities in the Middle East and in Spain during the height of the Arab empire (seventh to thirteenth centuries). During this time, Arab poetry and literature, science, astronomy, geography, and mathematics were the most sophisticated in the world. Islamic civilizations also preserved the learning of the ancient Greeks and Romans and, eventually, passed this knowledge to Europeans. For those Westerners who simplistically and erroneously equate Islam with violence and terrorism, this history is a sobering corrective.

Finally, Islam exhibits a set of gender norms that are often at variance with Western standards. Many Westerners believe that the Muslim treatment of women renders them second-class citizens, as in many (but not all) Muslim countries women enjoy fewer rights than do men. In Saudi Arabia, for example, women cannot drive. The Quran says a man may marry up to four wives, although in practice, only a few wealthy Muslim men do so. Adult Muslim women tend to live lives segregated from males with the exception of their husbands. Frequently they do not leave the house unaccompanied by a man, and often, they remain covered by garments known variously as a *chador, abaya,* or *burkha* (Figure 4.12). Typically, Muslim women would not want to be seen by men other than their husband. Covering women is seen as a way to protect them from male lust. However, treatment of Muslim women varies considerably, depending on how religious, conservative, or Westernized the country is. In some countries, such as Malaysia or Turkey, women enjoy near equality with men, whereas in more conservative ones, particularly the Arab world and Iran, women tend to hold inferior social roles.

By the time of Muhammad's death in 632 CE, Islam had diffused throughout Arabia, spreading rapidly as Arab armies carried their faith with them. Soon thereafter, however, it began to split. The problem was that Muhammad left no son or specific heir. One group, the party of Ali (*shia Ali)*, believed that Muhammad had designated his cousin and son-in-law Ali to lead the Muslim community, or *umma.* However, Ali died in battle, as did Ali's son Hassan, and the problem of succession grew in political importance, acquiring religious overtones and becoming the basis for an ideological schism. Shiites, as this group came to be known, account for roughly 10% of all Muslims today. Although they form a minority of Muslims worldwide, Shiites make up the majority in three countries (Iran, Iraq, and Bahrain) and form minorities of varying sizes in many other Muslim states, including Kuwait, Saudi Arabia, Yemen, and Pakistan. The other group, the Sunni, which accepted the succession of leaders following Mohammed's death as a historical reality, constitute about 90% of Muslims worldwide today, and with the previously noted exceptions, dominate the vast majority of most Muslim states, ranging from Morocco to Indonesia, including Pakistan and Bangladesh (Figure 4.13). The break between these two groups, formalized in 680 CE, thus revolved around the question of human leadership of a religious community.

For most of the history of Islam, the Shiite–Sunni divide has been peaceful, at least partly because the two groups usually occupied different places. More recently, the division has

Figure 4.12: Among more traditional Muslims, women cover their hair, bodies, or faces, depending on their degree of religiosity. Less traditional women often go uncovered.

acquired geopolitical significance. For example, predominantly Sunni Saudi Arabia is engaged in a protracted rivalry with mostly Shiite Iran for influence in the Middle East. In Iraq, Sunnis and Shiites have engaged in widespread violence against one another since the U.S. invasion that began in 2003; Iraq is mostly Shiite, a group that suffered under the former Iraqi leader Saddam Hussein and has returned the favor upon coming to power following the American invasion in 2003. Tensions, which involve occasional violence, remain high between Sunnis and Shia in many other Muslim states today, including Syria, Lebanon, Bahrain, and Pakistan. Groups such as Al-Qaeda and the Islamic State arose as militant Sunni activists opposed to Shiites everywhere.

From the seventh to the thirteenth centuries, Islam spread far and wide. Indeed, it is one of the world's great proselytizing religions, actively seeking converts. As Arab armies overran the Middle East and North Africa, they carried the Muslim faith with them. In 732 CE, they crossed the straits of Gibraltar (derived from the *gabal al-Tariq*, or "rock of Tariq," the general who led the army there) into Spain, which the Arab Moors ruled for 700 years. During the period of the Arab conquest, Spanish culture acquired many influences from the Arabs. The country served as a conduit for Arab goods and innovations and those they had acquired from Asia to enter into Europe, such as oranges, papermaking, algebra, the compass, and others. The Moors were finally driven out of Spain by Christians in a long process that finally ended in 1492 CE. The Arabs also conquered non-Arab neighbors, such as Persia (now Iran), and carried Islam into Central Asia. At its height from the seventh to the thirteenth centuries, the empire stretched from Spain to Uzbekistan (Figure 4.14).

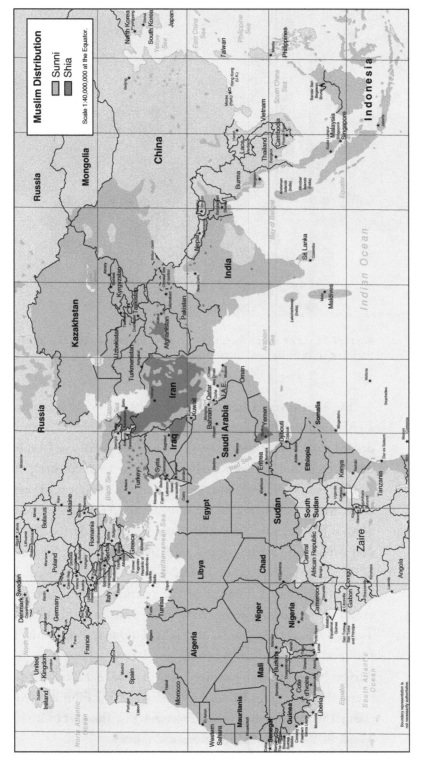

Figure 4.13: Islam is divided into two major branches, Sunni and Shiite. Although 90% of the world's Muslims are Sunnis, Shiites are concentrated in Iran, southern Iraq, and Bahrain, but pockets of Shiites are found elsewhere, in countries such as Lebanon, Afghanistan, and Pakistan.

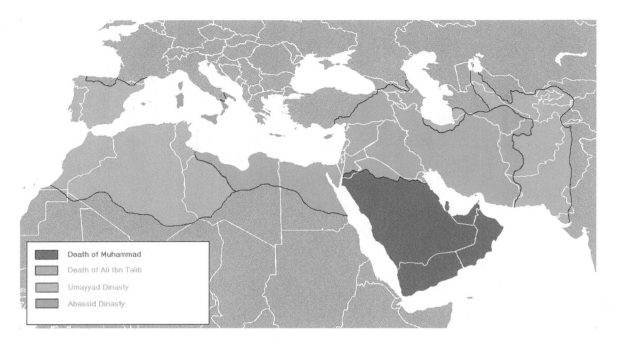

Figure 4.14: At its height from the seventh to thirteenth centuries, the Arab empire stretched from Spain to Central Asia.

The expansion of Islam beyond the Arab world reminds us that not all Muslims are Arabs, and not all Arabs are Muslims. Many Westerners do not realize that these terms are not synonymous. "**Arab**" refers to a distinct culture (or person belonging to that culture); Arabs speak Arabic, eat Arab food, and may identify as members of one of a dozen or so Arabic countries. Most, but not all, Arabs are Muslims; roughly 90% of the Arab world, which includes North Africa and the Arabian Peninsula, follows the Islamic faith. Minorities of Christian Arabs live in Iraq, Syria, Lebanon, Israel, Egypt, and elsewhere. In contrast, "Muslim" means a follower of the Islamic faith; it is a religious, not ethnic or linguistic, term. Non-Arab Muslims include, for example, Turks, Iranians, Afghanis, Kazakhs, Uighurs, Pakistanis, Malaysians, Indonesians, Muslim Bosnians, and many people in sub-Saharan Africa. None of these peoples speak Arabic, eat Arab food, or think of themselves as Arabs. Indeed, of the world's 1.6 billion Muslims, only about 180 million are Arabs (Figure 4.15). The vast bulk of the world's Muslims (90%), therefore, are not Arabic.

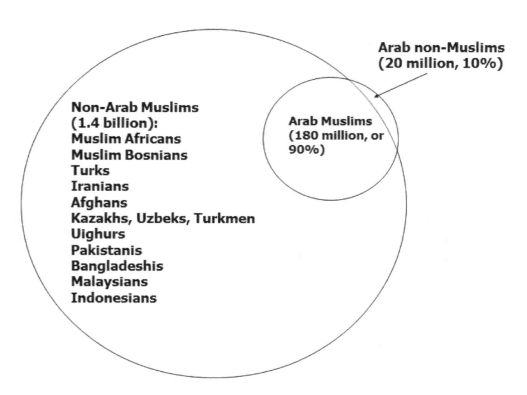

Figure 4.15: The vast majority of the world's Muslims are not Arabs, and a small minority of Arabs are not Muslim.

Islam eventually traveled far beyond the Arab world. The Turks, an Asian people speaking an Altaic language (Chapter 3), migrated to the Anatolian Peninsula in the eighth and ninth centuries and converted to Islam. They founded the great Ottoman Empire and spread the Muslim faith into parts of southeastern Europe, which they occupied for centuries. Albania, for example, is predominantly Muslim, and there is a Muslim plurality in Bosnia (i.e., Muslims make up about 40% of the population, not the majority but the largest religious/ethnic group). In West Africa, Islam was carried over the trans-Saharan trade routes (Chapter 2), and the northern parts of many West African countries today, such as Nigeria, are mostly Muslim. In East Africa, Arab traders brought Islam down the coast from Somalia to Muslim communities as far south as South Africa. Carried by merchants, mercenaries, and missionaries, Islam spread into Central Asia. It remains the dominant faith in countries such as Kazakhstan, Uzbekistan, and Turkmenistan, as well as among the Uighurs, a Turkic minority in western China. Islam was brought into South Asia by the Mughals, becoming the dominant religion of Pakistan and Bangladesh. The Muslim Mughal Empire dominated the Hindu majority in India for centuries until British colonialism arrived in the seventeenth century (Chapter 5). Islam briefly penetrated Indochina, as evidenced by the minority Cham people of Vietnam. Carried by merchants traveling by ship, Islam entered Malaysia and Indonesia (the world's largest Muslim country) in the thirteenth century, never going farther east than the island of Bali but penetrating the southern part of the Philippines (e.g., the Moro people in Mindanao).

As a result of this historical geography, Islam is the world's second-largest religion today, occupying a vast number of countries in Africa, the Middle East, and Asia (Figure 4.1). The proportion of the populations of different countries who follow the Muslim faith varies considerably, although it is close to 100% in some Middle Eastern states. In absolute terms, the largest populations of Muslims reside in Asia (Figure 4.16), notably Indonesia, Pakistan, Bangladesh, and India. Although they make up only about 9% of the population of India, more than 100 million Muslims live there. Tensions between India's Hindus and Muslims at times have erupted into horrific violence.

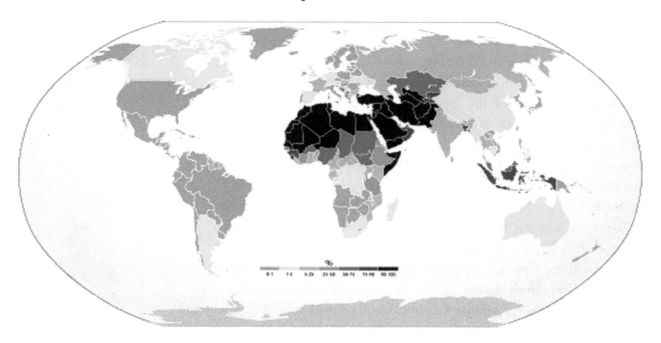

Figure 4.16: The geography of contemporary Islam.

HINDUISM

In South Asia, the world's third-largest faith, Hinduism, emerged in the Indo-Gangetic hearth—a lowland area of northern India that drains into the Indus and Ganges Rivers—around 2000 BCE. Some scholars argue that Hinduism arose around the time of the Indo-European invasion of the subcontinent, as reflected in the fact that Hindu holy scriptures, the Vedic texts, such as the Bhagavad-Gita, are written in Sanskrit, an Indo-European language (Chapter 3). Hinduism spread east down the Ganges River and, then, south through India, eventually dominating the entire region for the next 4,000 years. As it diffused during the last few centuries BCE and much of the first millennium CE through an already diverse cultural landscape, Hinduism absorbed and blended with other local religious beliefs. The diffusion and ongoing evolution of Hinduism means it might be better described as a variety of related religious beliefs, rather than as a homogeneous religion. Indeed, variations of Hinduism are evident even at the scale of local villages, in which different gods assume different levels of significance.

Figure 4.17: Unlike the Western monotheistic faiths, such as Judaism, Christianity, and Islam, Hinduism is polytheistic with thousands of deities. Among the most important is Brahma the Creator.

Hinduism has no authoritative doctrines and only a loosely defined philosophy compared with other religions. It comprises a complex set of beliefs, with many variations and thousands of different deities. Unlike the three Abrahamic religions, it is **polytheistic** (Figure 4.17). Some of the most important gods include Brahma the Creator, who gave rise to the universe; Vishnu the Preserver, who is sometimes also represented as Krishna, the god of love; Shiva the Destroyer; and

Ganesh (or Ganesha), associated with teaching and learning (often represented as having the head of an elephant). Hinduism holds that the individual soul (atma) periodically sheds its earthly body, only to be reincarnated in a new form. One's soul may reappear, therefore, as an animal or another human being, its status reflecting on the good and bad deeds (*karma*) one has done during a previous lifetime. Cows are regarded as holy animals in Hinduism (Figure 4.18) and are never eaten, because they may be the souls of the dead come back to life. The Ganges, the great river of northern India, is regarded as a holy body of water, particularly in cities such as Varanasi, and bathing in it increases the likelihood one's soul will reappear in a more advanced and enlightened form in the next life. However, whereas Westerners regard reincarnation in positive terms as a kind of immortality, among Hindus, reincarnation is a sign of being trapped in the circle of life, or *samsara*, that is, a condemnation to continued suffering. They strive to reach life's blissful end point, **nirvana**, in which pain and suffering finally cease. Nirvana does not directly equate to the Western notion of

Figure 4.18: Cows are regarded as sacred in Hinduism, so Hindus never eat beef. In India, beef slaughter and consumption are performed by Muslims.

heaven; rather, it indicates achieving "nothingness," escaping the endless cycle of earthly suffering.

Hinduism also includes the rigid social order of **castes.** These social strata determine almost everything about a person's life: educational and work opportunities, as well as prospective marriage partners. One is born into a particular caste and cannot escape it; caste standing reflects one's karma in a previous life. Unlike class, a socioeconomic category with some degree of mobility determined largely by economic status, caste is a religious category. Hinduism has a variety of castes, but several major groupings are the most important. At the top of the social order, Brahmins, those closest to Brahma, tend to dominate prestigious jobs, such as teacher, judge, and cook. At the other end of the social spectrum, the *Dalit*, formerly called untouchables, have such low status that they do not even belong to a caste. Traditionally, they were confined to the worst professions, such as digging graves and latrines or caring for lepers. Although the Indian constitution makes discrimination by caste illegal, particularly regarding the untouchables, in practice, they still suffer from widespread maltreatment.

From India, Hinduism spread to nearby regions. For a time, it dominated in Indochina, where a Hindu-Buddhist culture in Cambodia built the great temple complex of Angkor Wat. For this reason, Sanskrit became a popular form of writing in non–Indo-European languages, such as those in Thailand, Cambodia, and Laos. Later, Buddhism displaced Hinduism entirely in Indochina. Merchants also carried Hinduism to Indonesia, where, with Buddhism, it played a major role prior to the arrival of Islam in the thirteenth century. The temple of Borobudur in central Java stands as evidence. Today, the only place Hinduism survives in significant numbers outside of India is the island of Bali, which never converted to Islam. Thus, Hinduism is mostly a religion associated with a single country, India (Figure 4.19).

Similar to many other religions, Hinduism has spawned numerous offshoots. Jainism, developed from the teachings of a sixth-century BCE holy teacher, rejects Hindu rituals but shares many basic tenets of Hinduism, including the belief in reincarnation and *ahimsa* (the ethical doctrine that humans ought to avoid hurting any living creature). A more recent offshoot is Sikhism, a hybrid of Hinduism and Islam that arose about 500 years ago in the Punjab region of India. However, only a small minority practices Sikhism; Punjab remains more than 80% Hindu.

BUDDHISM

Also founded in the Indo-Gangetic hearth, Buddhism arose as an offshoot of Hinduism, much as Christianity emerged from Judaism. The central figure of Buddhism was Prince Siddhartha

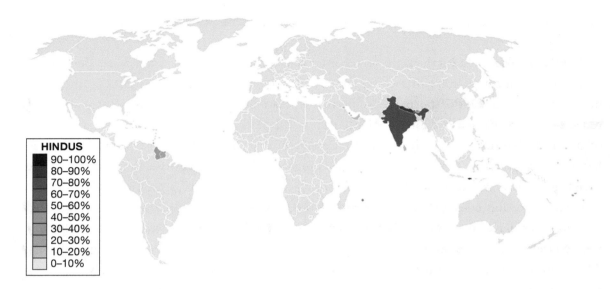

Figure 4.19: Global distribution of Hindus

Gautama, born in 644 BCE near present-day Nepal. Although raised in luxury, Siddhartha traveled widely as a young man and was appalled by the pain and suffering he witnessed. After years of study and meditation, he achieved enlightenment, or divine inspiration, becoming the Buddha ("Enlightened One") (Figure 4.20).

Buddhism incorporated certain elements of Hinduism and rejected others. For example, from Hinduism, Buddhists retained the notion of karma, reincarnation, and nirvana, preaching that the soul is endlessly recycled in the wheel of suffering (samsara) until it achieves enlightenment. However, Buddhism rejected the multiple gods of Hinduism; although some forms of Buddhism have metaphysical deities (such as Bodhisattvas, or enlightened beings who postpone their entry into nirvana to help others), it is largely a secular philosophy of life, rather than a true religion in the Western sense. Most forms of Buddhism lack a god or supernatural deity, or a sense of afterlife, which are common in Western faiths. Buddhism also became consciously egalitarian, rejecting the Hindu order of castes and preaching equality among all human beings. Buddhism centers on the Four Noble Truths that Siddhartha preached. These are the following:

1. There is suffering in the world.
2. Suffering has distinct causes.
3. Suffering can cease, that is, it is not inevitable.
4. There is a path out of suffering, meaning that it can be overcome through correct thought and action.

The core of the Buddhist worldview is that desire—wanting things to be different, rather than accepting them as they are—is the cause of suffering. To escape suffering, Buddhists follow the eightfold path of proper speech, understanding, livelihood, aspiration, behavior, effort, absorption of good influences, and mindfulness. Buddhism is largely about the art of selflessness, of escaping the ego that traps us in a world of desire and pain. To escape the self is, therefore, to escape want and frustration. This is no easy task, and it takes enlightenment to accept the world without wanting to change it, to realize that everything is temporary and ephemeral. The most important Buddhist virtues are compassion, empathy, and kindness. For example, many Buddhists practice the doctrine of **ahimsa,** or nonviolence toward all living things. For many Buddhists, this includes vegetarianism.

Figure 4.20: Upon achieving enlightenment, Siddhartha Gautama became the Buddha ("Enlightened One") and spread a doctrine that has touched the lives of billions of people for the past 2,500 years.

Buddhism also advocates an emphasis on living life in the present moment rather than regretting the past or planning for the future. Buddha emphasized the transitory nature of being (i.e., that nothing lasts forever); to miss the present moment is to miss one's appointment with life. For example, where, exactly, is the past? Where is the future? From a Buddhist perspective, past and future only exist in relation to the present. Thus, if Western culture typically views time as a line stretching from past to present to future, Buddhists see time is a point. Achieving such insights takes great work, and Buddhists strive to gain mindfulness by meditation: sitting quietly and emptying the mind of all thought (Figure 4.21).

After Siddhartha Gautama's death, Buddhism was spread by missionary monks into other parts of India, briefly becoming the state religion before being displaced by Hinduism. However, although Buddhism is almost extinct in India today, it became widespread throughout much of East and Southeast Asia. Indeed, Buddhism is as close as East and Southeast Asia have ever come to exhibiting a pan-Asian philosophy that crosses national, ethnic, and linguistic boundaries. Similar to Hinduism, Buddhism has numerous local variations that result from blending with preexisting local beliefs.

Figure 4.21: A Buddhist monk meditating. Meditation, or practicing mindfulness, is key to the Buddhist path to enlightenment. By focusing on the here and now, Buddhists seek to escape the prison of desire and ego. Emptying the mind of all thought sounds easy but isn't. Try it sometime.

Three major strains of Buddhism emerged over time, each carried to different parts of the Asian continent (Figure 4.22). Mahayana ("Greater Vehicle") Buddhism traveled via the Silk Road routes through Afghanistan and Central Asia and, then, into China, where Buddhist devotees decorated caves and erected many famous temples and monuments. In China, where Buddhism arrived in the first century CE, Mahayana Buddhism met initial opposition from people of indigenous Chinese faiths, such as Confucianism. However, Mahayana Buddhism coexisted and mixed with these earlier beliefs. Today, despite decades of hostility from the government of China, Buddhism continues to thrive there. From China, Mahayana Buddhism traveled east into Korea and Japan and south into Vietnam during a long period of Chinese occupation of that country from the first to the tenth centuries. In Japan, Mahayana Buddhism entered from China in the sixth century as Zen Buddhism, whose ascetic simplicity and emphasis on self-control were popular among the Japanese aristocracy.

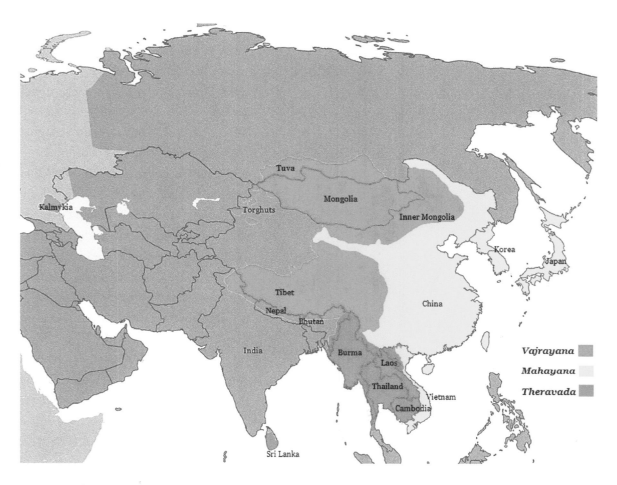

Figure 4.22: Three strains of Buddhism were carried out of India. Mahayana Buddhism was carried through Afghanistan and, via the Silk Road, into China, from which it spread to Korea, Japan, and Vietnam. Vajrayana Buddhism leaped over the Himalaya Mountains into Tibet and, later, came to include the Mongols (some regard this branch as a variant of Mahayana). Theravada Buddhism, starting in Sri Lanka, was carried to Burma, Thailand, Laos, and Cambodia.

The second strain, Vajrayana Buddhism, traveled from India over the Himalaya Mountains and became the state religion of Tibet, where it has played an enormously influential role for centuries (Figure 4.23). Tibet was long an essentially Buddhist religious state ruled by lamas, including the Dalai Lama, who was regarded as the reincarnation of the Buddha himself. Tibetan culture exhibited exceptionally rich Buddhist influences in its art, music, and architecture (such as the Dalai Lama's Potola Palace in Lhasa, the Tibetan capital). Tibetan Buddhists also created *mandalas*, or religious designs made in colorful sand that are swept away when completed

Figure 4.23: Tibetan Buddhism plays an enormous role in the cultural life of that nation.

to illustrate the ephemerality of life. Buddhist monks and monasteries played a hugely influential role in structuring daily life and beliefs. In the thirteenth century, when the Mongols conquered and occupied China, they adopted Vajrayana Buddhism, as well. In 1959, the Chinese government invaded and occupied Tibet. They tried to eradicate much of Buddhism's influence there, killing countless monks and destroying thousands of temples and monasteries. The Dalai Lama fled to India, and China occupies Tibet to this day.

The third strain of Buddhism, the conservative Theravada ("lesser vehicle") version, is the oldest and does not seek to reconcile with other belief systems. It began on the island of Sri Lanka and moved into Southeast Asia, including Myanmar (formerly Burma), Thailand, Cambodia, and Laos. Buddhism has had a particularly profound influence on these cultures, including the landscapes, social structures, and everyday life (Figure 4.24).

DAOISM

In China, the oldest indigenous faith is Daoism (or Taoism); unlike Buddhism, which was, at first, regarded as a foreign import, Daoism originated in China. Daoism incorporates nature-focused beliefs that probably originated in the Chinese Neolithic period (Chapter 2), but the philosopher Lao Tzu codified it and made it explicit in his book, the *Dao De Ching* (the *Book of Changes*). Like Buddhism or Confucianism, Daoism is not so much an explicit explanation of the universe (with deities and such) as it is a set of beliefs about how people should act.

Figure 4.24: A Buddhist temple in Thailand reflects the profound importance of that faith to Thai culture and history.

At its core lies the notion of the Dao, the way of nature or of the universe itself. The Dao is not a deity, but rather a mysterious, pervasive force that supposedly gives the world both order and disorder. Those who practice Daoism seek to act in accordance with the Dao. The Dao cannot be understood through direct observation or experimentation, but must be appreciated through contemplation and meditation. In the *Dao De Ching*, for example, Lao Tzu writes:

> He who follows the Dao
> is at one with the Dao.
> He who is virtuous
> experiences virtue.
> He who loses the way is lost.
> When you are at one with the Dao,
> the Dao welcomes you.

In its mysterious quality and emphasis on harmony and meditation, Daoism resembles Buddhism in some respects. Indeed, the similarities between the two faiths meant that Daoism helped to pave the way for Buddhism's diffusion into China.

Daoism is essentially an animistic, or nature-worshipping, religion, seeing the Dao in plants, animals, stones, rivers, and mountains. The Dao is manifested in the flow of **chi**, or life force, through people and other living beings. The Daoist emphasis on chi appears, for example, in practices such as *tai chi*, a kind of slow dance that has been likened to "moving meditation" (Figure 4.25) and is widely practiced today in much of China and some neighboring countries. Likewise, the Daoist practice of acupuncture, using very fine needles to control pain, is based on the view that the body exhibits a distinct flow of chi through various pressure points.

Perhaps the most famous expression of Daoism is its symbolism of *yin* and *yang*, complementary terms that represent opposite, interwoven sides of the Dao. The representation of yin and yang as two forces united in a circle, each with a dot carrying elements of the other, is world renowned (Figure 4.26). Yin and yang represent equally important facets of the Dao. Thus, classically, yin represents that which is negative, evil, darkness, the moon, the Earth, weakness, femininity, and emotions, whereas yang is associated with that which is positive,

Figure 4.25: *Tai chi*, sometimes called moving meditation, is part of the beliefs and practices of Daoism.

goodness, light, the sun, heaven, strength, masculinity, and logic. The balance of yin and yang creates stability and harmony in the universe; if one side dominates the other, instability and disorder follow.

Although it intermixed with, and was often overshadowed by, Buddhism in China, Daoism retains followers there to this day. There are numerous Daoist temples, for example, where people may go to pray to their ancestors and for good luck.

Figure 4.26: The symbols for yin and yang reflect Daoism's emphasis on the unity of opposites. Nowhere can the circle be divided as to show only one color.

CONFUCIANISM

The third major ideology (not really a faith) in China, Confucianism, has had a tremendous influence on the cultures of much of East Asia. This set of teachings (not really a religion at all in the Western sense) began with the renowned philosopher Kung Fu Tzu (known to Westerners as Confucius), who lived and taught in the sixth century BC (Figure 4.27). Confucius himself never wrote anything; rather, his sayings were collected later by his disciple, Meng Tzu (Latinized as Mencius), in a volume called the *Analects of Confucius*.

Rather than the afterlife, Confucianism concerns itself with a pragmatic understanding of how to behave in this world. Essentially, it forms a guide for proper behavior, ranging from the individual to the administration of the kingdom of China. Confucianism, thus, places enormous emphasis on what it views as correct ethics and etiquette; if each person knows his or her place in the world and abides by the rules associated with it, harmony and peace will follow.

Confucianism is a highly conservative doctrine that aims to preserve the land's established order. It views the world as stratified between five tiers of superior and inferiors. At the lowest level is the relation between parents and children; parents, as superiors, should show their children kindness and understanding; children, in return, must honor their parents, alive or dead, with obedience, devotion, and respect. Thus, Confucianism advocates the

Figure 4.27: Kung Fu Tzu, or Confucius, was an enormously influential philosopher whose ideas shaped East Asian societies for millennia.

notion of filial piety, in which children are expected to take care of their elders in their waning days. The second tier consists of relations between older and younger people: Confucianism places a great value on age, and the elderly show the young their wisdom in return for care and respect. Unlike in Western cultures, in which people tend to value youth, in Confucianism, age evokes the most respect.

The third tier concerns the relation between males and females, particularly between husbands and wives. Confucianism holds that men are superior to women, who must show complete devotion and obedience. Confucius said, for example, that every woman has three masters: first, her father; then her husband; and finally, her sons. This view also underpinned the classical Chinese practice of foot binding, in which the feet of wealthy girls were tightly wrapped to the point of disfiguring them (Figure 4.28). This made these women unable to perform farm labor and indicated to the world that the family could afford to support them. Women with bound feet had difficulty walking and served as little more than adornments. Small feet were also viewed as sexually attractive.

The fourth tier applies to the relations between the emperor and his subjects: The emperor is regarded as the nation's elder parent and should rule wisely in return for obedience from the masses (Figure 4.29). In classical China, the emperor had complete power, but also complete responsibility, for managing the affairs of state. If an emperor ruled well, it meant that he enjoyed the **"mandate of heaven"** (i.e., the gods looked well upon him), and thus, there was peace and prosperity. If things went badly, however, such as with drought, famine, plague, or invasion, it indicated that the gods were displeased and that the emperor had lost the mandate of heaven. Chinese history is replete with famous peasant uprisings against emperors who were believed to have lost divine favor.

Figure 4.28: Foot binding was a common practice among wealthy women in Confucian China. Small feet symbolized that the woman did not work in the fields, and reveal that even sexuality is socially constructed.

The fifth and last tier of Confucianism reflected China's relations with its neighbors. In this view, China was *Zhong Guo*, the

Middle Kingdom, and the emperor's throne was the exact center of the universe. China played the father figure or elder brother to smaller countries on its periphery, such as Mongolia, Korea, Japan, Vietnam, and Tibet. Rulers of other lands were expected to accept China's superiority and pay tribute to the emperor. When the Europeans first began to enter China in the sixteenth century, the Chinese expected them to abide by these rules.

Note that, in each of these cases, similar rules apply: Superiors should exhibit wisdom, patience, and kindness; inferiors should accept without questioning and show deference and obedience.

Confucianism became institutionalized as the state ideology of China in the third century BC during the Ch'in dynasty and has played a remarkably influential role ever since. Much of Chinese statecraft, including foreign policy and the politics of the imperial court, revolved around Confucian principles. Only with the Communist revolution of 1949 was Confucianism finally displaced as the dominant belief system of government. The ideology was exported to Korea and Japan, along with Buddhism, and played an important role there, too. Confucianism also shaped daily life in important ways: its disdain for manual labor and stress on intellectual work, for example, helped to promote an emphasis on reading and studying that continues to shape life today for hundreds of millions of Asian people. Its sexism has encouraged countless couples to seek abortions of female fetuses, especially in China, where most couples were limited to one child until very recently (Chapter 8). Some observers have asserted that the docility of the labor force (e.g., infrequent strikes) in countries

Figure 4.29: In the Confucian system of governance, the rule of the emperor of China was legitimized by the Mandate of Heaven, with which he enjoyed absolute authority, but also absolute responsibility, to govern.

such as Japan, Korea, Taiwan, and Singapore is in part due to deep-seated Confucian principles.

SHINTOISM

Shintoism arose in Japan's early history as an indigenous religious system. This ideology has long been fused with Japanese nationalism, that is, its sense of itself as a nation superior to other states. Shinto practitioners, for example, trace Japan's creation to the sun goddess Amaterasu, whose tears formed the principal islands of Honshu, Hokkaido, Shikoku, and Kyushu. Nicknamed the "land of the rising sun," Japan adopted the red circle on its national flag.

Shintoism is largely an animistic faith, finding meaning in nature. Thus Shintoists revere Mt. Fuji as a sacred location (Figure 4.30), and

Figure 4.30: Mt. Fuji, Japan, is a sacred place in Shinto ideology.

Japanese culture has long been infused with a deep appreciation for the beauty of the natural world (Figure 4.31). In this context, death and beauty were often fused: In feudal Japan, for example, the cherry blossoms, or *yayesakura*, were celebrated as being the most beautiful just as they fell from the tree. Japanese samurai died honorably by committing the act of *sepuku*, or *hara-kiri*, ritual self-disembowelment: to die an honorable death was considered beautiful. (This notion was later adopted by the *kamikaze* pilots of World War II as they crashed their planes into American battleships).

Central to Shinto mythology is the role of the emperor, who is regarded as a direct descendent of Amaterasu. Unlike China, in which Confucianism permitted uprisings against emperors who had lost the mandate of heaven, in Japan the emperor was held to be a deity. Devotion to the emperor was considered a sacred duty. Thus, Japanese history is characterized by one very long line of rule by the royal family; secular rulers may come and go, but the imperial throne remained sacrosanct. The Japanese population had never even heard, let alone seen, the emperor, until Emperor Hirohito's surrender at the end of World War II. Today in Japan, the imperial throne has

Figure 4.31: A Shinto shrine in Japan.

lost its sacred overtones, but the tradition of obedience and devotion remains important to Japanese culture and society.

Shintoism coexisted in Japan with both Mahayana Buddhism (notably Zen) and Confucianism, which were imported from China. Indeed, the Japanese landscape contains large numbers of both Buddhist and Shinto temples and shrines. The fact that it is possible to belong to all three faiths simultaneously reflects their ability to blend and their lack of exclusivity. Even today, Japanese companies will, at times, ask Shinto priests to pray for a tree that is to be cut down or for sales of a new model of automobile.

4.2 GEOGRAPHIES OF SECULARISM

The study of the geographical dimensions of religious beliefs also includes those who do not adhere to any religion. Ever since the Enlightenment, religion in the West has suffered a long, slow, and steady retreat. On the heels of ascendant capitalism, European culture witnessed the separation of church and state and the rise of an alternative, secular set of institutions. The printing press, rising literacy levels, and the scientific breakthroughs of the Enlightenment all contributed mightily to undermining the legitimacy of religion, the intellectual monopoly of the church, and the rise of secular alternatives, such as universities. Organized religion, which held a monopoly over cultural, legal, and political power in Europe for a millennium during the medieval era, has seen a gradual decline in the number of believers, church attendance, its public visibility, and the depth of religiosity in everyday life. As Christianity has incrementally lost much of its ideological power and political clout, it has opened a space for secular and scientific thought. From the Copernican revolution to Darwin and the flourishing of secular social sciences and humanities in the twentieth century and today, alternatives to religious worldviews abound in power and popularity.

The most famous explanation of the growth of secularism originates in the works of famed sociologist Max Weber. Weber's argument began with his observations that the Protestant ethic, which he held up as the motivating cause of capitalism, enticed believers into working hard, delaying gratification, and accumulating savings (i.e., capital) as a sign of God's grace (i.e., the likelihood of entering heaven). To succeed as incipient capitalists, adherents had to become increasingly "rational" in outlook and behavior, adopting logical, even scientific norms. Rational secularization was accelerated by the growth of the market, with its emphasis on profit maximization, and through the impersonal bureaucratization of politics and collective decision making. In a secular capitalist culture the commodity displaces God as the object of holy attention, a process Weber called the "disenchantment of the world." The very rise of capitalism, he maintained, gradually lowered an "iron cage" of rationalism over the very culture that created it, squeezing religion into the domain of the irrational. The commodification of daily life and the adoption of explicit, impersonal rules of bureaucratic behavior entailed the steady decline in the appeal of religious authority.

Atheists are those who deny the existence of God. In this respect, they differ from agnostics, who hold open the possibility of a deity. For most people, the religions of others are simply not true, and they tend to believe that their specific theological interpretation alone has a monopoly over divine truth; atheists simply extend this line of thought to include all metaphysical beliefs. Atheists should not be confused with misotheists, who actively hate God; to hate God, one must believe that he/she/it exists. Atheists often believe that religions have adverse social effects, including the subordination of women, hostility to science, interreligious conflicts, and the naturalization of inequality.

Although there were occasional atheists from the Renaissance to the nineteenth century (e.g., Nietzsche, Marx, Freud), atheism is largely a product of the twentieth century. Twentieth-century atheism is frequently discussed in terms of the triumph of Communism in the Soviet Union, China, eastern Europe, Mongolia, North Korea, Vietnam, and Cuba. State-enforced atheism is quite a different animal from its intellectual counterpart. Viewing religion as bourgeois decadence and a challenge to the power and authority of the secular state, Communists in the Union of Soviet Socialist Republics (USSR) and China attempted the most wholesale liquidation of religious faith in world history. For seven decades, atheism reigned as state ideology in the Soviet Union and client states, imposing incalculable human suffering and loss of cultural heritage. Under Soviet leader Joseph Stalin, from the 1930s to the 1950s, the atrocities against people of faith are too numerous to recall: religious instruction was prohibited; numerous places of worship were closed; tens of thousands of priests and nuns were imprisoned, tortured, or executed; entire religious communities were terrorized, moved to distant regions, and/or annihilated; schools spewed forth antireligious propaganda; and the populace was virtually forced to give up its religious beliefs under the threat of severe sanctions. Likewise, in China, state-sponsored atheism was manifested in the suppression of its Buddhist heritage, particularly in Tibet, including the murder of more than one million people and the ransacking or obliteration of nearly all of its 6,000 monasteries during the Great Leap Forward and the Cultural Revolution. The failure of state-sponsored atheism in the communist world, as reflected in the recent resurgence of the Orthodox Church in Russia and Buddhism in China, points to the cruelty of Communism's rulers more than it does to the intellectual legitimacy (or lack thereof) of atheism as a worldview. It is as inaccurate to burden all atheists with the crimes of the USSR and China as it is to paint all Christians as anti-Semites or bloodthirsty crusaders:

Today, roughly one billion people, or 15% of the planet, may accurately be described as nonreligious, although not all of them are atheists (Figure 4.32). The geography of atheism reflects several important world historical forces, including the lasting influence of Communist state-enforced atheism in North Korea (56%), Russia (26%), and China (40%), where it reflects decades of enforcement by politically repressive regimes. In addition, global geographies of wealth and poverty are also important here; countries with severe social problems are closely linked to high degrees of religiosity. Thus secularism is least present in most of Africa, South Asia, Indonesia, and the poorest parts of South America, as societies plagued by chronic insecurity, hunger, unemployment, poverty, and disease are the most religious on Earth.

Europe, with an historical legacy of social democracy and the enduring legacy of traumatic religious strife, has essentially become a secular continent, in marked contrast to the United States, which remains the world's only industrial yet religious country. In western Europe, Christianity has faced a steady diminution of its power, popularity, and respectability since the Enlightenment. European churches, supported by their governments and facing little competition, have become often incapable of attracting a widespread audience. As a

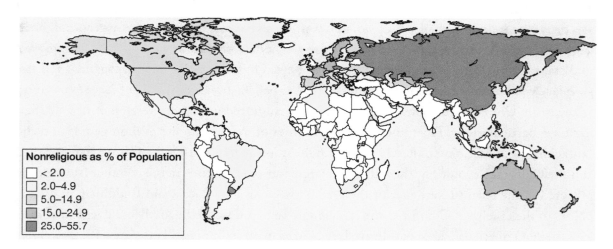

Figure 4.32: In large parts of the world, nonreligious people, including atheists and agnostics, make up a substantial share of the population.

result, European politics generally lack the religious dimensions found in the United States, the latter of which includes an aggressive evangelical movement.

In contrast, religion is deeply woven into American culture and politics. The United States is, perhaps, the most religiously diverse society in the world. In contrast to Europe, the constitutionally enshrined separation of church and state in the United States created a relatively free arena in which different faiths compete for adherents. Thus, the country is not only religiously diverse, it exhibits high levels of religiosity. Roughly 90% of Americans say they believe in God, 44% of Americans attend church weekly, 9 out of 10 Americans say they pray at least some of the time, and 58% claim they pray daily. However, even in the United States, organized religion is waning. Weekly church attendance, for example, has declined gradually. Since 1990, the percentage of Americans who do not identify with any religion has doubled to 15%, and the proportion self-identifying as atheists rose from 4% in 2003 to 10% in 2008. Only 53% of Americans born after 1981 say they believe in God.

■ KEY TERMS

Ahimsa: the Buddhist doctrine of nonviolence toward all living things

Arab: a person who speaks Arabic and practices Arab culture

Castes: groups that form the basis of the religious order of Hinduism

Chi: the Daoist notion of a life force that flows through the universe, including people

Dalit: the Hindu untouchables, so low that they do not even have a caste

Ethnic religions: those practiced predominantly by one ethnic group

Haj: Arab term for a pilgrimage to Mecca

Mandate of heaven: the Confucian notion that the emperor rules on behalf of the gods

Muslim: a person who belongs to the Islamic faith

Nirvana: Hindu and Buddhist term for the end of existence and escape from desire and suffering

Polytheistic: religions that follow more than one god

Quran or Koran: the holy book of Muslims reflecting the word of God

Ramadan: the ninth month of the Muslim calendar devoted to daylight prayer and fasting

Sharia: a body of Muslim law derived from Koranic principles

Universalizing religions: those that actively proselytize and seek converts

Zionism: the political ideology that advocates for a Jewish homeland, i.e., Israel

■ STUDY QUESTIONS

1. What is a religion and what social functions do religions perform?
2. Where in the world is each of the six largest religions found?
3. What was the first monotheistic religion?
4. What was the Diaspora?
5. What is the Torah? The Talmud?
6. What is Zionism?
7. Who are the Ashkenazim and Sephardim? Describe their differences.
8. In what languages were the Old and New Testaments written?
9. How did Christianity triumph in the Roman Empire and, later, in the rest of Europe?
10. Where is the epicenter of Catholicism located? Why?
11. Where in Europe are Catholicism, Protestantism, and Orthodox Christianity to be found?
12. What are the three "Abrahamic" religions?
13. What is the Ka'aba?
14. What is the Quran? Who wrote it?
15. What are the two holiest cities of Islam?
16. What are the Five Pillars of Islam?
17. What is sharia?
18. Who are the Shiites and Sunnis? What is their relative proportion of Muslims? Where are they found?
19. What is the difference between an Arab and a Muslim? How do they overlap?
20. What is Sanskrit, and what is its significance in regard to Hinduism?
21. What are the four most important gods of Hinduism?
22. What are karma, reincarnation, and nirvana?
23. What is a caste in Hinduism?
24. Who was Siddhartha Gautama?
25. How does Buddhism resemble and differ from Hinduism?
26. What are the Four Noble Truths?

27. What is the point of meditation in Buddhism?

28. What are the three major forms of Buddhism, and where are they located?

29. What and where is Zen Buddhism?

30. Who is the Dalai Lama?

31. Who was Lao Tzu, and what did he write?

32. What is the Dao? Yin and yang?

33. Who was Kung Fu Tzu?

34. Describe the Confucian stratification based on familial relations, age, gender, imperial status, and China's relations to its neighbors.

35. What and where is Shintoism?

36. What are three current examples of conflicts between Christianity and Islam?

37. Give three examples of sacred spaces from different religions.

38. How are globalization and religious fundamentalism intertwined?

39. Give an example of religious fundamentalism as currently found in Christianity, Islam, Judaism, and Hinduism.

40. What forces explain the global distribution of secular societies, that is, populations that do not adhere to religion?

41. Why is Europe a relatively secular region of the world?

42. What explains the fact that the United States is the world's only wealthy yet religious country?

BIBLIOGRAPHY

Bruce, S. (2002). God Is Dead: Secularization in the West. *Oxford: Blackwell.*

Dawkins, R. (2006). The God Felusion. *New York: Houghton Mifflin.*

Eck, D. (2001). A New Religious America: The World's Most Religiously Diverse Nation. *New York: HarperCollins.*

Freeman, C. (2002). The Closing of the Western Mind: The Rise of Faith and the Fall of Reason. *New York: Vintage Books.*

Harris, S. (2006). Letter to a Christian Nation. *New York: Vintage Books.*

Hervieu-Léger, D. (2002). Space and religion: New approaches to religious spatiality in modernity. International Journal of Urban and Regional Research, 26, *99–105.*

Hitchens, C. (2007). God Is Not Great: How Religion Poisons Everything. *New York: Hatchette Book Group.*

Hout, G., & Fischer, C. (2002). Why More Americans Have No Religious Preferences: Politics and Generations. American Sociological Review, 67, *165–190.*

Kong, L. (2010). Global shifts, theoretical shifts: Changing geographies of religion. Progress in Human Geography, 34*(6),* 755–776.

Lewis, B. (2003). What Went Wrong? The Clash Between Islam and Modernity in the Middle East. *New York: HarperCollins.*

Stark, R., & Finke, R. (2000). Acts of Faith: Explaining the Human Side of Religion. *Berkeley: University of California Press*

Stump, R. (2000). Boundaries of Faith: Geographical Perspectives on Religious Fundamentalism. *Summit, PA: Rowman and Littlefield.*

Taylor, C. (2007). A Secular Age. *Cambridge, MA: Harvard University Press.*

Tiryakian, E. (1993). American religious exceptionalism: A re-consideration. The Annals, 527, *40–54.*

Warf, B., & Winsberg, M. (2008). The geography of religious diversity in the United States. Professional Geographer, 60, *413–424.*

Warf, B., & Vincent, P. (2007). Religious diversity across the globe: A geographic exploration. Social and Cultural Geography, 8, *597–613.*

Zelinsky, W. (2001). The uniqueness of the American religious landscape. Geographical Review, 91, *565–585.*

IMAGE CREDITS

- Figure 4.2: Copyright © Richardprins (CC BY-SA 3.0) at http://commons.wikimedia.org/wiki/File:Kingdoms_of_Israel_and_Judah_map_830.svg.
- Figure 4.4: Clinton C. Gardner, http://commons.wikimedia.org/wiki/File:Buchenwald_Corpses.jpg. Copyright in the Public Domain.
- Figure 4.5: Copyright © Anwar Saadat (CC BY-SA 3.0) at http://commons.wikimedia.org/wiki/File:Jewry.PNG.
- Figure 4.6: Smith, "Map Expansion Christianity," http://commons.wikimedia.org/wiki/File:Expansion_of_christianity.jpg. Copyright in the Public Domain.
- Figure 4.7: Copyright © Ra Boe (CC BY-SA 3.0) at http://commons.wikimedia.org/wiki/File:Reims_07_(RaBoe).jpg.
- Figure 4.8: Copyright © San Jose (CC BY-SA 3.0) at http://commons.wikimedia.org/wiki/File:Europe_religion_map_en.png.
- Figure 4.9: Verrai, http://commons.wikimedia.org/wiki/File:Religions_of_the_US.PNG. Copyright in the Public Domain.
- Figure 4.10: Copyright © Antonio Melina (CC BY 3.0 BR) at http://commons.wikimedia.org/wiki/File:Mosque.jpg.
- Figure 4.11: Copyright © Depositphotos/Zurijeta.
- Figure 4.12: Copyright © THX_9151bs (CC by 2.0) at https://commons.wikimedia.org/wiki/File:Saudi_in_niqap.jpg.
- Figure 4.13: https://www.flickr.com/photos/telcomintl/14630104417/. Copyright in the Public Domain.
- Figure 4.14: Copyright © Busterof666 (CC BY-SA 4.0) at http://commons.wikimedia.org/wiki/File:Arabempire_isoch.png.
- Figure 4.16: Copyright © The Emirr (CC BY-SA 3.0) at http://commons.wikimedia.org/wiki/File:Map_of_Islam.svg.
- Figure 4.17: Ravi Varma, http://commons.wikimedia.org/wiki/File:Murugan_by_Raja_Ravi_Varma.jpg?fastcci_from=20043968. Copyright in the Public Domain.
- Figure 4.18: Copyright © John Hill (CC BY-SA 3.0) at https://en.wikipedia.org/wiki/Cattle_in_religion_and_mythology#/media/File:Well-loved_cow,_Delhi.jpg.
- Figure 4.19: ThrashedParanoid, http://commons.wikimedia.org/wiki/File:Hinduism_By_Country_Percent.png. Copyright in the Public Domain.
- Figure 4.20: Copyright © Dirk Beyer (CC BY-SA 3.0) at https://commons.wikimedia.org/wiki/File:Kamakura_Budda_Daibutsu_front_1885.jpg.
- Figure 4.21: Copyright © Tevaprapas (CC by 3.0) at https://commons.wikimedia.org/wiki/File:Abbot_of_Watkungtaphao_in_Phu_Soidao_Waterfall.jpg.
- Figure 4.22: Copyright © Javierfv1212 (CC BY-SA 3.0) at http://commons.wikimedia.org/wiki/File:Buddhist_sects.png.
- Figure 4.23: Copyright © Antoine Taveneaux (CC BY-SA 3.0) at https://commons.wikimedia.org/wiki/File:Buddhist_monks_of_Tibet7.jpg.
- Figure 4.24: Copyright © Depositphotos/anekoho.

CHAPTER FOUR: THE GEOGRAPHY OF RELIGIONS   183

- Figure 4.25: Copyright © Jakub Hałun (CC BY-SA 4.0) at http://commons.wikimedia.org/wiki/File:20091004_tai_chi_Hong_Kong_Kowloon_6887.jpg.
- Figure 4.26: Copyright © JohnLangdon (CC BY-SA 3.0) at https://commons.wikimedia.org/wiki/File:Black_and_White_Yin_Yang_Symbol.png.
- Figure 4.27: E.T.C. Werner, "Confucius - Project Gutenberg," https://commons.wikimedia.org/wiki/File:Confucius_-_Project_Gutenberg_eText_15250.jpg. Copyright in the Public Domain.
- Figure 4.28: Copyright © Woller (CC BY-SA 3.0) at https://commons.wikimedia.org/wiki/Category:Foot_binding#/media/File:Lotosfuesse-2010.jpg.
- Figure 4.29: http://commons.wikimedia.org/wiki/File:Chinese_Emperor_Fu_Hsi,_wearing_traditional_costume,_Wellcome_V0018487.jpg. Copyright in the Public Domain.
- Figure 4.30: Copyright © Midori (CC by 3.0) at http://commons.wikimedia.org/wiki/File:Lake_Kawaguchiko_Sakura_Mount_Fuji_3.JPG.
- Figure 4.31: Copyright © Rdsmith4 (CC BY-SA 2.5) at https://commons.wikimedia.org/wiki/File:Itsukushima_floating_shrine.jpg.

Chapter Five

THE RISE OF GLOBAL CAPITALISM

MAIN POINTS

1. Capitalism arose within the context of feudalism, which dominated Europe for a thousand years. Feudalism was marked by the dominance of the church, a small land-owning elite, vast numbers of serfs and peasants, and a stable, tradition-oriented culture.

2. Following the disastrous epidemic of bubonic plague in the mid-fourteenth century, capitalism gradually began to take shape in Europe. Capitalism was characterized by the dominance of markets, an urbanized merchant class, production for profit, uneven development, and a long series of cultural and scientific changes that gradually undermined the monopoly of the church.

3. In the eighteenth century, the Industrial Revolution led to new energy sources, enormous technological changes, massive productivity increases, and the growth of cities throughout Europe, North America, and Japan.

4. As capitalism became dominant in Europe, it spilled across the world in the form of colonial empires. These held sway over the planet for almost 500 years with enormous consequences for the global economy and the peoples and places within it.

As we have seen, geographies are not created overnight. The spatial distributions of peoples, cultures, and activities reflect processes that take years, even centuries, to unfold. Because the present is produced out of the past and shaped by it in countless ways, any serious understanding of the world's human geography must include an appreciation of how the contemporary world came to be. An historical appreciation also reminds us that the landscapes of the present are constantly changing (reflecting the theme of historical context introduced in Chapter 1).

A major development underlying current cultural and economic geographies is the rise of what may loosely be called the "West," which, today, refers primarily to Europe and North America. As we shall see, the rise of global capitalism and the creation of the "West" as a distinct entity were two sides of one coin. As capitalism emerged in Europe, it radically changed the continent's societies, cultures, politics, and landscapes; with colonialism, Europe conquered vast parts of the world, dramatically reconfiguring the world's distribution of peoples and their languages, religions, economies, social relations, and landscapes. Indeed, the division between "the West and the Rest" remains one of the most important and enduring features of the planet's geography. However, many writers often oversimplify and exaggerate this division, as if the West did not contain fundamental internal differences, as well as similarities to the rest of the world. Societies, such as Japan, for example, arose outside of Western colonialism only to become firmly entrenched within the West economically and politically.

This chapter provides a historical appreciation of the rise of global capitalism in several ways. Global capitalism has been the motor driving the formation of the world's geographies for five centuries, and it continues to reshape the world's peoples, places, and cultures along primarily Western lines. Thus, to understand contemporary human geographies, we must comprehend how capitalism emerged and conquered the world. First, this chapter delves into the context in which capitalist economies and societies were born and developed, particularly feudalism. Second, it explores the unique characteristics of capitalism, the dominant form of production and consumption around the world today. Third, this chapter turns to the Industrial Revolution, which marked an exponential increase in the scale and speed of human activities, radically reconfiguring the worlds of work, daily life, and transportation, creating huge cities and new landscapes as a result. Finally, it addresses the issue of colonialism, the process by which capitalism "went global," spilling out of Europe and conquering the rest of the globe.

5.1 FEUDALISM AND THE BIRTH OF CAPITALISM

Prior to capitalism, the prevailing form of economic and social relations in Europe was **feudalism,** which lasted from roughly the sixth to the sixteenth centuries (persisting later in some places, even into the early twentieth century in parts of eastern Europe). After the Roman Empire's collapse in the fifth century, many aspects of feudalism emerged from the Frankish Empire that dominated present-day France, Germany, and Italy from the seventh to the tenth centuries. Feudal society remained relatively stable for a long period, sometimes called the Middle Ages. Feudalism was not unique to Europe, but existed in similar form in Japan, China, and (to some extent) India.

Politically, this type of system manifested in Europe as a changing set of empires, including the Frankish Empire, the Norman Empire, the Holy Roman Empire, tsarist Russia, the Austro-Hungarian Empire. Imperial Russia and the Austro-Hungarian Empire survived until World War I (Figure 5.1). A major difference between Europe and the United States is the impact of feudalism in Europe. In North America, capitalism emerged on a landscape that had not been shaped by more than a millennium of feudalism, as was Europe, including its land use and property systems, cities, and class and gender relations.

CHARACTERISTICS OF FEUDALISM

Feudalism's political, cultural, and economic characteristics made it qualitatively different from capitalism. Compared with the dynamic, ever-changing world of market-based societies, feudalism produced a remarkably stable and conservative world that changed slowly. Most people's lives—working the land in a cycle of endless drudgery—were the same from one generation to another.

However, the feudal era was not completely static. The later feudal period (after about 1000 CE) saw significant changes: populations grew, universities developed, new types of farming emerged, people drained wetlands and cleared forests, and new technologies arose. Plagues and diseases caused massive dislocation, and political and religious conflicts led to enduring changes, such as the split between the Catholic and Orthodox branches of Christianity (see Chapter 4).

In feudal Europe, the Catholic church was, by far, the predominant political/ideological institution, having risen to power as the Roman Empire dissolved (Chapter 4). Everyone was deeply religious, and their belief in God informed every aspect of their life and worldview. In most towns, the cathedral was the largest and most impressive building, its size testimony to

Figure 5.1: The Holy Roman Empire exemplified the political geography of feudalism, which was based on multiethnic empires rather than nation-states. It contained numerous ethnic and linguistic groups, including Germans, French, Swiss, Italians, Poles, and others.

the wealth and power of the church. The Pope in Rome owned land, controlled armies, and held great sway over kings and emperors (Figure 5.2). Local priests, who were often the only ones who could read and write, served as judges, teachers, and officiates at weddings and funerals. In Rome, the Pope exerted great power over kings and nobles throughout the continent, often appointing leaders and threatening to excommunicate those who did not obey. The church owned farmland and hunting estates, raised taxes, and even had its own armies.

The ruling class of feudalism was an aristocratic nobility whose power lay in the ownership of land. Many tiers existed within this ruling class, including lords, dukes, earls, barons, and others. In an overwhelmingly rural society, in which agriculture produced a relatively small amount of food per unit of land, peasants and farmers formed the vast majority of the population. Aristocrats typically owned vast estates of farmland. In the manorial system, the extraction of wealth from peasants and farmers occurred through the payments of rent, in which tenant farmers paid a large share, perhaps half, of their output to the local lord. The aristocracy controlled the government, including the military and penal system. Knights and the military existed to enforce the rule of aristocratic law and to protect communities from brigands, robbers, and invaders.

Figure 5.2: The Pope was arguably the most powerful force in feudal Europe, controlling enormous resources and exerting huge political and religious authority.

Farming under feudalism was based on **animate** sources of energy, that is, human and animal muscle power. Peasants and draft animals worked the fields, collected firewood, drew water from wells, and performed innumerable other tasks (Figure 5.3). Because child labor was widely used on the farms, birth rates were high (meaning women were pregnant most of the time). The population lived in small hamlets and villages, producing their own food, clothing, and other necessities. Peasants were illiterate, ignorant of events even a few miles away, and unaware of what century they lived in.

Figure 5.3: Although many people think of the medieval era in terms of kings and knights, the vast majority of the population consisted of peasants and farmers. Serfs were a particularly oppressed group bound to the land by law and custom and obligated to provide unpaid work on their lord's lands for several days a week.

A large portion of the rural population consisted of **serfs**. A serf was not a slave, that is, he or she was not owned by a master. Rather, serfs were bound to the land by feudal law and custom and forced to give several days of work to their master each week. In addition to serfs, the farming population included pools of freemen, independent peasants with their own land. Serfs and freemen lived a monotonous life in which each day was identical to the one before. They did the same chores, ate the same food, and saw the same people. The standard of living for virtually everyone except aristocrats and some merchants was very low. Diets were typically inadequate, and malnutrition was common. Famines broke out every few years. Life expectancy in feudal Europe was typically under 40 years. Many women died in childbirth, and infant mortality rates were high. Water supplies were often infected by bacteria, and diseases such as cholera, plague, and tuberculosis took an enormous toll in human lives and suffering.

Agricultural work was organized around the rhythms of the seasons, with different tasks for the spring, summer, fall, and winter. Winters, for example, might be spent indoors, weaving or fixing farm implements. Spring was a time of planting. Summers involved tending to the crops and livestock. The fall was the time of the harvest, for many the central event of the year. Most people lived in extended families; grandparents, parents, nieces and nephews, and cousins lived in crowded rooms, often without paved floors. Families often slept together in one bed, sometimes including dogs and pigs for warmth.

Markets existed under feudalism but only the wealthy had the income required to buy luxury goods. Typically, markets consisted of seasonal fairs where itinerant merchants (often Jews) sold metal goods, silks, or jewelry. Barter was common. Thus, under feudalism, the state, rather than markets, allocated most resources.

Although feudal society was predominantly rural and agricultural, a few cities and towns existed. Urban areas under feudalism differed from those of today in many ways. Because agricultural productivity rates were low and, thus, farmers' ability to support city dwellers, cities were small (Figure 5.4). Most hamlets did not exceed 200 or

Figure 5.4: Carcassonne, France, exemplifies the feudal city: small, dense, and surrounded by concentric walls of fortifications. The vast majority of the population in the medieval era lived and worked on farms; the relatively small agricultural surplus generated by feudal agriculture, which relied on animate (or living) sources of energy, meant that farmers could sustain only a small proportion of people in cities.

300 people, and cities greater than 10,000 people were rare. Feudal cities were densely populated, with the inhabitants crowded together, often in very unsanitary conditions. The streets were often covered with mud and animal waste. The centers of feudal cities frequently consisted of a walled fortress, often with a palace located within, where the local lord lived, surrounded by concentric rings of protective walls. Because land was not a commodity to be bought and sold, but allocated on the basis of power and status, there was little differentiation among land uses; without land markets to separate land uses according to their profitability, commercial and residential land uses were mixed together.

Within the cities, feudal **guilds**, or associations of craft workers and artisans, produced a variety of goods. The primary functions of guilds were to limit competition by keeping imports out, as well as to train new generations of labor. Guilds consisted of skilled workers with years of experience, and were organized by the type of good they produced. There were, for example, blacksmiths' guilds, weavers' guilds, goldsmiths' guilds, and guilds for bakers, leather workers, paper makers, glass workers, and shoemakers. Young men who were lucky enough to be chosen to work in the guild escaped a lifetime of drudgery in the fields, and spent years as apprentices learning the trade before eventually becoming craftsmen in their own right.

THE END OF FEUDALISM

The feudal era in Europe saw relatively little change from the sixth to the sixteenth centuries. However, the late medieval period, starting roughly around the eleventh century, saw a gradual agricultural revolution based on the introduction of the heavy plow (which facilitated farming in the thick soils of northern Europe), waterwheels, the horseshoe, stirrups, the **three-field system** of farming (in which one field was left unplanted periodically to raise its fertility). Several other innovations from more advanced societies, such as the Arabs, Indians, and Chinese included cotton, the compass, sugar, rice, silk, paper, printing, the needle, the zero, oranges, and the windmill. The Arab conquest of Spain made the Iberian Peninsula a primary point of entry for new ideas and technologies.

Feudal Europe was the western terminus of a much larger world system that stretched across the Mediterranean, the Middle East, the Indian Ocean, and into eastern Asia (Figure 5.5), connecting most of the Old World. The Arab world's strategic location striding Asia, Africa, and Europe placed it at the center of the enormous trade networks that linked together places as far-flung as China, Mozambique, and Belgium. Straddling the center of the feudal world system, the Islamic rulers, centered in Damascus and Baghdad, guaranteed safe passage between two critical worlds—the Mediterranean and the Indian Ocean—that had been separated since the collapse of Rome. China formed the eastern end of this system. The major overland

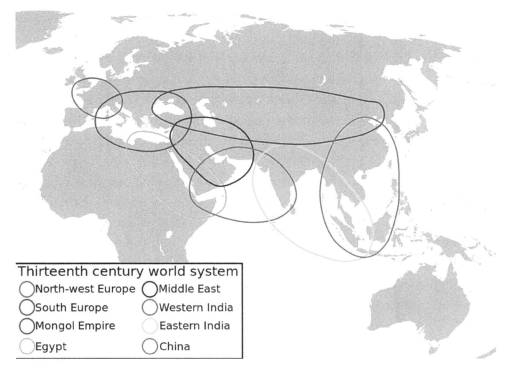

Figure 5.5: The thirteenth- and fourteenth-century world system consisted of a series of towns, kingdoms, and principalities that stretched across Eurasia and included parts of Africa. This network was tied together by trade over oceanic routes on the Mediterranean Sea and Indian Ocean and land routes, such as the Silk Road. It linked peoples and places as diverse as feudal Europe, the Arab world, western Africa, Zimbabwe, India, Mongol-dominated China, and Indonesia.

connection was the Silk Road, the umbilical cord that connected Europe, the Middle East, Central Asia, South Asia, Tibet, and China with ceaselessly flowing caravans bringing goods, innovations, ideas, merchants, and missionaries. Few individuals traveled the entire distance of the Silk Road; rather, it was served by networks of intermediaries. For 2,000 years, Silk Road caravans linked ports, trading cities, oases, and innumerable different cultures. From China came jade, paper, the compass, gunpowder, printing, porcelain, lacquer ware, silk, pearls, peaches, apricots, citrus fruits, cherries, and almonds; moving in the other direction, China acquired horses, hides, furs, dyes, amber, pistachios, saffron, sesame, peas, onions, coriander, cucumbers, grapes, sugar beets, ivory, and, oddly, the chair. So too did religions flow along this highway, including Buddhism and Islam (Chapter 4).

The Silk Road was, therefore, as important for the flow of cultures and ideas it was for the flow of goods.

New innovations imported via the Silk Road improved European agricultural productivity. The supply of agricultural land expanded as peasants and farmers cut down forests and drained swamps. Likewise, the heavy plow opened up the thick soils of northern Europe to farming. These factors led to a gradual increase in the urban population. By the fifteenth century, much of western Europe was carpeted with a growing network of cities, called *newtowns* in Britain and *villanovas* in Spain.

However, compared with much of the rest of the world, feudal Europe remained relatively primitive. Standards of living and rates of innovation in Europe from the sixth to the sixteenth centuries were much lower than the wealthier, more powerful, and more sophisticated societies of the Arab world, India under the Mughals, or China during the Sung and Ming dynasties.

Among the other things introduced to Europe via the Silk Road trade routes, a deadly disease commonly known as bubonic plague or the Black Death for the dark, swollen lymph glands it produced. In Asia, this disease was common in rodent populations in the steppes, or grasslands. In 1347, ships carrying plague-infected rats from Asia landed in Italy, Spain, and southern France, all of which were part of the expanding trade network between Europe, the Middle East, and Asia. Within 4 years, one-quarter of Europe's population—more than 25 million people—was dead (some estimates range as high as 50%). In some places two-thirds of the population died. The plague raced through the crowded, unsanitary cities, annihilating the majority of inhabitants in many areas. From southern Europe, it spread north, to Germany, Britain, Scandinavia, and Russia (Figure 5.6). Because the notion of germs was nonexistent, feudal Europe had no means of understanding or controlling the plague; frequently used "remedies," such as burning Jews or witches or whipping themselves with chains, did not halt its spread.

Historians have speculated that the plague played a major role in destabilizing feudal Europe and opening the door to a new type of society. Within a few years, much of the continent experienced severe labor shortages; Europe went from being land-poor and people-rich to being people-poor and land-rich. The chaos allowed serfs to run away without fear of being caught and returned. Labor shortages caused wages to rise, which improved standards of living for the survivors, and in many places serfs revolted, such as the English Peasants' Rebellion of 1381.

However, other historians have argued that feudalism was suffering from numerous problems anyway and would have collapsed without the plague. For example, in some cities, such as Florence, Italy, capitalist social institutions, such as banks, were already emerging before the fourteenth century. In any case, following other changes, including a mini Ice Age and the Hundred Years War between France and England, which bankrupted both powers, feudalism in Europe began to crumble. These intertwined climatic, ecological, political, and

Figure 5.6: The Black Death, or bubonic plague, which swept through Europe in the midfourteenth century, eliminated at least one-quarter of the population, causing social and economic havoc, leading to labor shortages, fueling peasant uprisings, and possibly helping to pave the way for the emergence of capitalism.

economic crises bankrupted several kingdoms and caused massive social disruption, including persistent crop failures and the exhaustion of state treasuries. Coupled with the devastation of the plague and the rise of a new merchant class, feudal societies faced a series of challenges that they simply could not overcome. Out of the ashes of feudalism a new economic, political, and social system emerged: capitalism.

5.2 THE EMERGENCE AND NATURE OF CAPITALISM

From the fifteenth to the nineteenth centuries, feudalism in Europe was gradually replaced by a new kind of society, capitalism, a different social, economic, political, and cultural system.

If capitalism can be said to have a birthplace, it would most likely be in northern Italy. The city-states of this peninsula, such as Florence, Venice, Pisa, and Genoa, played a key role in fomenting the new kind of society. They had large groups of wealthy merchants and flourishing trade networks across the Mediterranean, including with the Arabs in Egypt and the Middle East (Figure 5.7). The famous Medici family of Florence had vast holdings in silver mines, silk production, and banking. In northern Europe, a network of cities formed the Hanseatic League, which stretched from Russia and Scandinavia across northern Germany and into the North Sea to Britain from the thirteenth to the seventeenth centuries (Figure 5.8). In these cities, groups of *burghers*, or merchants, accumulated wealth and power that would make them the dominant figures in the new social formation. The Hanseatic League eventually collapsed as trade shifted to the Atlantic Ocean following the discovery of the New World.

Like feudalism, capitalism possesses a distinct set of characteristics, including markets, class relations, finance, territorial and geographic changes, long distance trade, and new ideologies.

MARKETS

Unlike under feudalism, in which the state allocates most resources, such as land and labor, under capitalism, the **market** plays this role. Markets consist of buyers and sellers of **commodities**, which are goods and services bought and sold for a price. Not everything under capitalism is a commodity (for example, air). Only goods that command a price are sold on the market and can generate a profit for producers can be classified as commodities. The expansion of market societies saw the steady commodification of different goods and services, including food, housing, clothing, transportation, education, entertainment, health care, and other domains. Thus, things and services that used to be produced on a subsistence basis or bartered increasingly came to be made for sale on a market, their value determined by the amount of money they commanded.

There are a huge variety of markets based on the type of commodity being produced and consumed, as well as the amount and nature of competition. Markets range from being freewheeling and highly competitive, with many small producers who must accept the price the market hands them, to large ones dominated by a few major producers who can affect the market price. In market-based societies, ownership of private property is a key requirement to production; after all, a good cannot be sold unless the seller owns it. The incentive of producers to sell

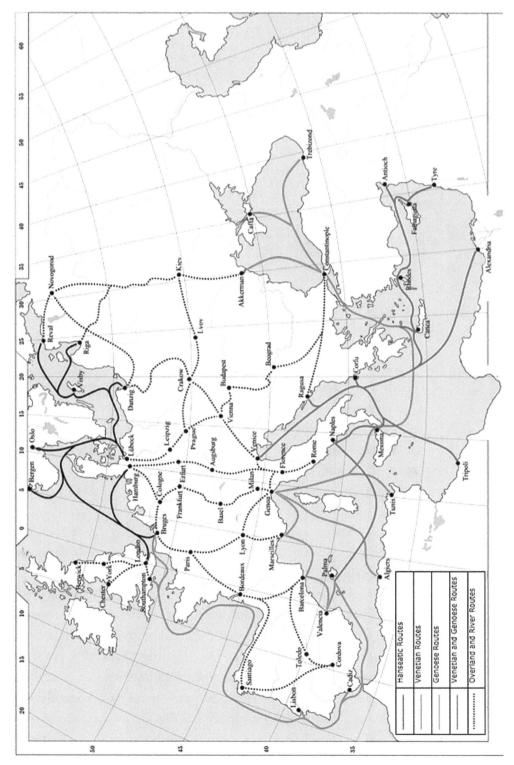

Figure 5.7: Medieval Europe exhibited growing trade networks across the Mediterranean, including with the Arab world, which helped to propel the northern Italian city-states to prominence in the early years of capitalism.

Figure 5.8: Dutch map of the different small and large Hanseatic Leage cities and trade routes.

goods and services is *profit*, the difference between payments received and production and transportation costs. Thus, markets involve production for sale, rather than subsistence or use.

Because markets involve competition among different producers, there is a strong incentive to produce goods and services cheaply and efficiently. Generally, the more competitive a market is, the more dynamism and innovation it exhibits. This powerful incentive to innovate is largely responsible for making capitalism a dynamic society. In this sense, capitalism differs considerably from noncapitalist societies, in that it rewards innovation and risk taking.

Markets are not unique to capitalism. For example, markets existed in slave-based economies, such as Rome. In feudal Europe, occasional markets, in the forms of annual fairs and festivals, brought traveling merchants and local populations together. Typically precapitalist markets were small and involved relatively few types of goods, mainly luxury items for the elites. However, markets are unique in their importance to capitalism. Only under capitalism are markets the major way in which resources are allocated. It is worth emphasizing that markets are not the *only* way in which resources are distributed. Even in supposedly "free market" societies, the government plays a key role, including protecting property rights, building the **infrastructure**, providing public services, and protecting firms from foreign competition.

CLASS RELATIONS

Although some observers depict capitalist societies as classless (as if they consisted only of individual buyers and sellers and no other type of social relation), market-based societies, in fact, do have social classes. In contrast to feudalism, the class system of capitalism reflected a broad-based shift from a hierarchy based on tradition to one based on money, from born rank to earned status. Historically, this process saw the ascendancy of the merchant class, or what is sometimes called the **bourgeoisie** (the middle class of feudalism, the ruling class of capitalism). This group gained enormous wealth through trade and banking, including loans to feudal kings, then came to own the means by which production was organized. Simultaneously, they steadily increased their political power. Merchants, thus, both reflected and produced the changes that accompanied rising capitalism.

The merchants and middle-class urban tradespeople of Europe steadily gained wealth, power, and prestige in the fifteenth through eighteenth centuries. Essentially, the merchant class rose to become the capitalist class, that is, the owners of **capital** (that is, money, investments, and other assets used to produce goods and services). The feudal aristocracy correctly perceived the newcomers to be a lethal threat to its centuries of rule; aristocrats held merchants in low regard as money-grubbers who were not motivated by ancient (and increasingly quaint) notions of honor. Conflicts between the merchant class and the feudal elite steadily increased in frequency and intensity. These disputes revolved around issues such as levels of taxation, the freedom to open markets at particular hours, and the terms under which merchants would lend money to kings (largely for purposes of waging war). The demise of the feudal aristocracy came gradually in some places, such as England (e.g., during the civil wars of the seventeenth century between Royalists and Parliamentarians), and suddenly in others, such as France, where the aristocracy was beheaded during the Revolution of 1789.

In addition to anointing a new ruling class, capitalism changed the role and nature of workers. Labor itself became a commodity, that is, bought and sold for a price (wages) in labor markets. Over several centuries, the peasants and serfs of Europe gradually shifted from subsistence farming to working for wages. The working class emerged as more and more rural workers had to leave the land and learn new norms, such as working the hours dictated to them by their new bosses, the capitalists.

FINANCE

The growth of capitalism saw money become the measure of all worth. In noncapitalist societies, barter plays a major role in economic relations: Goods may be traded for one another or labor traded for goods. In barter-based economies, money is relegated to a relatively small role. Obviously, money existed before capitalism, but under market-based societies money assumed a new level of importance. As the cash system replaced barter, money became standardized and ubiquitous as a measure of value. Human labor became measured in monetary terms, as workers were paid an hourly wage. Wealth and power were increasingly defined economically rather than politically. Many traditional social roles that were once structured around lines of kinship, religion, and friendship became depersonalized and mediated by money. For example, rather than conduct business on the basis of informal (typically oral) agreements among people who knew each other, in which trust and reputation were critical, economic transactions became formalized among strangers using written contracts backed by the power of the state. Money, thus, has important social, as well as economic roles.

Large, complex capitalist societies cannot function without well-established financial systems, which not only reflect production systems, but also shape them. The size and power of banking and credit markets often derives from the wealth of the sectors to which they lend funds. Conversely, as banking grew in influence, it could also shape those same markets by controlling the allocation of capital. The turnover rate of money—the pace with which it changes hands—is important to the process of capital accumulation and wealth creation. Because it is an abstract measure of value, money allows transactions to occur much more rapidly than if they are conducted through barter. Banking arose primarily with the goldsmiths of feudal Europe, who stored gold for their customers and then lent it out to borrowers for a price (interest). Modern banking arose to become a huge and complex industry linking savers and borrowers of different types and with different needs. By the seventeenth century, commercial credit (loans to companies) became widespread, and with it, different types of banks and insurance firms. Starting with small savers who pooled their funds to purchase ships to trade with Asia, joint-stock companies spread the risks of large investments over many small investors. These companies became the foundation of modern stock markets. Accounting became an important profession, and double-entry accounting the norm by which corporate assets, revenues, and costs were measured. By the nineteenth century, financial systems were increasingly regulated by the state through central banks, which sought to control money supplies and, thus, interest, inflation, and exchange rates. Most industrial economies had created state-run central banks by the turn of the twentieth century to regulate the amount of money in their economies, which, in turn, affects the price of money (interest rates) and the cost of doing business, including, for example, taking out loans.

TERRITORIAL AND GEOGRAPHIC CHANGES

If capitalism fundamentally changed the rules of societies, it also reshaped how they were organized geographically. Feudal geographies gave way to new, capitalist ones. Because the spatial structure of a society is fundamental to how it is organized, the geographies of capitalism are central to its structure and how it changed over time.

Geographers have noted that capitalism creates varying levels of economic growth, wealth, and poverty in different locations. Uneven development is reflected in the simultaneous existence of rich and poor places; those with high and low unemployment; regions and countries with happy, prosperous residents and those with large numbers of impoverished, hungry ones. Uneven development occurs through patterns of capital investment (an influx of wealth into an area, generating jobs and raising standards of living) and disinvestment (a departure of wealth, leaving a region impoverished). As capital seeks out the highest rate of profit, it flows into some regions and out of others, simultaneously creating prosperous places and in others creating economic decline in the process. For example, in the United States, capital has steadily abandoned many communities in the industrialized Northeast and Midwest (see Chapter 11), turning it into a Rustbelt, and flowed to the Sunbelt states of the South and West. In this way, wealthy regions and poverty-stricken ones are intimately connected; as capital flows from declining areas to growing ones in search of new opportunities to generate profit, it recreates inequalities over space.

The production of uneven development occurs at global, regional, national, and local spatial levels. All of these scales represent different versions of the same process.

At the global level, capitalism, through the colonial empires that Europe established across the planet, created a worldwide economic system of commodity production and consumption. The capitalist world economy positioned Europe at the center, with rapidly rising wealth and power, and its colonies on the global periphery as a series of impoverished resource-producing areas. Essentially, Europe became wealthy, in part due to its extraction of cheap raw materials from its colonies in the Americas, Asia, and Africa. Prior to colonialism, Europe was poor and underdeveloped compared with competing powers, such as China, India, and the Arabs. By the sixteenth century, when colonialism began in earnest, Europe had begun to surpass many other regions of the planet in wealth and technology. Thus, early capitalist globalization in the form of colonialism had highly uneven impacts—positive and negative—on different parts of the world.

At a second scale, within Europe, capitalism created a division between a relatively prosperous northwestern part and less wealthy southern and eastern parts. Historically, southern Europe, including Greece, Italy, and Spain, had been the wealthiest and most powerful part of the continent, reflecting the earlier system that revolved around the Mediterranean Sea. With the ascendancy of capitalism and the opening of the Atlantic to trade and commerce, northwestern Europe became much more economically advanced. To this day, the societies of northwestern Europe remain wealthier than those elsewhere in that continent.

At a third scale, within the individual countries of Europe and other emerging nation-states eventually dominated by capitalism, there arose a pronounced division in wealth between cities and the countryside. Under feudalism, with its tiny cities, there were relatively few differences in standards of living between urban and rural areas. Under capitalism, as rural areas were reshaped by the commodification of agriculture (reorganization of farming around profit) in the form of cash crops, large numbers of people migrated to urban areas in search of job opportunities. Given the maritime world economy, port cities, in particular, thrived. Today, almost everywhere, cities have higher incomes and more job opportunities than do rural areas.

Finally, within capitalist cities, urban land became a commodity, organized through land markets. The commodification of land meant that its price—rent—became the measure of its value. The rent that landowners could charge became the basis of profit derived from land, as well as its price (price is the sum of potential rents over time). As land markets grew, profit became the mechanism for separating different land uses, including a division between work and home, or areas of production and housing, respectively. Gradually, as cities grew larger, particularly under the Industrial Revolution in the nineteenth century (discussed later in this chapter), the distances between home and work increased to a point at which workers engaged in mass commuting.

LONG-DISTANCE TRADE

The ability to buy and sell goods over long distances is a fundamental part of capitalist societies. Trade reflects the geographic organization of production and exchange, that is, spatial differences between producers and consumers, linking sellers and buyers who rarely see one another. As capitalism took hold and became entrenched across the European continent, trade networks proliferated in diversity and extent. Of course, there was trade prior to capitalism. In feudal Europe, trade with the Muslim world, and through the Silk Road, with East and

South Asia, allowed the influx of many goods that Europeans did not produce for themselves. Indeed, as we have seen, Europe was the western endpoint of a much larger fourteenth-century trading system that stretched across the Middle East, India, and into Southeast Asia and China (Figure 5.5). Prior to capitalism, trade was largely confined to luxury goods, such as spices, silks, porcelain, and precious metals. Only aristocrats had the means to purchase such luxuries.

Although long-distance trade was peripheral to feudalism, it is central to capitalism. In market-based societies, trade occurs in all sorts of goods, from luxuries to everyday items. The desire to expand trade networks was the major incentive that led to the formation of land and sea routes that tied different parts of Europe together and to the rest of the world. Within Europe, the sixteenth to nineteenth centuries saw a vast expansion in roads, canals, and eventually railroads, which sutured places together into an increasingly interdependent system. As Chapter 1 noted, one of the major analytical themes used to understand geographies is that places are always connected to one another. Geographers use the term **time-space compression** to describe the increased speed with which people, capital, goods, and information circulated among places. New ships allowed Europeans to sail long distances relatively quickly, and in the process, they created new maps and charts for nautical navigation, learning the behavior of winds and currents all over the planet.

If trade reflects differences among places in production, it also helps to shape those places. In economics and economic geography, this idea is reflected in the concept of **comparative advantage**, the difference in production that grows when places begin to trade extensively with one another. Trade allows people in some places to consume goods and services that they cannot produce for themselves (or do not produce in sufficient quantities) while selling their excess output to people in other places. The growth of long-distance trade within Europe, and between Europe and its colonies around the world, increased competition with producers who had long relied only on local markets to sell their output. This increased competition fueled declines in production costs and prices, as well as associated increases in standards of living. By the seventeenth century, for example, an upper-middle-class family in Britain could purchase salted cod from Newfoundland, furs from Russia, timber from Scandinavia, wines from France, blown glass from eastern Europe, and olive oil and citrus fruits from Spain or Greece.

NEW IDEOLOGIES

Capitalism is not simply economic, political, cultural, or geographic in nature, it is all of these simultaneously. Just as feudal society was permeated with an ideology dominated by religion, so too does capitalism exist within the domain of culture and ideology, or the ways in which people perceive and think about the world. Chapter 1 maintained that culture and human consciousness are fundamental parts of how human geographies are made. Thus, understanding how ideologies changed during the rise of capitalism is necessary for appreciating its complexity and power. If market-based systems affected the ways in which people were organized and interrelated, they also brought an array of ideological changes that revolutionized the ideas, science, and culture of the modern world.

After its invention in 1450, the printing press made the production of books much easier, faster, and cheaper (Figure 5.9). The invention of the movable-type printing press had huge impacts on European societies, allowing large quantities of materials to be produced cheaply and distributed quickly and accelerating the decline of the feudal order. Printing was the first major step in the mechanization of communication and accelerated the diffusion of information by packaging it conveniently for all who could read it. In doing so, it undermined the centrality of the clergy in the production of knowledge and broke the

Figure 5.9: The printing press, appearing at the dawn of the Renaissance, contributed enormously to the flow of ideas and information in early modern Europe. As books became cheap and literacy rates rose, printing had powerful effects on language, religion, science, and nationalism, particularly in the Protestant countries of northwestern Europe.

monopoly of learning held by monasteries. Thus, printing fomented the growth of an intellectually active class of lay people (the inteligentsia). Ideas of many sorts began to circulate around the continent, and larger numbers of people learned to read and write. With printing, readers gained access to people, places, and events traditionally far removed from them historically or geographically. Printing and the rising rates of literacy facilitated the Renaissance, the Protestant Reformation, European expansionism, and the rise of modern capitalism and science.

In the sixteenth and seventeenth centuries, starting in Italy, Europe witnessed the explosion of artistic and scientific knowledge that we know as the Renaissance (a term coined in the nineteenth century). Leading intellectuals, such as Leonardo da Vinci, exemplified the rise of secular (nonreligious) knowledge, as did Erasmus and the Humanists in northern Europe, who put humans, not God, at the center of their worldview.

The sixteenth century also saw the Protestant Reformation, the second great schism in Christianity (Figure 5.10; Chapter 4). Starting with Martin Luther in Germany, Protestantism offered a different worldview than did Catholicism, emphasizing the role of the individual and his or her direct relation to God, bypassing priests as intermediaries. This elevation of the individual reflected the broader growth of individualism in different social spheres at this time. Protestantism spread with the growth of literacy and printing as people were encouraged to read the Bible in their vernaculars.

Figure 5.10: In many respects, the second schism in Christianity, between Protestants (in blue) and the Catholic Church, mirrored the social and geographic divide between northwestern Europe, in which capitalism was becoming predominant, and the remainder of the continent, in which feudal social relations persisted for much longer.

The famous sociologist Max Weber studied the relations between Protestantism and the development of industrial capitalism in Northwest Europe. Weber argued that the "Protestant ethic," which stressed delayed gratification, savings, and material success as a sign of one's potential entry into heaven, was instrumental to the development of capitalism. He held that Protestantism elevated work into a moral obligation, making profit a reward rather than a sin, and paved the way for the accumulation of wealth. Other scholars have challenged this perspective, maintaining that Protestantism followed in the wake of market relations rather than causing them or that Protestantism and capitalism coevolved. Some others note that the origins of capitalism in Catholic Italy undermine the claim that it began in predominantly Protestant northern Europe.

In the sixteenth and seventeenth centuries, science played a critical role in restructuring how people viewed the world and their place in it. Evidence came to be the central criterion of knowledge, not faith. The Copernican revolution, empirically confirmed by Galileo and the telescope, led to a view of the universe in which Earth revolved around the sun, replacing the older geocentric perspective endorsed by the Catholic Church. Modern science was born as scientists such as Francis Bacon, Isaac Newton, Boyle, Pascal, Linnaeus, and Lavoisier made enormous strides in the understanding of physics, astronomy, and chemistry and their applications to gravity, optics, and other fields.

In the seventeenth and eighteenth centuries, Western societies underwent the Enlightenment, or Age of Reason, an explosion of science and secular political thought. In Britain, France, Germany, and elsewhere, advances were made in geology, chemistry, physics, and biology, including the discovery of atoms, electromagnetism, and bacteria, leading to the germ theory of disease. The Darwinian revolution of the nineteenth century introduced the concept of evolution in direct challenge to Christian doctrine. These discoveries demystified nature and subjected it to scientific law, a process that was accelerated by the proliferation of universities and institutionalized learning through academic societies. In political thought, thinkers such as John Locke, David Hume, and Adam Smith developed a worldview that stressed secularism, individualism, rationality, progress, and democracy. Indeed, the last, articulated by Thomas Jefferson, among others, may be seen as a largely American contribution to the Enlightenment.

5.3 THE INDUSTRIAL REVOLUTION

The pace of economic, technological, social, and geographic change accelerated greatly in the eighteenth and nineteenth centuries during the Industrial Revolution. Note that that the Industrial Revolution occurred long *after* capitalism began; indeed, for most of capitalism's history, it involved preindustrial forms, labor-intensive artisanal and household-based, of production. However, starting in the mid-1800s, an explosive increase in the speed and productivity of capitalist production in Europe, North America, and Japan transformed the worlds of work, everyday life, and the global economy. In the twentieth century, industrialization spread to eastern Europe and the Soviet Union, and since the 1970s, it has spread to selected parts of the developing world, particularly East Asia (Chapter 11).

Three elements of industrialization are particularly important: the use of inanimate energy, technological changes, and productivity growth.

INANIMATE ENERGY

If preindustrial societies relied on animate sources of energy (i.e., human and animal muscle power) to get things done, industrialization can be defined loosely as the harnessing of inanimate sources of energy. This process marked a major milestone in human economics. Several types of inanimate energy were tapped historically. The first involves running water moving from higher to lower elevations. This source had been used even in the late medieval era in mills to grind grains into flour. Running water was a major source of energy in the earliest stages of the Industrial Revolution, but it constrained firms to locations near streams and rivers. Many textile plants in Britain and New England, for example, used this strategy. There were severe drawbacks to this approach: Many streams dry up in the summers, and locating on one may put the producer inconveniently far away from the market.

A more efficient source of **inanimate energy** involves the steam engine, for which the designs were created by James Newcomen in 1712 (Figure 5.11); the first operating model was built by the Scottish engineer James Watt in 1769. This technology marked a turning point

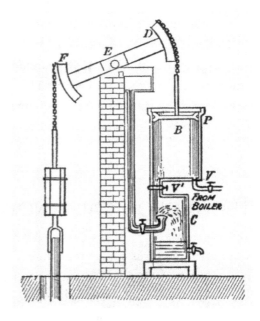

Figure 5.11: The steam engine was the decisive technology of the nineteenth century and the motor of the Industrial Revolution. Introduced in 1769, it revolutionized both production, helping to create the factory system and new forms of transportation such as the railroad and steamship.

in the process of industrialization. Originally designed to pump seawater out of underwater coal mine shafts that penetrated coastal coal layers that extended under the ocean, the steam engine could do the work of dozens of men far more efficiently. Wood provided the first major source of fuel for this invention, which required heating water into steam to drive the engine's pistons. As the demand for wood led to forests in Britain being cut down in large numbers, wood supplies dwindled, and the rising cost eroded profits.

As wood became scarce, producers switched to coal, which was mined in large quantities. Thus, as Britain industrialized, several areas became major coal-producing centers, including Wales and Newcastle. As the Industrial Revolution spread across Europe, the large coal deposits of the northern European lowlands became increasingly important, including in northern France, Belgium, the Ruhr region in Germany, and Silesia in southern Poland. This relation exemplifies how nature helps to shape the formation of geographies, one of the five analytical themes introduced in Chapter 1. In the United States, coal deposits in Appalachia played a key role in the nation's industrialization.

In the nineteenth century, coal was joined by other fossil fuels, particularly petroleum and, to a lesser extent, natural gas. The abundance of cheap energy was the lifeblood of industrialization, and production processes became increasingly energy-intensive as a result. This substitution of inanimate for animate energy both freed tens of millions of people from drudgery and allowed large numbers to live relatively comfortable lives in the expanding middle class that industrialization produced.

TECHNOLOGICAL INNOVATION

Capitalism involves constant change. Firms, under the lure of profits and threat of ruin, engage in innovation as a way to reduce costs and increase revenues. Although innovations certainly emerged prior to industrialization, the Industrial Revolution witnessed a dramatic jump in the number, diversity, and applications of new technologies. A technology is a means of converting inputs to outputs. These can range from extremely simple (e.g., a digging stick) to highly sophisticated (e.g., a computer). As industrialization produced an increasingly complex division of labor based on mass production, new inventions sprouted up rapidly (Table 5.1).

During the Industrial Revolution, a major reorganization in the nature of labor occurred with the development of the factory system. Prior to this era, production was organized on a small-scale basis, including home-based work. By the late eighteenth century, firms in different industries began grouping large numbers of workers together under one roof, a process that effectively created the industrial working class. Never before in human history had so many

TABLE 5.1: SOME MAJOR INNOVATIONS OF THE INDUSTRIAL REVOLUTION

1708	mechanical seed sower
1712	steam engine
1758	threshing machine
1765	spinning jenny
1787	power loom
1793	cotton gin
1807	steamboat
1828	railroad
1831	electric generator
1834	reaper
1839	photography, vulcanized rubber
1844	telegraph
1846	pneumatic tire
1849	reinforced concrete
1850	refined gasoline
1851	refrigeration; sewing machine
1857	pasteurization
1859	gasoline engine
1866	open hearth furnace
1867	dynamite
1873	typewriter
1876	telephone
1877	phonograph
1878	microphone
1879	electric light bulb
1884	rayon
1886	hydroelectric power plant
1888	camera; radio waves
1892	diesel engine
1895	X-rays
1896	wireless telegraphy
1899	aspirin
1900	zeppelin
1903	airplane
1906	vacuum tube
1925	television

workers been concentrated on a permanent basis, a feature that changed how they lived and viewed each other, and themselves. Inside factories, workers used many types of machines representing vast amounts of capital. The introduction of interchangeable parts, invented by American gun maker Eli Whitney, made machines more reliable and easier to fix (Whitney also invented the cotton gin). By the early twentieth century, Henry Ford introduced the moving conveyor belt, which further accelerated the tempo of work and the ability of workers to produce.

PRODUCTIVITY INCREASES

Because of the massive technological changes of the Industrial Revolution, productivity levels surged. Productivity refers to the level of output generated by a given volume of inputs. For example, a farmer may measure productivity in terms of bushels per acre (or metric tons per hectare), or a firm might measure it in terms of units produced per unit of labor hour. Productivity increases refer to rising levels of efficiency (i.e., greater levels of output per unit of input [e.g., labor hour or unit of land] or, conversely, fewer inputs per unit of output).

Productivity levels rose exponentially in the nineteenth century, generating several important effects. As the cost of producing goods declined, standards of living rose. Most workers labored long hours under horrific conditions and endured standards of living still quite low compared with those we enjoy today. But nonetheless, over several decades, industrialization saw many kinds of goods become increasingly affordable, meaning that the working class gradually became better off. Clothing, for example, which was scarce before the Industrial Revolution, became relatively cheap. As agriculture became mechanized, food supplies got cheaper and more varied, malnutrition declined, and famines in Europe essentially disappeared (except for the Irish Potato Famine of the 1840s).

THE GEOGRAPHY OF THE INDUSTRIAL REVOLUTION

Like all major social processes, the Industrial Revolution unfolded unevenly over time and space. Whereas capitalism had its origins in Italy, industrialization was very much a product of northwestern Europe. Some observers put the first textile factories in Belgium in cities such as Liege and Flanders. However, Britain became the world's first industrialized nation, that is, the first to utilize inanimate energy on a mass basis and engage in widespread machine-based production. Britain had already enjoyed a network of long-distance trade relations with its colonies in North America and India. The commodification of agriculture (i.e., the process of organizing farming around markets and profit) in Britain was advanced compared with the European continent. Britain also benefited from large deposits of coal and was the locale where the steam engine was invented.

By the end of the eighteenth century, Britain stood virtually alone as the world's only industrial economy, a fact that gave it an enormous advantage over its rivals. Britain's industrial base, for example, allowed it to triumph over France in their eighteenth-century rivalry for global dominance and to flood its competitors' markets with cheap textiles. Cities in the Midlands of Britain, such as Leeds and Manchester, were known as the workhouses of the world for their high concentrations of workers, capital, and output, becoming centers of the British textile and metal working industries. Other cities, such as London, Glasgow, and Liverpool, became centers for shipbuilding, a major industry in a maritime-based world economy. In many cities, networks of producers of garments, guns, watches, shoes, cutlery, and other goods formed dense industrial districts of small firms.

A half century after it began in Britain, the Industrial Revolution diffused to the European continent. France saw the formation of industrial complexes in the lower Seine River and Paris. In northern Italy, the Po River Valley became a major producer of textiles and shoes. In Scandinavia, cities such as Stockholm became major shipbuilders. After Germany became a unified nation in 1871, the Ruhr region developed into a global center of steel, automobile, and petrochemicals firms.

By the early nineteenth century, the Industrial Revolution began to spread worldwide. It leapfrogged across the Atlantic as the textile industry arose in southern New England and, later, spread with the formation of the Manufacturing Belt. Starting in the 1870s, Japan became the first non-Western country to industrialize as the old feudal order there had collapsed in the mid-nineteenth century. Russia did not industrialize until the 1920s, when the Soviet Union leaped from being a backward, agrarian society to become the world's second-largest economy in the span of a decade. In the twentieth century, the process of industrialization diffused to many developing countries, particularly in East Asia, where it has had profound consequences for hundreds of millions of people. In a sense, then, the industrialization of the developing world, which remains partial and incomplete, continues a long-standing historical process. The large centers of industrialization in Europe, North America, and East Asia formed by the diffusion of the Industrial Revolution remain highly important to the global economy today.

CYCLES OF INDUSTRIALIZATION

The nature and form of industrialization varied in successive historical periods. Capitalism is prone to large, periodic changes in its industries, technologies, products, labor markets, and geographies, which take the form of business cycles and periodic economic crises. Frequently, the historical development of industrialization is divided into five major periods of change, each roughly 50 to 75 years long, which involved the rise of different industries, technologies, products, and types of labor organization.

The first wave of the Industrial Revolution (1770s–1820s) centered on the textile industry. Easy to enter, with few requirements of capital investment (i.e., machinery and equipment) or labor skills, textiles was the first sector to industrialize in Britain, the rest of Europe, North America, Japan, and the developing world. Because this initial wave of industrialization centered in Britain, it made that nation the world's leading economic power.

The second wave (1820s–1880s) was a period of heavy industry, or large, capital-intensive firms. In the nineteenth century, sectors, such as shipbuilding and iron plants, were critical. Such sectors differed markedly from the light industry of textiles. They required massive capital investments and, thus, were difficult for new firms to enter. U.S. manufacturing originated during this period, although most of its growth occurred after the Civil War of the 1860s.

The third wave of industrialization (1880s–1930s) saw numerous other heavy industries appear, including steel, rubber, glass, and automobiles. This was a period of massive technological change, including the automation of work, as well as economic changes. Aided by the railroads and telegraph, local markets gave way to national markets. As a result, firms became national in scope, and ownership became increasingly concentrated in a few large firms (what economists call an oligopoly). Many companies became multiestablishment corporations, with factories and offices located in several cities and serving multiple local markets. Not surprisingly, this wave saw the rise of robber barons, such as Andrew Carnegie (steel), John D. Rockefeller (oil), James Duke (tobacco), Eleuthere Dupont (chemicals), J. P. Morgan (banking), and John Deere (agricultural machinery).

In the fourth wave of industrialization (from the Depression of the 1930s until the 1970s), the primary growth sectors were petrochemicals (including plastics), automobiles, and aerospace. With a relatively stable global economy, this era saw the economic and political domination of the world system by the United States, which produced a huge share of the planet's industrial output.

The fifth wave of industrialization, often held to begin after the oil shocks of the 1970s, has been led by the electronics industry, which was powered by the microelectronics revolution and computers. The digitization of information enabled the widespread adoption of computer-based technologies in manufacturing and other sectors of the economy. Much of the growth of this sector was centered in East Asia, including the rapid rise of China. This wave also witnessed the explosive growth of producer services and telecommunications (Chapter 12).

It is important to note that during each era, the major propulsive industries were commonly featured as the "high-tech" sectors of their day. Thus, just as electronics is often celebrated at this historical moment for its innovativeness and ability to sustain national competitiveness by lowering costs and raising productivity, so too were the textile industry in the eighteenth century and steel industry in the nineteenth century associated with rising wages and standards of living. However, because capitalism is ceaselessly innovative, what was a leading industry at one moment became a lagging one at the next historical epoch. The rise and fall of different industries saw high-wage, high-value-added sectors, which are innovative and productive, replace low-wage ones in the world's wealthy countries and low-wage, low-value-added ones, such as assembly plants, disperse to the world's poorer countries.

CONSEQUENCES OF THE INDUSTRIAL REVOLUTION

The Industrial Revolution permanently changed the social and spatial fabric of the world, particularly in the societies that now make up the economically developed world. No part of their social systems, economy, technology, culture, or everyday life was left untouched. Within a century of its inception, industrialization changed a series of rural, poverty-stricken societies into relatively prosperous, urbanized, and cosmopolitan ones. Some of the major changes included the following.

CREATION OF A WORKING CLASS

As noted earlier, a significant part of industrialization was the reorganization of work along the lines of the factory system (Figure 5.12). For the first time in history, large numbers of workers labored together using machines. These conditions were quite different from those facing agricultural workers, who were dispersed over large spaces and relied on animate sources of energy. Industrialization gave rise to organized labor markets in which workers were paid by the hour, day, or week.

As firms created a new form of labor, they created a new form of laborer, the industrial working class. Working conditions were typically brutal: workers typically labored for 10, 12, even 14 hours per day, 6 days per week, for relatively low wages. (The 8-hour day and Saturdays free from work were products of workers' movements in the 1930s.) Often, work was unsanitary and dangerous,

Figure 5.12: The factory system brought large numbers of workers together with machines for the first time in human history, initiating a new division of labor with unprecedented productivity. In forging the industrial working class, the factory system also brought with it enormous human suffering. Working conditions in many East Asian factories today are not remarkably different from those that characterized Europe and the United States in the nineteenth and early twentieth centuries.

even lethal, and workers were subjected to accidents, poor lighting, and poor air quality. Child labor was also common, subjecting those as young as four or five to horrendous and exploitative conditions, such as those now found in the developing world.

As a result of this process, time—like space, and so much else—became a commodity, something bought and sold as employers paid workers for their hours worked. Prior to the Industrial Revolution, people experienced time seasonally and rarely felt the need to be conscious of it. Time was simply lived, without worry about the precise beginnings and endings of events. With industrialization, however, time was measured and divided into discrete units, as signaled by the factory whistle, bell, and stop watch. For hourly wage earners, time was literally money. This change marked the commodification of time through the labor market.

Industrialization produced both a working class and labor unions. The first resistance to employers included the British Luddites in the eighteenth century, who blamed their miserable working conditions on the machines they used, and often destroyed them in attempts to halt their exploitation. In France, workers wearing large wooden shoes, or "sabots," jammed them into the machinery in an act of sabotage. By the late nineteenth century, workers' movements created a number of labor unions, which, in the United States, included the Knights of Labor, the American Federation of Labor, and in the twentieth century, the Industrial Workers of the World and the Congress of Industrial Organizations. Industrialization, thus, often produced considerable class conflict.

URBANIZATION

Geographically, the Industrial Revolution was closely associated with the growth of cities. Almost everywhere, industrialization and **urbanization** have been virtually simultaneous processes. Manufacturing firms concentrated in cities, especially large ones, such as the British Midlands or throughout the U.S. Manufacturing Belt (e.g., Pittsburgh, Cleveland, Detroit, Chicago, and Milwaukee).

The reasons why firms concentrated in cities are important. Often, there is a tendency to attribute this phenomenon to the presence of workers in urban areas. Which came first, firms or workers? Cities were clearly centers of capital investment, as much as they were centers of labor. Yet, cities were very small when the Industrial Revolution began, and through agricultural mechanization (which reduced rural job opportunities) and rural-to-urban migration, the urban labor supply was produced.

Firms have powerful reasons to concentrate, or **agglomerate**, in cities. Most firms benefit substantially by having close proximity to other firms, including suppliers of parts and ancillary services. Concentration allows firms to reduce transportation costs, share an infrastructure, obtain specialized information, and generate and develop a labor force with necessary skills for the field. By locating near one another, firms become more productive than if they are alone. The cities of the nineteenth century were composed of dense webs of industrial firms, with intricate input and output relations tying them together. Companies in these cities bought parts, services, and information from, and sold their products to, other companies located nearby.

Industrialization changed societies from predominantly rural to increasingly urban in character. In Europe, North America, and Japan, the majority of people lived in cities for the first time in history. The growth of cities in industrial societies is frequently depicted using an urbanization curve, which illustrates the percentage of people living in urban areas over time. In the United States, for example, the first national census of 1790 showed that 95% of Americans lived in rural areas. This proportion decreased throughout the nineteenth century, and by 1920,

50% of the nation's population lived in cities. Today, it is roughly 85%. This pattern is similar in every other country that has industrialized.

POPULATION EFFECTS

Industrialization changed more than just the geographic distribution of people (i.e., in cities); it also shaped the growth rates and demographic composition of societies (Chapter 8).

At the eve of the Industrial Revolution, the famous theorist Thomas Malthus predicted that rapid population growth would create widespread poverty and famine. Yet, Malthus was soon proven wrong, at least in the short run. The industrialization of agriculture generated productivity increases greater than the rate of population growth, and the creation of a stable and better food supply improved most people's diets. As a result, life expectancy rose. Industrialization also lowered death rates, particularly as malnutrition declined and infant mortality rates dropped. Eventually, public health measures and cleaner water helped to control the spread of most infectious diseases. As death rates dropped, the populations of industrializing countries increased dramatically. This change was also accompanied by a shift from the extended to the nuclear family. Eventually, industrialization also led to a drop in the birth rate, and growth rates declined.

GROWTH OF GLOBAL MARKETS AND INTERNATIONAL TRADE

Yet another impact of the Industrial Revolution concerned the global economy. Capitalism had formed a loose network of international trade well before the eighteenth century, including extensive linkages across Eurasia and the Indian Ocean in the early fourteenth century. The harnessing of inanimate energy for transportation, however, dramatically accelerated the speed of land and water transportation, forming a significant round of **time-space compression**, or shrinking of the relative distances among places when measured in terms of travel time. Whereas both sailing ships and horse-drawn transport traveled at roughly 10 miles per hour (mph), for example, steamships could reach speeds of 40 mph and railroads more than 65 mph. Moreover, the new, industrialized forms of transportation were also cheaper, resulting in declines in the transport costs required to move goods from one location to another.

These changes dramatically accelerated the volume of international trade, and imports and exports soared. Europe, starting with Britain, could import unprocessed raw materials, including cotton, sugar, timber, and metal ores, and export high-value-added finished goods, a process that generated large numbers of jobs in Europe and contributed to a steady rise in the standard of living. Britain became an ardent proponent of free trade, which opened up other countries' markets to its goods, especially textiles. At this historical moment, classical political economists, such as Adam Smith and David Ricardo, began to demolish the philosophy of mercantilism, which preached state protection against imports.

Finally, the industrial world economy saw an explosion of international finance. British banks, largely concentrated in London, for example, began to extend their activities on an international basis, lending to clients overseas. For example, much of the capital that financed the American railroad network came from Britain. The globalization of production was, thus, accompanied by the steady globalization of money and credit.

The timing of industrialization was significant to individual nations. There existed an important difference between early and late industrializers in this regard. Early industrializers (e.g., Britain, the United States) faced little competition internationally. Thus, their light industries (i.e., textiles) associated with the first wave of industrialization, which placed few demands on the government, such as an infrastructure, could develop with little government intervention. This reality shaped national political climates characterized by minimal government involvement in trade and the economy.

In contrast, relatively late industrializers faced a significantly different international climate, one dominated by early industrializers. Countries such as Germany and Japan, which did not begin industrializing until the late nineteenth century—long after Britain and the United States—faced significant competition in industries, such as textiles. Consequently, these nations tended to move rapidly into heavier sectors. Germany, for example, developed steel, armaments, and automobile industries. In countries in which heavy industry dominates and places significantly higher demands on the state for labor training, infrastructure, and trade protection, national political cultures that look favorably upon state intervention are more likely to develop. This is true of newly industrializing countries today. Thus, the internal political culture within countries was strongly affected by the timing of their industrialization.

5.4 COLONIALISM: CAPITALISM ON A GLOBAL SCALE

The development of capitalism in Europe was closely linked to Europe's conquest of the rest of the world. Capitalism was both the motor of European expansion and the beneficiary, as colonial powers in the West benefited from their conquests of other parts of the planet. Between the fifteenth and twentieth centuries, a small group of European powers came to dominate most of the planet. This process can be viewed as the expansion of capitalism on a global scale. The geographies of capitalism are typified by uneven economic development in different areas, and colonialism created uneven development on a global scale with Europe at the center and its colonies on the world periphery. This theme appears in many theories of world development (Chapter 7).

Colonialism had cultural dimensions, as well as political and economic ones. It led to the distinction between the "West" and the "rest." In conquering the "Orient," a term meaning "East" that Europeans applied to most of Eurasia, Europeans came to see themselves as Europeans, and later, as Westerners. They formed this self-representation in contrast to the way they saw the people they conquered, whom they represented in frequently racist, simplistic, and highly erroneous terms (e.g., as backward, primitive, decadent, corrupt, and unchanging).

Although colonized people around the word fought back fiercely against colonial rule, Europeans brought technology and diseases that their opponents could not withstand. Examples of resistance include the Inca rebellions against the Spanish, Zulu attacks on the Dutch Boers, the Indian Sepoy uprising of 1857, and China's Boxer Rebellion from 1899 to 1901. Yet Western powers, armed with guns, ships, and cannon, effectively dominated the entire planet by the midnineteenth century. Although a few countries were nominally independent, the only one to escape colonialism substantively was Japan, which closed itself off from the world until 1868.

Colonialism had profound implications for both the colonizers and the colonized. Globally, colonialism produced the division between developed and less developed countries, a theme explored in detail in Chapter 7. Colonialism changed European states too, deepening the formation of capitalist social relations and markets in western Europe. Prior to colonialism, Europe was relatively poor and powerless compared with the Muslim world, India, or China. Afterward, Europe became the most powerful group of societies on the planet.

Colonialism did not happen in the same way in different historical moments and geographic places. Historically, two major waves of colonialism occurred (Figure 5.13). One sprang from the preindustrial, mercantile era and the other one from the Industrial Revolution. From the sixteenth century, when colonialism began, until the early nineteenth century, the primary colonial powers were Spain and Portugal, and their primary colonies were in the New World and parts of Africa, such as Mozambique and Angola. By the midseventeenth century, the British began to exert dominance worldwide, surpassing not only the Spanish but also the Dutch. With the gradual onset of the Industrial Revolution, Britain, the world's first (and for a while, the only) industrial country, defeated its major rival, France, which had established control over Quebec. In North America, Britain lost its American colonies in the late eighteenth century but retained control over Canada until the nineteenth.

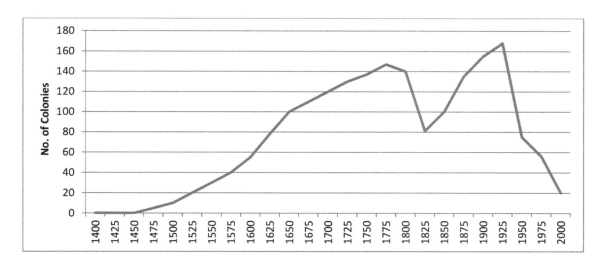

Figure 5.13: Temporally, colonialism consisted of two waves of European domination over two eras, one before the Industrial Revolution and the other during it. The first, preindustrial phase consisted primarily of Spanish and Portuguese conquests in the New World, colonies that became independent following the Napoleonic Wars of the early nineteenth century. The second, industrial phase consisted mostly of British, French, and other countries' conquests of Africa and Asia.

Following the Napoleonic wars, which ended in the Treaty of Vienna in 1815, European powers were relatively weak. This situation allowed nationalists in Latin America, led by Simon Bolivar, to throw off their Spanish colonial overlords and form independent countries. Thus, the number of colonies declined sharply in the early nineteenth century.

During the second phase of industrialization, national economic policy was increasingly characterized by the ideology of free trade or the removal of barriers to imports. This served British interests nicely as they expanded their investments and sales of goods (notably textiles) throughout the world. As Europeans relentlessly sought out more markets for their goods, the number of colonies grew again. Britain emerged as the world's premier geopolitical power, and along with France, colonized large parts of Africa and Asia. The nineteenth century marks the Pax Britannica, when Britain ruled the world's oceans and the pound sterling was the most important currency. Finally, as Figure 5.13 illustrates, the number of colonies declined rapidly after World War II amid the era of decolonization.

Spatially, colonialism was also uneven. Different colonial empires had widely varying geographies, as illustrated by the distribution of empires at their peak in 1914, on the eve of World War I (Figure 5.14). The British Empire, which encompassed one-quarter of the world's land surface, stretched across every continent of the globe, including large parts of western Africa, the Indian subcontinent, and Malaysia and Singapore. The French ruled Indochina (present-day Vietnam, Laos, and Cambodia) as well as much of Central and western Africa. The Portuguese had Brazil, large chunks of Africa, Goa in India, Timor in Indonesia, and Macau in China. The Belgians possessed Congo in Africa. Portugal controlled parts of Africa and outposts in East Asia. Italy also had African colonies (Libya and Ethiopia). Even Germany, late to unify and industrialize, controlled parts of Africa (Togo, Namibia, Tanganyika) and New Guinea, although it lost these following its defeat in World War I.

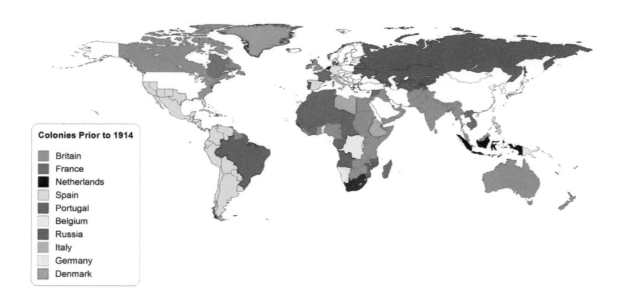

Figure 5.14: Colonialism was unevenly distributed as European powers dominated different parts of the world at varying times. Particularly notable are the British Empire, which covered 25% of the world's land surface, the Spanish and Portuguese rule in Latin America, and large French holdings in western Africa and Indochina. Minor colonial powers included Belgium, the Netherlands, whose major colony was Indonesia.

WHY THE WEST?

What allowed Europe to conquer the rest of the planet? The answers to this question are not simple. For much of history, Europe was relatively weaker and poorer than the countries it conquered. By the sixteenth century, however, Europe did come to possess several technological and military advantages over its rivals.

Jared Diamond, in his Pulitzer Prize–winning book *Guns, Germs, and Steel,* maintains that Western societies enjoyed a long series of advantages by virtue of geographical accident. Agriculture in the West, centered on wheat, was productive and could sustain large, dense populations. In contrast to the Americas, Old World societies were often stretched across vast East-West axes, or regions with common growing seasons (Chapter 2). There was a long Western history of metalworking, which not only increased economic productivity but also allowed for weapons, such as guns and cannons. By the Renaissance, Europeans had become highly skilled at building ships and navigating the oceans. During the eighteenth century, the West had discovered inanimate energy (such as steam power), which offered numerous economic and military advantages. In addition, the Europeans unintentionally unleashed devastating diseases, particularly smallpox and measles, on people with no natural immunities to those diseases. Epidemics killed far more than armed combat ever did.

Others maintain that the West's advantages were not simply technological, but political. The Western "rational" legal and economic system stressed secular laws and the importance of property rights. Unlike the Arabs, Mughal India, or China, Europe was never united politically. Indeed, every time one European power attempted to conquer the others, it was defeated, as exemplified by Napoleonic France in the early nineteenth century and Germany twice in the twentieth. The lack of centralized political authority created a climate in which dissent and critical scholarship were tolerated. For example, the French Huguenots, Protestants in a predominantly Catholic country, could flee persecution by moving to Switzerland, where they started the Jura district watch industry. Similarly, when Columbus failed to obtain financing for his voyages from the Italians, he went to Spain, whose king ultimately consented.

HISTORY AND GEOGRAPHY OF CONQUEST

To appreciate colonialism, it is necessary to understand how it worked in varying historical and geographical circumstances. This short overview reveals that colonialism meant quite different things in the context of different times and places.

LATIN AMERICA

Home to wealthy and sophisticated civilizations, such as the Incas, Mayans, and Aztecs, Latin America was the first major region to be taken over by Europeans. Two years after Columbus arrived, Spain and Portugal struggled over who owned the New World, a contest settled by the Pope with the Treaty of Tordesillas in 1494, which divided the world along the 46th degree of longitude into a western part, to be colonized by Spain, and an eastern part, to be colonized by Portugal. The conquistadors who conquered Mexico and Peru annihilated the Aztecs and Incan civilizations, respectively. In large part, this feat was accomplished through the introduction of smallpox, which killed 50 to 80 million people within a century of Columbus's arrival, the greatest act of genocide in human history.

Figure 5.15: Whereas Portugal colonized Brazil (purple), the Spanish empire in the Western Hemisphere included western South America, Central America, parts of the Caribbean, and much of what would later become the western United States (red denotes Spanish colonies, pink denotes disputed Spanish influence).

The Spanish retrieved enormous quantities of silver from the New World. The silver mines in central Mexico were among the largest in the world, and two million Aymara Indians perished in the Potosi mines in Bolivia. Argentina takes its name from the Latin word for silver and is home to the Rio Plata, literally "river of silver" in Spanish. Most of this metal was taken back to Spain and provided an enormous base of capital that financed economic activities throughout Europe.

The conquest of Latin America initiated a vast flow of plants, animals, and viruses between the Old and New Worlds in a process called the Columbian Exchange (Table 5.2). Both hemispheres contributed to this process, which had profound ecological impacts and changed agricultural practices and diets in both Europe and the Americas. For example, Europeans acquired the potato, which became a staple crop in countries ranging from Ireland to Russia, as well as chocolate and

TABLE 5.2: THE COLUMBIAN EXCHANGE

FROM NEW WORLD	FROM OLD WORLD
Corn	horse
Potatoes	cattle
Sunflowers	sheep
Tomatoes	pigs
Peppers	chickens
Peanuts	wheat
Pineapples	citrus fruits
Blueberries	bananas
Cocoa	sugar
Tulips	olives
Tobacco	okra
Syphilis	smallpox

the tomato. The Columbian Exchange reminds us that human and physical geographies are deeply interconnected, one of the five themes raised in Chapter 1. They shape one another over time, so that the world's biological landscapes today are anything but "natural."

Spain also introduced the land-grant system into the New World through large landed estates known as *encomiendas*. As a small landed aristocracy consolidated its hold, the distribution of farmland became highly uneven, with a few wealthy landowners and large numbers of landless *campesinos* (farmers). This pattern continues in the present day, indicating how the legacy of colonialism still shapes the geographies of the contemporary world. It also reminds us of the need to view human geographies historically.

The Spanish empire in the New World largely ended with the independence movements of the 1820s that followed the Napoleonic Wars, although Spain did not lose its last colonies until the Spanish-American War of 1898. Upon independence, the Spanish empire broke up into a series of independent countries stretching from Mexico to Argentina. The Portuguese colony of Brazil did not fragment in the same way, leaving that nation the giant of Latin America.

Figure 5.16: Unlike Latin America, in which the Spanish and Portuguese dominated, a variety of colonial powers were present in North America. Of these, the English would eventually triumph to become the continent's dominant power.

NORTH AMERICA

In North America, several colonial powers vied for territory. The Spanish were active in Florida and in the Southwest (Figure 5.16). A century before the Pilgrims arrived at Plymouth Rock, the Spanish controlled what is now Texas and California. The French took over Quebec and the St. Lawrence River Valley, only to lose it to Britain in the eighteenth century. France also settled the Mississippi River Valley, with the key port of New Orleans, only to sell it to the United States in the Louisiana Purchase of 1803. The Russians crossed the Bering Straits and seized Alaska, but sold it to the United States in 1867. The Dutch established the colony of New Amsterdam, but the British captured it in 1664 and renamed it New York. Britain emerged as the dominant power in North America, controlling New England and the Piedmont states along the eastern seaboard. Canada was colonized largely through Britain's Hudson Bay Company, a chartered monopoly that controlled the fur and fish trade.

British colonialism in North America began with a series of port cities on the east, typically at the mouths of rivers (e.g., Boston, New York, Philadelphia). After the independence of these colonies, settlers moved west, across the Appalachians in the early nineteenth century and across the Great Plains and Rocky Mountains somewhat later. Railroads opened up this region to the east coast cities and markets. As in Latin America, this process involved the wholesale eradication of Native American peoples and theft of their land.

SUB-SAHARAN AFRICA

Before Europeans settled in Africa and took control of its peoples, they engaged extensively in the slave trade. This process included the kidnapping of roughly 20 million people for export to the New World (Figure 5.17), typically under brutally inhumane conditions. The African slaves were used to compensate for the labor shortages brought on by the decimation of Native Americans by smallpox. The capture of slaves robbed African societies of young adults in their prime working years and sometimes occurred with the assistance of local kings who profited from the trade. Slaves were generally taken from western Africa, and the largest numbers were brought to work the sugar plantations in Brazil and the Caribbean. Others were brought to labor in the cotton and tobacco plantations of the American South.

Africa is rich in minerals, and colonialists were quick to seize upon that fact. In the nineteenth century, European powers penetrated from

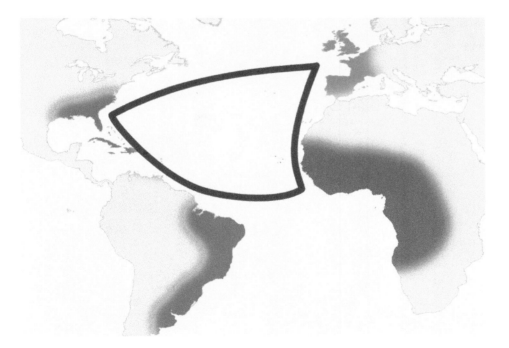

Figure 5.17: The slave trade, which saw an estimated 20 million Africans kidnapped and forced to labor in the sugar, cotton, and tobacco fields of the Americas, reflected the severe labor shortages in the Western Hemisphere brought on by the enormous smallpox epidemics that decimated the native populations. In this way, and others, global capitalism unleashed a massive reshuffling of the world's peoples.

the coastal areas into the interiors and mined for copper, gold, and diamonds, often using slave labor and building railroads to do so. These mines remain the basis of many African economies today.

Perhaps the most important impact of colonialism on Africa was the political geography that Europe constructed. At the Berlin Conference of 1884, European powers drew maps demarcating their respective areas of influence. The borders of the colonies that resulted bore no resemblance whatsoever to the distribution of indigenous peoples. Roughly 1,000 tribes were collapsed into about 50 colonies (Figure 5.18). Some tribes were

Figure 5.18: Among the massive effects on the African continent that colonialism initiated was an extensive redrawing of its political geography, in which roughly 1,000 different tribes were collapsed into 50 colonies, which eventually became independent states. Because these were often peoples of very different languages, ethnicities, and economies, lumping them together contributed significantly to the persistent turmoil, secessionist movements, tribal conflicts, and civil wars that have plagued Africa for decades.

separated by colonial boundaries; many others, with widely different cultures and economic bases, were lumped together. Not surprisingly, because these colonies became independent countries in the 1950s and 1960s, African states have been wracked by numerous civil wars and tribal conflicts in which tens of millions have perished. Conflicts have occurred in Angola, Congo, Rwanda, Liberia, Sierra Leone, Ethiopia, Somalia, and Sudan.

THE ARAB WORLD

The Arab world, which was one of the most powerful and sophisticated centers of world culture from the seventh to the fourteenth centuries, had been colonized long before the Europeans arrived. The expansion of the Ottoman Empire over several centuries saw Muslim Turks (who are not Arabs) dominate the Muslim Arab peoples of the Middle East and North Africa (Figure 5.19). The Ottomans posed a major threat to

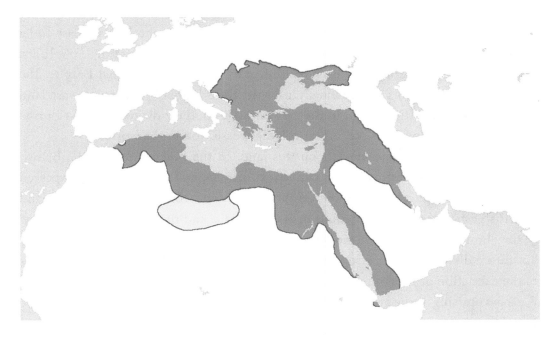

Figure 5.19: The Ottoman Empire represents a rare form of non-European colonialism. Turks, who are not Arabs (Chapter 4), established a formidable empire that dominated the eastern Mediterranean Sea, the Middle East, and North Africa for centuries. Essentially this saw one group of Muslims, Turks, colonize another, the Arabs. The empire finally collapsed during World War I. Green: Ottoman Empire. Light Green: Ottoman vassal, Fezzan.

Europe. Their control over the Middle East was one reason the Europeans explored the Atlantic in search of alternative routes to the Orient. Europe and the Ottomans fought a series of wars during the sixteenth, seventeenth, and eighteenth centuries.

Starting in the nineteenth century, European powers gradually encroached on this vast domain. In 1798, the French seized Egypt from the Ottomans, only to lose it to Britain 4 years later. The British used Egypt as a source of cotton, growing large plantations along the Nile River and building the Suez Canal between the Mediterranean Sea and the Indian Ocean in 1869. In the nineteenth century, France seized Morocco, Algeria, and Tunisia.

After World War I, the Ottoman Empire collapsed, and the French and British seized its Arab colonies in the Middle East. The French took Syria and Lebanon. The British assumed control over Palestine, much of which became Israel in 1948, as well as Iraq and the sheikdoms of the Persian Gulf. Many Arabs who initially welcomed the Europeans as liberators quickly learned that they had traded one set of foreign oppressors for another.

SOUTH ASIA

The Indian subcontinent was known as the jewel in the crown of the British Empire because of its enormous size and riches. Starting in the seventeenth century, the British East India Company established footholds in this domain, founding the city of Calcutta in 1603. India, which is predominantly Hindu, had long been controlled by the Muslim Mughal Empire. The English exploited these divisions and gradually usurped the Mughals' position. A vast land that stretched from Muslim Afghanistan in the east through Hindu India to Buddhist Burma (Figure 5.20), the Indian colony was the largest colonial possession in the world.

Britain had enormous economic impacts on this land. In the nineteenth century, British textile imports flooded India, destroying the native textile industry. Indian labor was exported throughout the British Empire, including the Caribbean, Eastern Africa, and the Pacific Island of Fiji. The British established tea plantations in Assam, India, and the island of Ceylon (now Sri Lanka); they brought Tamil Hindu workers into a predominantly Buddhist island, sowing the seeds for future discord. Britain also built the world's most extensive network of railroads to facilitate the extraction of Indian resources.

In 1857, a mass uprising against the British took place, known as the Sepoy Rebellion (Figure 5.21), which claimed the lives of tens of thousands of Indians when it was brutally crushed by the colonial authorities. Although it failed, the action forced the British crown to assume direct control over this land, rather than administer it through the East India Company, which was disbanded as a result. The Sepoy Rebellion was an important moment in the long struggle for Indian independence, which eventually succeeded in 1947 when it was led by Mahatma

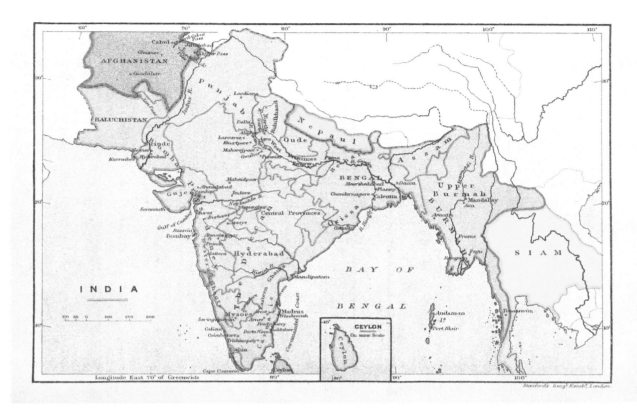

Figure 5.20: The British colony of India included the entirety of South Asia, including what would later become the countries of Pakistan, India, and Bangladesh. The British also colonized Ceylon, later renamed Sri Lanka.

Mohandas Ghandi. It reminds us that colonialism was always resisted and contested.

EAST ASIA

East Asia, comprised of China, Korea, and Japan, also had a unique colonial trajectory. Japan, as noted earlier, was never colonized. When it emerged from a long period of isolation that began in the early seventeenth century and lasted until 1868, Japan rapidly Westernized and industrialized. It became the only non-Western power to build a Western-style colonial empire, gradually expanding its power through Asia. Korea was taken over by Japan in 1895, annexed in 1905, and occupied until the end of World War II. Similarly, Japan occupied Taiwan from 1895 to 1945.

Figure 5.21: The Sepoy Rebellion in India did not succeed in driving the British out, but it did cause the collapse of the British East India Company and sowed the seeds of later nationalist movements. It illustrates that colonialism was always resisted by colonized peoples.

China was a very different story. The Manchus, from Manchuria to the north, an Altaic people (Chapter 3), ruled China as the Chi'n dynasty from 1644 until the nationalist revolution of 1911. Under the Manchus, the Chinese government was weak and corrupt. Except for a few cities along the coast, China was never formally colonized; rather, European control operated through a pliant Manchu government. Chinese laborers emigrated to British colonies in Southeast Asia and to the United States, where they built railroads in the West. British, French, German, and American trade interests purchased vast amounts of Chinese tea, silks, spices, and porcelain. In fact, Britain bought more from China than it sold to China (a negative trade balance) in the eighteenth and early nineteenth centuries. The situation changed when the British began introducing large amounts of opium to China. By the 1830s, opium addiction was widespread in China, causing severe social disruptions. Nonetheless, profits were more important than people, and the British trade balance was restored.

In a rare moment of defiance, the Manchu government resisted opium imports. This action led to two conflicts between Britain and China, the Opium Wars of the 1840s. The British won easily and seized several coastal cities, including Hong Kong and Shanghai (Figure 5.22). In these "treaty ports," only Western, not Chinese, law applied. Britain held Hong Kong until 1997.

Chinese resentment against the Manchus culminated in the Taiping Rebellion, a huge uprising in the southern part of the country led by Chinese Christians (Figure 5.23). Lasting from 1851 to 1864, this rebellion led to the deaths of more than 20 million people but was ultimately crushed by the Chinese government with Western backing. The shorter Boxer Rebellion of 1899 to 1901 was more explicitly anti-Western. These revolts set the stage for the successful nationalist revolution of 1911, which ended Manchu rule.

SOUTHEAST ASIA

The peninsula of Indochina and the islands of Southeast Asia, long home to a diverse series of peoples and civilizations, were conquered by a variety of different European powers (Figure 5.24). The Philippines, named after King Philip II, was Spain's only Asian colony and served as the western endpoint of the trans-Pacific galleon trade. Spanish rule in the Philippines included sugar plantations, spread of the Spanish language, and made the Philippines becoming the only predominantly Christian country in Asia. The United States took the Philippines from Spain in 1898 during the Spanish-American war. The Philippines became independent shortly after World War II.

Starting in the midnineteenth century, the French controlled much of Indochina, present-day Vietnam, Laos, and Cambodia. French rule shaped the design and architecture of cities, such as Saigon, introduced rubber plantations, and spread Catholicism in Vietnam, although it never displaced indigenous Buddhism. French domination did not end until 1954, when a Communist insurgency defeated them in the Battle of Dien Bien Phu. This event laid the foundations for the American military involvement in Vietnam, which lasted until 1975.

Elsewhere in Southeast Asia, Britain ruled Burma (today's Myanmar) and heavily influenced the economy of Thailand. These areas became rice exporters for other parts of the British Empire. The British controlled the colony of Malaya (later Malaysia), including the

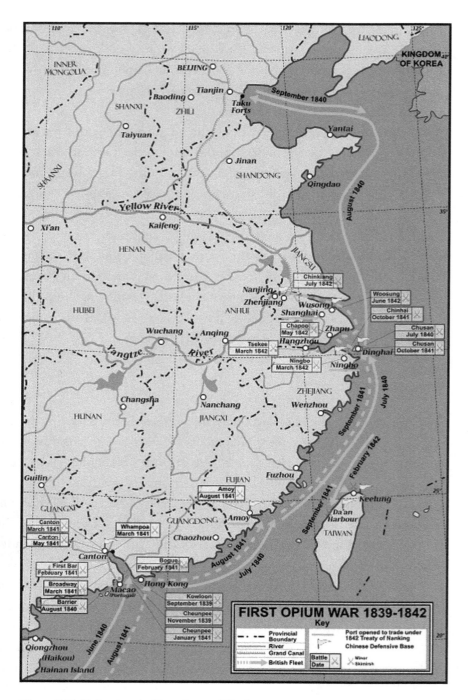

Figure 5.22: Following the First Opium War from 1839 to 1842, Europeans held several cities along China's coast. These "treaty ports" were the only places in China that Europeans ruled directly. Elsewhere in China, colonialism acted through the corrupt government of the Manchu dynasty, turning the country into a vast supplier of cheap labor and raw materials and a market for Western goods and bringing about untold levels of human misery.

Figure 5.23: The Taiping Rebellion was a massive uprising against the Chin or Manchu dynasty that led to millions of deaths. It was ultimately crushed by Western powers, including the United States.

strategically critical Malacca Straits linking the Indian and Pacific Oceans. The British founded the city of Singapore as a naval station and commercial center to exert control over this region. Malaya, like other colonies in the area, became a major producer of rubber products, as well as timber and tin. Europeans brought large numbers of Chinese people to Southeast Asia as craft workers, bankers, and other professionals. These Chinese migrants, known as "overseas Chinese," became the majority in Singapore. Because they were concentrated in skilled professions, they formed a small but wealthy ethnic minority in many other areas, and they have often been the targets of attacks by radical Muslims there.

Indonesia, now the fourth most populous country in the world, was dominated by the Dutch for several hundred years. Dutch rule, starting with the founding of the Batavia colony on Java in the eighteenth century, gradually expanded to include the other islands. The primary

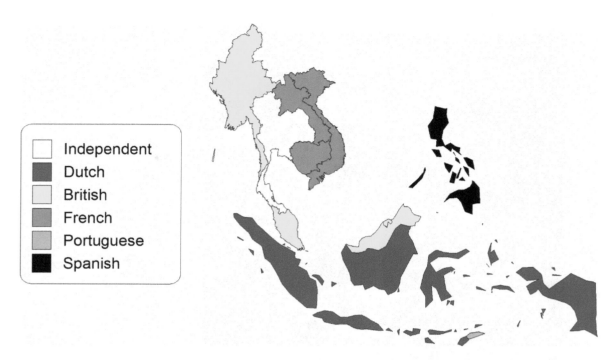

Figure 5.24: As in several regions of the world, several European powers established colonies in Southeast Asia. The Dutch colony of Indonesia was the largest. This region became an important exporter of rice, spices, tin, and rubber.

institution involved was the Dutch East Indies Company (known by its Dutch initials VOC), a chartered crown monopoly similar to the British East India and British West Indies Companies. Indonesia became a significant source of spices, tropical hardwoods, rubber, cotton, and palm oils, all of which reflect the broader colonial process of turning colonies into resource-exporting areas.

OCEANIA

Australia, New Zealand, and the Pacific Ocean islands, which are commonly grouped as the region of Oceania, formed yet another domain of colonialism. Australia was inhabited for thousands of years by indigenous aborigines, most of whom were eradicated by the colonizing British. Like the native populations of the Americas, the aborigines of Australia lacked many domestic animals and were susceptible to diseases such as smallpox. The native population on the

Island of Tasmania, south of Australia, was exterminated, often through acts of genocide. Today, aboriginals make up only 1% of Australia's population. The continent originally served as a penal colony for criminals. In the nineteenth century, it became a significant exporter of wheat and beef.

Britain also colonized New Zealand, but the much larger native Maori population survived and continues to form a significant presence there. The introduction of refrigerated shipping in the late nineteenth century turned this island nation into a major producer of lamb and dairy products.

Finally, the countless islands of the Pacific Ocean, home to varying groups of Polynesians, Micronesians, and Melanesians, were conquered by the British and French. Captain Cook sailed through in the seventeenth century, paving the way for followers. French colonies, such as Tahiti, became well known in Europe as lush paradises. Fishing and whaling interests used these islands to refuel in the nineteenth century. Following World War II, when the United States drove Japan out of the Pacific, America became the leading political force in the area.

THE EFFECTS OF COLONIALISM

Colonialism obviously had enormous, lasting effects on colonized places, which are summarized below.

ANNIHILATION OF INDIGENOUS PEOPLES

Often, colonialism involved the destruction of the people who were conquered. At times, this involved open genocide, such as in Australia, in which more than 90% of the aboriginal population was exterminated. In the New World, disease led to the deaths of tens of millions, almost depopulating the Western Hemisphere. The African slave trade devastated tribal societies on that continent. Although colonizers did not always assert control so bluntly, the European conquest of the world often involved violence.

RESTRUCTURING ECONOMIES AROUND RAW MATERIALS

The incorporation of colonies into a worldwide division of labor led to their development as exporters of raw materials, including logging, fishing, mining, and agriculture. Plantations in Africa, Southeast Asia, and the Americas grew cash crops for sale abroad, including sugar, cotton, fruits, rubber, tea, coffee, jute, and tobacco. Silver, tin, gold, copper, and other metallic ores were mined using slave labor or peasants working in slave-like conditions. For this reason, many developing countries today export raw materials and must import finished goods.

FORMATION OF A DUAL SOCIETY

Colonialism produced great inequality in colonized societies. Often, colonial powers used a small native elite drawn from a privileged ethnic minority to assist them in governing the colonies, part of a colonial strategy of "divide and conquer." For example, the French utilized the Alawites in Syria, an offshoot of Shiite Islam (Chapter 4), which continues to control the government today. The Germans and Belgians favored lighter-skinned Tutsi over darker-skinned Hutu in Rwanda and Burundi, setting the state for the tribal bloodbath there in 1994. The British relied on the Muslim Mughal rulers to govern Hindu India. The overseas Chinese effectively ran the economies of many Southeast Asian colonies.

For the bulk of the population, colonialism entailed declining economic opportunities. Colonial powers disrupted traditional patterns of agriculture, often with disastrous effects. Europeans often commandeered the best farmland for themselves. People living in dry climates in western Africa coped well with drought until the British and French forced them to grow cash crops that needed more moisture than did traditional crops. Huge famines shook India, Egypt, and China in the nineteenth century because people grew cash crops for export instead of food for their own consumption.

POLARIZED GEOGRAPHIES

As colonial societies became more unequal, their spaces became more polarized. Because Europeans relied on maritime trade and transportation, ports became important centers of commerce. Their growth often came at the expense of traditional capitals further inland. For example, in Peru, Lima displaced the Incan capital of Cuzco; in western Africa, the famous trading city of Timbuktu declined as new maritime routes flourished (Chapter 2); in India, coastal cities, such as Calcutta, Mumbai (formerly Bombay), and Chennai (formerly Madras) displaced the Mughal capital of Delhi; in Myanmar (formerly Burma), Mandalay, in the northern interior, fell far behind coastal Rangoon; in Vietnam, the imperial capital Hue declined to the advantage of Saigon; and in Indonesia, the traditional center of Jogjakarta became marginalized by the Dutch port city of Batavia, later Jakarta. As cities grew and offered more opportunities, millions of people left the poorer rural areas in waves of rural-to-urban migration.

From the coasts, railroads extended colonial control into the interior, often reaching into mineral-rich regions or plantations. A long-term consequence was that the road and rail networks of developing countries often bear little resemblance to the distribution and needs of the population that lives there. Rather, these transit systems were constructed to facilitate the export of raw materials to the colonizing country, and today, to the global economy.

COLONIAL POLITICAL GEOGRAPHIES

Europeans imposed a new, and often disastrous, set of boundaries on their colonies. Unlike in Europe, where countries were centered on some degree of ethnic similarity, those that emerged from colonial boundaries bore little resemblance to the distributions of different ethnic, linguistic, and religious groups. In Africa, where this process was the most notorious following the Berlin Conference of 1884, colonialism led to highly artificial borders that are still the source of considerable ethnic strife today. Similarly, Burma and Afghanistan were creations of the British. Even India, which is more culturally diverse than all of Europe, was stitched into one country; surprisingly, upon independence, it divided only into three (including Pakistan and Bangladesh). In such diverse societies, where local religious, ethnic, and tribal loyalties take precedence over allegiance to the state, political conflicts greatly impair economic growth and development.

CULTURAL WESTERNIZATION

As noted previously, colonialism is not simply an economic or political process, but also a cultural and ideological one. Western economic and political control was accompanied by the imposition of Western culture. Missionaries, for example, spread Christianity throughout the colonial world, sometimes successfully (as in parts of Africa, Latin America, and the Philippines) and sometimes less so (as in most of Asia) (Chapter 4). Often religion justified colonialism as an expression of divine will, and conversion was often forced upon native peoples. Similarly, colonial languages, such as English, Spanish, French, and Portuguese became spread worldwide (Chapter 3). School systems in colonies, set up to benefit the ruling elite, offered extensive instruction in the history and culture of the colonizing country, but little about the society in which the students lived.

More broadly, colonialism reflects the broader process by which global capitalism homogenizes lifestyles, values, and role models around the world. As European-style governments, schools, religions, and products became widespread, they changed how colonized peoples thought about themselves. In this way, capitalism became woven into the fabric of everyday life across the planet, displacing traditions and substituting in its place a culture based on markets, profit, and commodities.

THE END OF COLONIALISM

European colonial empires lasted for nearly 500 years, yet they ultimately collapsed. In Latin America, this process began after the Napoleonic wars of the early nineteenth century weakened Spain and Portugal sufficiently to allow the colonies to break away. In Africa and Asia, the end of colonialism came much later in a similar fashion, after World Wars I and II greatly diminished Europe's economic, military, and political power.

The international environment following World War II provided an ideal opening for various nationalist and independence movements in the Arab world, Africa, India, and Southeast Asia. Devastated by the war, Europeans had lost much of their ability to exert control over their colonies. The Japanese occupation of Southeast Asia destroyed the myth of European invincibility. Moreover, the Cold War rivalry between the United States and the Soviet Union (1945–1991) allowed political leaders in the developing world to play the superpowers off against one another. Although some independence movements succeeded peacefully, others had to resort to violence. The British left India in 1947 without an armed struggle in response to Mohandas Gandhi's campaign of nonviolent resistance. Some African colonies were granted independence without violence. However, the Vietnamese and Algerians waged protracted war on the French and defeated them in battles that claimed the lives of millions. In the Portuguese colonies of Angola and Mozambique, guerilla wars against the colonists lasted for decades, eventually succeeding in 1975.

Diverse groups struggled against colonial powers. Sometimes communists were involved; other groups included nationalists, students, workers' organizations, and religious movements. Often, independence movements were led by intellectuals educated in the West, such as Ghana's Kwame Nkrumah, Vietnam's Ho Chi Minh, and India's Gandhi.

As a result of this process of decolonization, the number of independent states multiplied rapidly in the 1950s, 1960s, and 1970s. Today, very few official colonies remain, including a few islands in the Caribbean and the Pacific Ocean, Puerto Rico, and French Guiana in South America. Whether colonialism is truly dead, however, is less certain. Chapter 7 discusses neo-colonialism: the idea that some countries and cultures exert cultural and economic control over others without ruling them politically. Because Western transnational corporations dominate the economies of many former colonies, many question whether colonialism is truly over or if it has simply taken a new name and new form.

■ KEY TERMS

Agglomerate: to cluster together; economically, this allows firms to reduce transport costs and to acquire specialized information and services

Animate energy: energy derived from living muscle, whether human or animal power.

Bourgeoisie: a Marxist term for the ruling class of capitalists (i.e., those who own capital)

Capital: funds and goods used to produce commodities, including financial sums, machinery, equipment, and buildings

Commodities: products produced for sale on the market to earn a profit

Comparative advantage: in market-based societies, the ability of a region to produce an output more efficiently or profitably than other regions

Feudalism: the social, political and economic system that dominated Europe for 1,000 years before the rise of capitalism; it was also found in Japan and other parts of the world

Guilds: feudal organizations for the production of handicraft goods and to train apprentices

Inanimate energy: energy that is not derived from living muscle, including fossil fuels, as well as nuclear, wind, and solar energy

Infrastructure: the transportation and communications network of a region, as well as systems to control water and generate electricity

Market: an institution that links buyers and sellers of commodities

Serfs: the largest group of agricultural workers under feudalism; unlike slaves, serfs at least owned themselves, but were bound by law and custom to work the land for aristocrats

Three-field system: a form of farming in which one-third of the fields are left fallow to enhance their productivity

Time-space compression: the "shrinking" of distances due to improved transportation and communications technologies

Urbanization: the process by which a growing share of a population comes to live in cities

■ STUDY QUESTIONS

1. Why should geographers study historical contexts?
2. Describe the key economic, social, and political aspects of feudalism.
3. What was the fourteenth-century bubonic plague, and what impacts did it have?
4. How does capitalism differ from feudalism economically and socially?
5. How does social class function under capitalism?
6. How does the role of markets in capitalism differ from that in noncapitalist societies?
7. What territorial changes accompanied the emergence of capitalism?
8. How did capitalism unleash new ideas and ways of looking at the world?
9. How is the nation-state related to the growth of capitalism?
10. When and where did capitalism begin?
11. What is industrialization?
12. When and where did the Industrial Revolution begin?
13. What are some major economic, social, and geographic impacts of the Industrial Revolution?
14. How did Europe manage to colonize the rest of the world?
15. How did colonialism differ among Latin America, Africa, and Asia?
16. What are five ways in which colonialism affected the societies and geographies of the colonies?
17. When did colonialism come to an end, and why?

BIBLIOGRAPHY

Abu-Lughod, J. (1989). Before European Hegemony: The World System A.D. 1250–1350. *New York: Oxford University Press.*

Berman, M. (1982). All That Is Solid Melts Into Air: The Experience of Modernity. *New York: Penguin Books.*

Blaut, J. (1993). The Colonizer's Model of the World: Geographical Diffusionism and Eurocentric History. *New York: Guilford Press.*

Boorstin, D. (1983). The Discoverers. *New York: Random House.*

Cipolla, C. (1965). Guns, Sails and Empires: Technological Innovation and the Early Phases of European Expansion, 1400–1700. *New York: Minerva.*

Crosby, A., (1986). Ecological Imperialism: The Biological Expansion of Europe, 900–1900, *Cambridge: Cambridge University Press.*

Diamond, J. (1999). Guns, Germs and Steel. *New York: Norton.*

Edgerton, S. (1975). The Renaissance Rediscovery of Linear Perspective. *New York: Icon.*

Eisenstein, E. (1979). The Printing Press as an Agent of Change. *New York: Cambridge University Press.*

Frank, A., & Gills, B. (Eds.). (1993). The World System: Five Hundred Years or Five Thousand? *London: Routledge.*

Headrick, D. (1981). The Tools of Empire: Technology and European Imperialism in the Nineteenth Century. *New York: Oxford University Press.*

Headrick, D. (1988). The Tentacles of Progress: Technology Transfer in the Age of Imperialism, 1850–1940. *New York: Oxford University Press.*

Hugill, P. (1993). World Trade Since 1431: Geography, Technology and Capitalism. *Baltimore: Johns Hopkins University Press.*

Hugill, P. (1999). Global Communications Since 1844: Geopolitics and Technology. *Baltimore: Johns Hopkins University Press.*

Johnson, P. (2002). The Renaissance: A Short History. *New York: Modern Library.*

Jones, E. (1981). The European Miracle: Environments, Economies, and Geopolitics in the History of Europe and Asia. *New York: Cambridge University Press.*

Kern, S. (1983). The Culture of Time and Space 1880–1918. *Cambridge, MA: Harvard University Press.*

Landes, D. (1969). The Unbound Prometheus: Technological Change and Industrial Development in Western Europe from 1750 to the Present. *New York: Cambridge University Press.*

Marks, R. (2007). The Origins of the Modern World: A Global and Ecological Narrative From the Fifteenth to the Twenty-first Century. *Lanham, MD: Rowman and Littlefield.*

McNeill, W. (1963). The Rise of the West. *Chicago: University of Chicago Press.*

McNeill, W. (1982). The Pursuit of Power: Technology, Armed Force, and Society Since A.D. 1000. *Chicago: University of Chicago Press.*

Mignolo, W. (1995). The Darker Side of the Renaissance: Literacy, Territoriality, and Colonization. *Ann Arbor: University of Michigan Press.*

Ong, W. (1982). Orality and Literacy. *London: Routledge.*

Pomeranz, K., & S. Topik. (1999). The World That Trade Created: Society, Culture, and the World Economy, 1400–the Present. *Armonk, NY: M.E. Sharpe.*

Sobel, D. (1998). Longitude. *London: Fourth Estate.*

Thompson, E. (1967). Time, work-discipline, and industrial capitalism. Past and Present, 38, *56–97.*

Tilly, C. (1990). Coercion, Capital, and European States, AD 990–1990. *Oxford: Blackwell.*

Tracy, J. (Ed.). (1991). The Political Economy of Merchant Empires. *Cambridge, UK: Cambridge University Press.*

Warf, B. (2008). Time-Space Compression: Historical Geographies. *London: Routledge.*

Wintle, M. (1999). Renaissance maps and the construction of the idea of Europe. Journal of Historical Geography, 25, *137–165.*

Withers, C. (2007). Placing the Enlightenment: Thinking Geographically about the Age of Reason. *Chicago: University of Chicago Press.*

Wolf, E. (1982). Europe and the People Without History. *Berkeley and Los Angeles: University of California Press.*

IMAGE CREDITS

- Figure 5.1: Copyright © OwenBlacker (CC BY-SA 3.0) at https://commons.wikimedia.org/wiki/File:Holy_Roman_Empire_11th_century_map-en.svg.
- Figure 5.2: Guido Reni, http://commons.wikimedia.org/wiki/File:Pope_Gregory_XV.jpg1. Copyright in the Public Domain.
- Figure 5.3: http://commons.wikimedia.org/wiki/File:Reeve_and_Serfs.jpg. Copyright in the Public Domain.
- Figure 5.4: Greudin, http://commons.wikimedia.org/wiki/File:Carcassonne-vignes.jpg. Cleared via GNU General Public License.

- Figure 5.5: Derfel73, http://commons.wikimedia.org/wiki/File:Archaic_globalization.svg. Copyright in the Public Domain.
- Figure 5.6: Copyright © Roger Zenner (CC BY-SA 3.0) at http://commons.wikimedia.org/wiki/File:Bubonic_plague_map.PNG.
- Figure 5.7: Lampman, http://commons.wikimedia.org/wiki/File:Late_Medieval_Trade_Routes.jpg. Copyright in the Public Domain.
- Figure 5.8: Copyright © Doc Brown (CC BY-SA 3.0) at http://commons.wikimedia.org/wiki/File:Kaart_Hanzesteden_en_handelsroutes.jpg.
- Figure 5.9: http://commons.wikimedia.org/wiki/File:Press_-Bettman.jpg. Copyright in the Public Domain.
- Figure 5.10: Andrei Nacu, http://commons.wikimedia.org/wiki/File:Reformation.gif. Copyright in the Public Domain.
- Figure 5.11: Newton Black and Harley Davis, https://commons.wikimedia.org/wiki/File:Watt7783.png, pp. 220. Copyright in the Public Domain.
- Figure 5.12: Karl Edouard Biermann, https://www.flickr.com/photos/blvesboy/2121487586/. Copyright in the Public Domain.
- Figure 5.15: Albrecht / Arthur Wellesley / XGustaX, http://commons.wikimedia.org/wiki/File:Spanish_colonization_of_the_Americas.png. Copyright in the Public Domain.
- Figure 5.16: Esemono, http://commons.wikimedia.org/wiki/File:Non-Native_American_Nations_Control_over_N_America_1810.png. Copyright in the Public Domain.
- Figure 5.17: Copyright © Sémhur (CC BY-SA 3.0) at https://commons.wikimedia.org/wiki/File:Triangular_trade.png.
- Figure 5.18: Isriya Paireepairit, https://www.flickr.com/photos/isriya/5227947131/. Copyright © by Isriya Paireepairit.
- Figure 5.19: Copyright © Gabagool (CC BY-SA 3.0) at http://commons.wikimedia.org/wiki/File:OttomanEmpire1600.png.
- Figure 5.20: Edgar Sanderson, History of England and the British Empire, 1893. Copyright in the Public Domain.
- Figure 5.21: Granger, http://commons.wikimedia.org/wiki/File:SepoyMutiny.jpg, ~1. Copyright in the Public Domain.
- Figure 5.22: Copyright © Philg88 (CC by 4.0) at http://commons.wikimedia.org/wiki/File:First_Opium_War_1839-42_Conflict_Overview_EN.svg.
- Figure 5.23: http://commons.wikimedia.org/wiki/File:Suppression_of_the_Taiping_Rebellion.jpg. Copyright in the Public Domain.

Chapter Six

GLOBALIZATION

MAIN POINTS

1. Globalization entails the growth in the volume, range, velocity, and impacts of international linkages.
2. International trade, which makes up about 40% of what the world produces, both reflects and, in turn, shapes geographic differences in production costs and productivity.
3. International movements of money through capital markets, including global banking, are enormous, and have profound effects on national economies.
4. Transnational corporations are some of the major actors in globalization and are responsible for a large share of international trade and foreign investment.
5. Around the world, groups of countries have formed regional trade associations to spur economic growth, such as the European Union and the North American Free Trade Agreement.
6. Tourism, the world's largest industry by employment, sees about 15% of the world's population visit another country every year and is a major source of revenue for many places.
7. Globalization also includes the internationalization of culture, which has largely involved the export of American culture to the rest of the world.

8. Numerous myths surround globalization, such as: (1) it is only an economic process; (2) it homogenizes local cultures; (3) it has only begun recently; (4) it means the end of the nation-state; (5) it is unstoppable; and (6) it is beneficial to everyone.

9. Globalization is resisted in many ways by many different groups. Their efforts range from peaceful grassroots protests to violent attacks against institutions perceived to be responsible for destroying traditional ways of living. By threatening traditional values, globalization has unleashed a worldwide torrent of religious fundamentalism.

We live in a globalized world, although many people may not appreciate what that means. Many firms in the United States are owned by foreigners. Our banks trade trillions of dollars on the foreign currency exchange market every day, and our economy is financed, in part, by Chinese purchases of U.S. Treasury Department bonds. In the grocery store, fruits from Latin America, meat from Argentina, wine from Australia, vinegar from Italy, and canned goods from France are stocked on the shelves by workers from Guatemala, El Salvador, or Vietnam. The vegetables we eat were picked by workers from Mexico. Our clothes are often made in China, and our cell phones assembled in Guangdong, a southern province of that country. "American" cars may be engineered in Germany and contain a combination of Korean, Mexican, Malaysian, and Canadian parts. They might be shipped to the United States by a Greek vessel, built in Korea, which employs Philippine sailors but is registered in Panama and insured in Britain. The gasoline we put in them comes from Saudi Arabia, Gabon, and Venezuela. Our textbooks may be copy edited in India and printed in China. The software on our laptops, assembled in Tijuana, Mexico, might also be written in India. In this and countless other ways, the global economy, through international flows of money, goods, people, and information, shapes our daily lives and our prospects for the future.

Globalization refers to a broad set of processes that span the world. There is no single process of globalization; rather, a diversity of intertwined processes reflects the persistent tendency of people, capital, goods, ideas, technologies, and diseases to move across the planet. An alternative definition is that globalization involves a massive compression of time and space. Globalization is, thus, a prime example of the theme of regional interdependence raised in Chapter 1.

This chapter examines several dimensions of globalization. It starts with a brief overview of international trade, one of the major mechanisms of globalization for centuries. Long-distance trade long preceded the rise of capitalism (Chapter 2), but it grew substantially during the centuries of colonialism and throughout the twentieth century. Next, it turns to the globalization of money and banking; international finance is one of the dominant forms of contemporary globalization. It then summarizes the important issue of foreign direct investment, especially as practiced by the world's major drivers of globalization, transnational corporations (firms that operate in more than one country). The fourth part turns to three influential organizations that shape global processes, the World Trade Organization, the International Monetary Fund, and the World Bank. Regional economic integration, including the European Union, the Association of Southeast Asian Nations, and the North American Free Trade Agreement, are examined in the next part. Tourism—the world's largest industry in terms of employment—is scrutinized. Globalization is not simply an "economic" process, but very much cultural and ideological, as depicted in the section concerning how it shapes lifestyles, viewpoints, and behaviors. The next section examines several inaccurate myths surrounding globalization—such as the idea that globalization is inevitable or that it benefits everyone. Finally, the chapter focuses on anti-globalization, or resistance to the encroachment of Western, especially American, influences. Opposition to globalization ranges from subtle contempt to violent terrorism.

6.1 INTERNATIONAL TRADE

Trade among countries has long been a central part of capitalism and a major factor linking various parts of the world together. Trade networks existed in prehistoric times (Chapter 2). From roughly the time of Jesus to the eighteenth century, Europe was linked to the Middle East and China through the Silk Road routes, along which flowed a huge variety of goods, as well as innovations, ideas, and diseases. But starting in the nineteenth century, particularly with the rise of the railroads and the steamship, international trade flourished dramatically, welding together different local economies into a single, integrated world system. In the twentieth century, as national barriers to trade began to decline, trade accelerated yet further. Today, trade makes up 40% of the world's output.

INTERNATIONAL TRADE AND SPECIALIZATION

Why are so many countries, large and small, rich and poor, deeply involved in international trade? One answer lies in the unequal distribution of productive resources among countries (i.e., uneven spatial development), which can be offset, to some extent, by trade. Countries trade the resources they have in abundance in exchange for the things they lack. However, whether a country can export successfully depends not only on its resources but also how its production system is organized, its efficiency and competitiveness, and related factors, such as government policies and the value of its currency, which affect the prices of imports and exports.

Production factors—labor, capital, technology, entrepreneurship, and land containing raw materials—have always varied enormously from country to country. Some countries, such as the United States, Japan, and Germany, have populations large enough to support large complexes of industry and domestic markets. Others, such as Singapore or Bahrain, do not, and are highly reliant on foreign trade. One country, such as China, may be home to workers adept at running modern machinery. Another, such as Japan, abounds with scientists and engineers specializing in products backed by extensive research. A third group has huge pools of unskilled workers; this includes regions such as Indonesia or the Philippines. The imbalance of natural and human-made resources accounts for much of the international interchange of production factors and the products and services. Numerous factors may reduce the ability of countries to best use their productive advantages, including inflation, exchange rates (the value of one currency in terms of other currencies), labor conditions, and government policies. Other countries are hobbled by the legacy of colonialism, drought, or political violence. Governments can encourage or discourage their export sectors through trade policies and public investments in infrastructure or education.

The production factors mentioned above have been identified in current international and economic circles as the keys to the international competitiveness of countries (Figure 6.1):

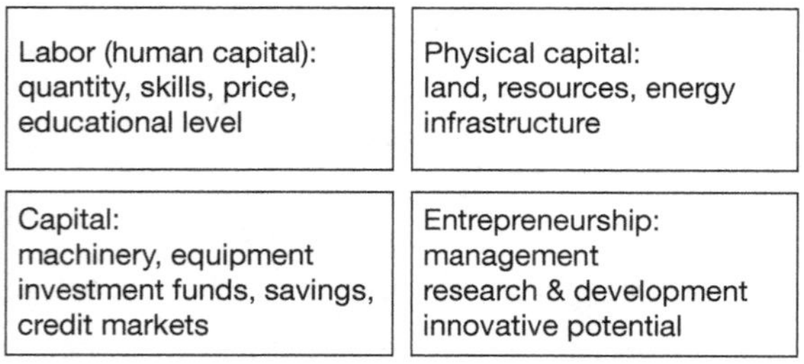

Labor (human capital): quantity, skills, price, educational level	Physical capital: land, resources, energy infrastructure
Capital: machinery, equipment investment funds, savings, credit markets	Entrepreneurship: management research & development innovative potential

Figure 6.1: Several intertwined factors of production, including labor, land, capital, and entrepreneurial ability, shape a country's economic potential, ability to innovate, productivity growth, and thus standards of living.

1. Human resources: the quantity of labor, skill, educational level, productivity, and cost of labor.
2. Physical resources: land; raw materials and their costs, relative location, and transport costs. Also in this category is infrastructure, such as transportation and communications systems, water management, electrical grids, ports, and airports.

3. Capital resources: all aspects of money supply and availability to finance the industry and trade from a specific country. These include the amount of investment capital available; the savings rate; the health of money markets and banking in the host country; government policies that affect interest and exchange rates and the money supply; levels of indebtedness; trade deficits; and public and international debt.
4. Entrepreneurship: knowledge-based resources, including management, the scientific and technical community, research and development institutions, and innovative capacity.

One of the hallmarks of capitalism is its tendency to generate uneven economic landscapes—that is, great differences in the types of economic activity from place to place, as well as the standards of living and life chances that those activities create. Different regions have long specialized in the production of different types of goods and services. During the Industrial Revolution, for example, Britain became a major producer of textiles, ships, and iron. France produced silks and wine. Spain, Portugal, and Greece generated citrus, wine, and olive oil. Germany, by the end of the nineteenth century, was a major exporter of heavy manufactured goods and chemicals. Czechs sold glass and linens. Scandinavia sold furs and timber. Iceland and Canada exported cod to the growing middle classes.

Within the United States, similarly, different places acquired advantages in some goods and not others. During the nineteenth century, New England was dominated by light industry, especially textiles. The Manufacturing Belt of the Midwest and Northeast became the centers of heavy industry. Appalachia developed a large coal industry to feed the furnaces of the industrial core. The South grew crops, such as cotton and tobacco. The Corn and Wheat Belt of the Midwest became the agricultural behemoth of the world. The Rocky Mountain states sold coal and copper. The Pacific Northwest was home to the expanding timber and lumber industry. Railroads tied these regions together into a national economy, which facilitated the shipment of resources from the West and Midwest to the urban centers of the East Coast.

When regions or countries specialize in the production and export of some goods or services, economic geographers say those regions or countries enjoy a **comparative advantage**. This notion was first introduced by the famous nineteenth-century economist David Ricardo, a contemporary of Thomas Malthus and one of most famous figures in the history of economics. Trade, Ricardo argued, forces producers to compete and become efficient: local producers that cannot compete with cheaper or better imports are driven out of business. Thus, trade not only reflects the uneven distribution of production factors around the world, but it also, in turn, helps to shape the distribution of those factors. A country often uses comparative advantage to justify free trade with other countries on the grounds that trade enlarges markets and makes the most efficient use of productive resources. Thus, international trade and the global division of labor are closely linked to one another.

INTERNATIONAL TRADE AND THE GLOBAL DIVISION OF LABOR

International trade is best understood from the standpoint of efficiency, but is it fair given the relationship of **unequal exchange** between developed and developing countries? This question is raised by theorists who view capitalism critically and for whom free trade is a weapon wielded by rich countries to take advantage of the markets of poor ones. Their argument is that an artificial division of labor has made it difficult for developing countries to earn revenues from free trade with developed countries.

The British were instrumental in creating a global division of labor in the eighteenth and nineteenth centuries. Implicit in the argument for free trade was the notion that what was good for Britain was good for the world. But, free trade was established within a framework of inequality among countries. At first, Britain—and later, the United States—used protectionism extensively. For example, Britain levied high tariffs (taxes on imports) against textiles imported from its colony in India. Britain found free trade and competition agreeable only after becoming established as the world's most technically advanced industrial nation.

Having gained an initial advantage over other countries, Britain then threw open its markets to the rest of the world in the mid-nineteenth century. Other countries were pressured to do the same. The pattern of specialization that resulted was obvious. Britain concentrated on producing manufactured goods, such as vehicles, engines, machine tools, paper, and textile yarns and fabrics, and exporting them in exchange for a variety of primary products, such as furs, wines, silks, and bulk imports, such as timber, grains, fruit, and meat. In this way, uneven spatial development was fostered and perpetuated. Although many countries gained from this division of labor, none gained more than Britain. Thus, globalization always involves social relations of power, one of the analytical themes raised in Chapter 1.

The only way other countries could break out of this division of labor was by interfering with free trade. The United States was highly protectionist in the nineteenth century, and tariffs on imports were a major source of federal government revenues. Germany, France, Japan and other countries with emerging industries did the same.

The global division of labor changed little until after World War II. At this point, a new global structure began to evolve, including the independence of many former colonies and the growth of transnational corporations. The basic trend was export-led industrialization concentrated in a few countries. For the best-off poor countries, industrial growth was geared toward the needs of the old imperial powers. Thus, the growth of manufacturing in these developing countries was not a sign of their emancipation from an unfair division of labor.

A common myth is that national competitiveness depends on cheap labor and abundant natural resources. Is cheap labor central to economic success? Cheap labor is, on a worldwide scale, virtually everywhere. The developing world has vast pools of unskilled workers. Countries that have succeeded, such as Germany or Japan, have done so with labor costs well above those of their competitors, because their labor force is productive and well educated. In contrast, countries with the cheapest labor—for example, most of Africa—have done poorly in the global economy. Cheap labor is usually unskilled and unproductive labor. Neither are abundant natural resources necessary for economic success. Japan, for example, has done well despite having virtually no resources, and many developing countries with resources are trapped in low-wage economies that export their raw materials to developed nations.

With these comments in mind, globalization has also had enormous environmental consequences. In opening up once-isolated regions to the world economy, foreign investment and international trade have accelerated the extraction of resources from many formerly pristine areas. For example, the global demand for tropical hardwoods has increased deforestation in countries such as Brazil, Thailand, and Indonesia, which are home to fragile ecosystems and innumerable species. In this respect, globalization raises the loss of biodiversity. The demand for petroleum has led to increased drilling in places such as equatorial Africa, Brazil, and the South China Sea, which, sooner or later, will generate oil spills. The globalization of mining has led companies to seek mineral ores in places such as Brazil, Australia, and Congo, which involves building roads to move equipment there. Rising global demand for fish (e.g., for sushi) has led to severely overfished oceans, parts of which have become almost unprofitable (such as off of New England). It is true that these examples all generate jobs and foreign revenues, but critics worry that this rapid pace of resource extraction creates a temporary set of benefits at the expense of long-term environmental stability. Moreover, the transportation and consumption of evermore goods on a worldwide scale increases the demand for fossil fuels, which are closely linked to concerns about human-induced climate change, and produces enormous amounts of trash. For example, plastic that is not recycled tends to accumulate in the oceans. In both the Pacific and Atlantic Oceans, currents have created artificial continents consisting of hundreds of millions of tons of floating plastic, which degrades under sunlight and wreaks havoc on marine ecosystems. In short, like all human geographies, the ones created by globalization cannot be understood independently of the natural environment, a theme raised in Chapter 1.

In a global economy, flows of oil, minerals, and foodstuffs are widely available. What, then, does determine economic success? The key is growth in productivity. Over the long run, rising productivity creates wealth for everyone, even if not equally. Creating productivity and maintaining its growth reflect many factors, including the education and skills of the labor force, available capital and technology, government policies enhancing competition, and infrastructure investment. The goal of national development strategies is to move into high-value-added, high-profit, high-wage industries as rapidly as possible. To accomplish this goal, firms and countries should seek to sell high-quality goods at premium prices. Quality is a key variable here. Countries often acquire reputations for producing high- or low-quality goods, earning (or not earning) brand loyalty as a result. Finland is well known for its production of cell phones, for example, just as South Korea is now becoming well regarded for its automobiles. By moving into high-value-added goods, nations should seek to automate low-wage, low-skill functions and retain knowledge-intensive ones.

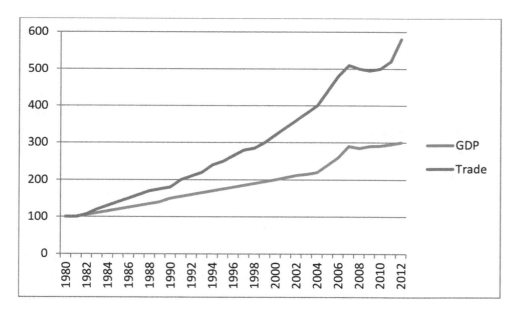

Figure 6.2: World trade has grown more rapidly than world output since 1950. Thus, national economies are more firmly linked to one another than ever before. Roughly 40% of the world's output crosses international borders.

World trade has jumped from $2 trillion annually in 1980 to more than $7 trillion today, or more than 40% of the gross world product (the sum of the value of everything produced worldwide), a clear sign of the increasing integration of national economies (Figure 6.2). Of these exports, agricultural goods make up 7.6%; mining ores, fuels, and minerals make up another 10.8%; all manufactured goods make up 61.3%; and all services make up 20.2%. Because a small group of economically developed countries produce the bulk of the world's tradable goods and have the disposable incomes to afford imports, global trade is largely confined to a "triad" consisting of Europe, North America, and East Asia.

6.2 INTERNATIONAL MONEY AND CAPITAL MARKETS

In addition to trade in goods and services, **capital markets**, or financial markets, form another dimension of globalization. *Capital* refers to goods used to make other goods, including machinery and equipment. In its liquid form—in which it can quickly be turned into cash—capital includes savings, stocks, bonds, loans, grants, and other financial instruments.

Capital takes two major forms. The first type involves lending and borrowing money. Lenders and borrowers may be in either the private or the public sector. The private sector includes banks, firms, and individuals, all of which may have accumulated deposits (banks) or savings (firms and individuals). The public sector includes governments or international institutions, such as the World Bank (which we will discuss below). The second type involves investment in the equity, or value, of companies, such as stocks and bonds. Capital is the result of historical development. It must be accumulated as a result of the willingness of a society to postpone consumption. Low-income countries have low capacities to generate investment capital; all the capital that they do generate is usually consumed domestically. High-income countries have a much greater capacity for generating investment capital, largely because they have higher productivity rates. They provide most of the world's private-sector capital, although a few fast-growing countries, such as those in Southeast Asia, are also capital exporters.

An important part of globalization is the internationalization of banking. International banks have existed for centuries (Chapter 5). For example, banking houses in the medieval period, such as those of the Medici in Florence, or the early modern era, such as the Rothschilds, helped to finance voyages of discovery, colonial operations, companies, and even governments and wars. The major banks of the two major colonial powers—Britain and France—have long been established overseas. American, Japanese, and other European banks became international much later. Major American banks moved into international banking in the 1960s, and the Japanese banks and their European counterparts followed in the 1970s.

Modern banks were enticed into international banking because of the explosion of foreign investment by transnational corporations in the 1950s and 1960s. The banks of different countries "followed the flag" of their domestic customers abroad. Once established overseas, many found international banking highly profitable. From their original focus on serving their domestic customers' international activities, banks evolved to service foreign customers as well, including foreign governments.

6.3 TRANSNATIONAL CORPORATIONS

Transnational corporations (TNCs) are the leading sources of foreign investment in the world and some of the most prominent actors involved in globalization. TNCs have a long history that stretches to the dawn of capitalism in the sixteenth century, when chartered monopolies, such as the Hudson's Bay Company, British East India Company, and Dutch East Indies Company, played a crucial role in European colonialism (Chapter 5). In the twentieth century, the numbers of TNCs grew exponentially (Table 6.1), as did their economic and political significance. Today, roughly 83,100 TNCs employ 120 million people worldwide, and generate

TABLE 6.1: ESTIMATED NUMBER OF TRANSNATIONAL CORPORATIONS SINCE 1700

YEAR	NUMBER OF TNCs
1700	1,000
1750	1,250
1800	1,500
1850	2,500
1914	3,000
1969	7,258
1988	18,500
1992	30,400
1997	53,100
1999	59,902
2005	63,000
2010	71,500
2015	83,100

the bulk of international trade and **foreign direct investment (FDI)**, or funds invested in buildings, equipment, and factories to produce goods and services overseas.

Most TNCs and FDI originate in economically developed countries that have the surplus capital to invest abroad. Despite common impressions that TNCs always invest in developing countries where labor costs are lower, the reality is that they concentrate most of their FDI in the developed world (Figure 6.3). American firms lead the world in FDI, but their share of the total is slipping. The rate of increase has been most rapid for companies from western Europe. However, some non-European countries have also increased their outflow of FDI, such as Brazil, Singapore, South Korea, and Taiwan. Investment in the developing world has focused mainly on a handful of countries—particularly China. Availability of natural resources, recent economic growth, and political and economic stability were among the factors that attracted foreign investment to less developed countries.

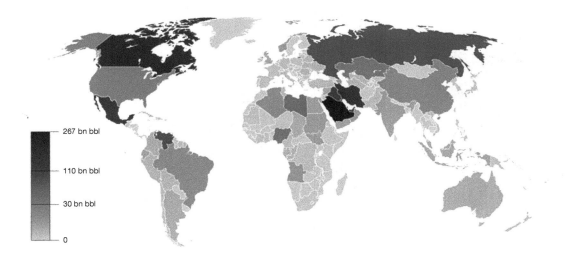

Figure 6.3: Contrary to popular myth, most of the world's FDI is channeled into economically developed countries, which have skilled labor, large markets, and well-developed infrastructures. China is the great exception to this trend; it is by far the largest recipient of FDI in the developing world.

6.4 THE WORLD TRADE ORGANIZATION, INTERNATIONAL MONETARY FUND, AND THE WORLD BANK

Globalization does not consist simply of firms and markets, but also of supranational organizations that play powerful roles in shaping trade, investment, and national policies around the world. Three especially important ones are the World Trade Organization, the International Monetary Fund, and the World Bank.

THE WORLD TRADE ORGANIZATION

In 1995, a permanent **World Trade Organization** (WTO) was established, to which most of the world's countries now belong (Figure 6.4). Before the existence of the WTO, countries with conflict over trade had to resolve their own problems. The WTO has the legal authority to settle conflicts between a pair of countries over trade disagreements.

Figure 6.4: WTO member states. Most of the world belongs to the WTO, which works to reduce trade barriers among member states.

Because its judgments are enforced through retaliatory trade sanctions by all other members of the WTO, each member country gives up a bit of its economic authority to this multilateral world organization. With the enactment of the WTO, a country loses some of its power to control flows across its borders. For some countries not abiding by the WTO trade agreements, this is a potentially devastating proposition.

The WTO attempts to reduce limits on imports, but many countries do not abide by its rules. For example, with this provision, Japan and South Korea restrict imports of rice, peanuts, dairy products, and sugar from the United States. The WTO also polices intellectual property rights, a topic of great interest to the United States, which has a strong competitive advantage in this arena. All signatories of the WTO are required to protect patents, copyrights, trade secrets, and trademarks. This measure is designed to end the wholesale pirating of computer programs, movies, television shows, musical recordings, books, and prescription drugs widely practiced in some developing countries—especially China. The WTO also prohibits members from requiring a certain proportion of content in products to be manufactured within their borders. This practice was widely employed as a device to limit the use of imported parts and components.

THE IMF AND THE WORLD BANK

The **International Monetary Fund (IMF)** and the International Bank for Reconstruction and Development, or **World Bank,** were established in 1945 as part of the Bretton-Woods agreement, one of the major pillars of the post–World War II global economy set up by the United States. Originally these institutions included only a handful of wealthy countries in Europe and North America. Today most countries in the world are members. These institutions offer significant sources of investment capital, especially aid for developing countries. The IMF was established to deal with short-term monetary issues, particularly exchange rate fluctuations, whereas the World Bank was established to finance long-term development projects.

The IMF provides short- to medium-term loans to member countries, and the World Bank provides longer-term loans for particular projects. Both institutions are supported by member governments, each paying a subscription or quota determined by the size of its economy. Because quotas determine a member's voting power, the banks are dominated by the most powerful economies—especially, by the United States.

The IMF and the World Bank were originally established to prevent a recurrence of the crisis of the 1930s: the Great Depression. For the first several decades, their primary concerns were preserving financial stability and promoting economic development. Starting in the 1980s, however, under pressure from the United States, these institutions adopted strong market-oriented positions and imposed them on the governments of less developed countries that needed their assistance, often at the cost of enormous human suffering. Loans from the IMF and the World Bank, therefore, tend to uphold the basis of U.S. economic and foreign policy. The **Washington Consensus** (so-called because of the influence of Washington, DC, policymakers) that these international organizations uphold involves, first, requiring less developed countries receiving loans to engage in tight monetary policy (restrict their national money supplies) to combat inflation. This condition serves investors well but raises interest rates for consumers. Second, recipients of aid must liberalize their financial markets, including a variety of deregulatory programs.

Thus, to acquire new loans from the IMF, governments of the less developed countries are often forced to adopt austerity measures such as reducing subsidies for the poor, including funds for public transportation, kerosene, or cooking oil. IMF-mandated privatization policies encourage or require governments to sell off public assets to private investors, often at reduced prices. All over the world, formerly public sector assets—for example, state-owned or state-operated power plants, bus routes, airlines, and telecommunications firms—are rapidly being sold to the private sector on the assumption that it is more efficient than the public sector. Yet public services often exist precisely, because unlike the market, they provide services to the poor. Similarly, IMF policies require trade liberalization, including the end of barriers, such as tariffs (import taxes), nontariff barriers (quotas and limits on the volume of imports), and subsidies of exports.

Critics note that such liberalization is often just a smokescreen for increased penetration of less developed countries' markets by firms from the United States and other Western countries and accuse the IMF of doing the dirty work of foreign capital. Even as less developed countries are forced to give up subsidies, the U.S. government lavishes subsidies on its farmers, giving them an unfair advantage in selling low-priced crops to foreign markets. In essence, the IMF imposes requirements on less developed countries that governments of developed countries would never accept.

By the IMF's own admission, its policies have sometimes worsened the problems of some countries. Tight monetary policies can generate recessions and lead to high unemployment. The economic models employed to buttress the IMF policies are often highly oversimplified and underestimate the complexity of the political and social contexts of the local countries. Many less developed countries lack a proper institutional environment for privatization to work successfully, including bankruptcy procedures, protection of property rights, and debt repayment programs. Moreover, the IMF increases inequality in less developed countries by protecting and rewarding investors at the expense of the poor. In pursuing policies that emphasize economic stabilization over job creation, the IMF tends to accommodate foreign banks over the impoverished masses.

6.5 REGIONAL ECONOMIC INTEGRATION

Regional economic integration is the international grouping of sovereign countries to form a single economic market that includes free trade among members and restrictions on trade with nonmembers. Many countries turned to regional integration schemes in the late twentieth century, such as the European Union and the Association of Southeast Asian Nations. Reasons for integration included a need to gain access to regional markets (which are larger than national ones), to obtain more bargaining power than they could if they adopted a "go-it-alone" policy, to strengthen their base for negotiating with multinational corporations, and to promote cohesive solidarity.

Five levels of economic integration are possible. At progressively higher levels, members must make more concessions and surrender more sovereignty (Figure 6.5). The lowest level is the *free trade area*, in which members agree to remove trade barriers among themselves but continue to retain their own trade practices with nonmembers. A *customs union* is the next higher level, in which members agree not only to eliminate trade barriers among themselves but also to impose a common set of trade barriers on nonmembers. The third type is the *common market*, which, like the customs union, eliminates internal trade

	REMOVAL OF TRADE RESTRICTIONS AMONG MEMBER STATES	COMMON EXTERNAL TRADE POLICY TOWARD NON-MEMBERS	FREE MOVEMENT OF CAPITAL AND LABOR AMONG MEMBER STATES	HARMONIZED ECONOMIC POLICIES, CENTRAL BANK, AND PERHAPS COMMON CURRENCY
Free trade area	*			
Customs union	*	*		
Common market	*	*	*	
Economic union	*	*	*	*

Figure 6.5: Several levels of regional economic integration exist today. These range from the relatively modest, in which only trade barriers are removed (e.g., NAFTA), to more highly integrated forms, in which movements of goods, people, and capital are unrestricted and economic policies are harmonized (e.g., the European Union).

barriers and imposes common external trade barriers. This regional grouping, however, permits the unfettered movement of capital and labor. At a still higher level is an *economic union*, which has the common-market characteristics, in addition to a common currency and a common international economic policy. The highest form of regional grouping is *full economic integration*, which requires the surrender of most of the economic sovereignty of its members, such as control over the money supply.

A variety of trade organizations exist throughout the world. These groups range from loosely integrated free trade areas, such as the Latin American Free Trade Association (LAFTA), to free trade regions, such as that created by the North American Free Trade Agreement (NAFTA), to common markets, such as the European Union (EU). Fully fledged interregional integration has yet to be achieved. Regional groups are more concerned with closer economic integration *within* regions than *among* regions.

THE EUROPEAN UNION

The most successful example of economic integration is the **European Union** (EU). It began as the European Economic Community (EEC) in 1957 with six nations: France, West Germany, Italy, Belgium, the Netherlands, and Luxembourg. As the EEC expanded over time, it deepened its level of economic integration, and in 1993, it became the European Union. Most western European countries subsequently joined, including the United Kingdom (which recently voted to leave), Ireland, Denmark, Greece, Spain, and Portugal. In addition, most eastern European nations have applied for membership in the EU, and several have been accepted. By 2015, the EU had 28 members and a total population of more than 500 million (Figure 6.6). Today, the EU is the largest single trade bloc in the world and accounts for 40% of international trade, which is three times its world share of population.

The intent of the EU was to give its members freer trade advantages while limiting the importation of goods from outside Europe. It called for

1. the establishment of a common system of tariffs applicable to imports from outside nations;
2. the removal of tariffs and import quotas on all products traded among the participating nations;
3. the establishment of common policies with regard to major economic matters, such as agriculture, transportation, and so forth;
4. free movement and access of capital, labor, and currency within the market countries;

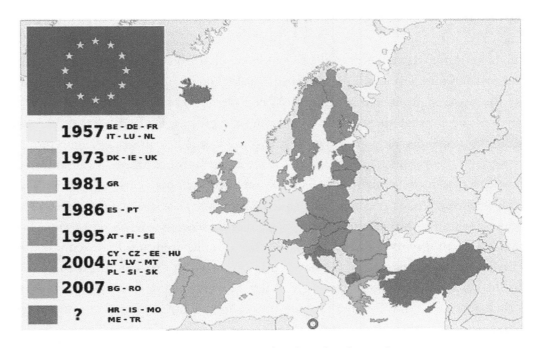

Figure 6.6: The European Union, with 28 members (as of 2016) and more than 500 million inhabitants, is the largest free trade area in the world. Member states have no barriers with one another to the flow of goods, capital, or labor. Many, but not all, EU members use the euro currency, as well.

5. transportation of goods and commodities across borders with no inspection or passport examination; and
6. a common European currency, the euro.

The EU has made tremendous progress toward its stated goals. The member nations have achieved more efficient, large-scale production because of potentially larger markets within the EU, permitting them to lower costs on manufactured goods. Americans are concerned about the economic power of the EU. Tariffs on their own products have been reduced to zero among EU nations, whereas tariffs on American-made products have been maintained. Consequently, importing goods to EU nations is relatively difficult. At the same time, increased prosperity among EU nations allows these countries to become potential customers for more exports from member states. EU integration also allows for the free movement of capital and labor among its members, giving the organization unprecedented economic integration. For example, workers from Portugal can look for jobs in Denmark, or a German dentist can open up shop in Greece.

In 1999, many (but not all) members of the EU adopted a unified currency, the euro. Britain, Denmark, and Sweden opted to retain their own currencies and lie outside of the "Eurozone," as do some of the newer states in eastern Europe. When members of the EU adopted a single currency, they were aware that there would be economic sacrifices. Each country effectively surrendered the right to independently balance its own budget and manage its own debt. As debt in many southern European countries, such as Greece and Italy has grown, these countries have been unable to react as they might have had they retained their own currencies. Countries with independent currencies can devalue them (lower a currency's value in relation to other currencies, which makes exports cheaper), an option not available to those using the euro. For example, Greece, with enormous debt and facing pressure from its creditors, has been in a dire economic crisis and cannot devalue the euro, which it shares with other countries. Therefore, opposition to the euro has grown, leading to calls for a "Grexit," or Greek exit from the currency union (but not the EU). In addition, many people in the EU are hostile to what they perceive as bureaucratic controls issued from its headquarters in Brussels. These tensions led to Britain's decision to leave the EU, or Brexit, which is discussed later.

NORTH AMERICAN FREE TRADE AGREEMENT

The economic pressure placed on the United States by the EU led the United States to promote freer trade through the U.S.–Canadian Free Trade Agreement, which phased out trade barriers between the world's two largest trading partners. In 1992, the **North American Free Trade Agreement** (NAFTA) was signed by the U.S. and Mexican presidents and the Canadian prime minister. In 1994, it went into effect.

NAFTA includes 330 million Americans, 35 million Canadians, and 110 million Mexicans. Unlike the EU, NAFTA includes a developing country. However, unlike the EU, NAFTA does not allow for the free movement of labor among its three member states, and they retain their own currencies.

NAFTA was not well received by all parties in North America. The main argument of its critics was that it transplants lower-skilled assembly and manufacturing jobs from the United States to Mexico, where labor costs are one-fifth to one-eighth as high. In addition, companies may flee America's more stringent regulation regarding pollution and workplace safety. Another argument was that NAFTA allowed the United States to flood Mexico with cheap, subsidized

American agribusiness exports such as corn and wheat. Indeed, since NAFTA began in 1994, two million Mexican farmers have gone bankrupt.

The principal argument in favor of NAFTA was that free trade would enhance U.S., Canadian, and Mexican comparative advantages by raising per capita incomes in Mexico and increase Mexican demand for goods from the United States and Canada. Another argument suggested that higher living standards in Mexico would help control the flow of undocumented aliens crossing the U.S. border, which was estimated at one million per year prior to the financial crisis that began in 2008 (since which it has declined dramatically to virtually zero). With free trade, wages should rise in Mexico. Therefore, potential undocumented aliens could stay home and work in their native country.

NAFTA turned out to free up financial and investment trade by reducing restrictions on subsidiaries of United States and Canadian financial services. Thus, NAFTA deregulated Canadian and United States banking, securities brokering, and insurance operations in Mexico. Truck transportation service allows free access to the Mexican market opened to Canadian and American trucking companies, as well as free access to the United States and Canada for Mexican truckers. Before NAFTA, a free trade zone 100 miles south of the border was in operation. NAFTA has, in effect, extended this *maquiladora* zone to all of Mexico, allowing imports from Mexican manufacturing plants to flow easily into the United States and imposing duties only on the value added by the manufacturer.

NAFTA has generated a loss of 600,000 U.S. jobs to Mexico, many of them in manufacturing—a small fraction of the 180 million jobs that make up the U.S. labor force. About 15 million of those are in manufacturing. Roughly an equal number of jobs have been created on the U.S. side of the border but are concentrated in lower-paying service sector occupations.

6.6 GLOBAL RESOURCE AND ENERGY FLOWS

To produce the goods and services people demand in today's global economy, we need to obtain natural resources. Natural resources include all substances of the biological and physical environment that people find useful under particular technological and socioeconomic conditions. Because these conditions are always subject to change, the definition of "useful" also changes. For example, petroleum was not considered a resource until the mid-nineteenth century, when the Industrial Revolution led to rising demand for fuels. Uranium, once a waste product of radium mining of the 1930s, now plays an key role in the generation of nuclear energy.

Resources are typically categorized as either nonrenewable and renewable. **Nonrenewable resources** consist of finite masses of material, such as fossil fuels (coal, petroleum, and natural gas) and metals, which cannot be used without depletion. Fossil fuels come from the fossilized remains of ancient plants and animals transformed by presure and heat over millions of years. They are, for all practical purposes, fixed in amount because they form very slowly over time. Consequently, their rate of use matters. Given that the world consumes many nonrenewable resources very quickly, most supplies in the world have been altered or depleted by use; petroleum is an example. **Renewable resources** can yield output indefinitely without impairing their productivity. They include water, sunlight, vegetation, fish, and animals. Renewal of such resources does not happen automatically, however. Resources can be depleted and permanently reduced by overuse. For example, overfishing can destroy productive fishing grounds.

Natural resources fall under the control of sovereign nation-states. Many wars in the twentieth century happened at least in part because of resources. For example, Japan invaded Korea and Taiwan in the 1890s largely to obtain arable land and coal. Concerns over oil supplies in the Persian Gulf played a role in both the Iraqi invasion of Kuwait in 1990 and the U.S. invasion of Iraq in 2003. In the Middle East, fierce national rivalries in a desert climate make water a potential source of conflict. Political tensions over the use of international rivers, lakes, and aquifers in the Middle East may escalate to war in the future.

MINERAL RESOURCES

A **mineral** refers to a naturally occurring inorganic substance in Earth's crust. Thus, silicon is a mineral, whereas petroleum is not, because the latter derives from the remains of ancient organisms. Although minerals abound in nature, many of them are insufficiently concentrated to be economically recoverable. Moreover, the richest deposits of metals are unevenly distributed and are being depleted worldwide. In contrast, nonmetallic minerals are plentiful and often widespread, including nitrogen, phosphorus, potash, sulfur for chemical fertilizer, or sand, gravel, or clay for building purposes. Except for iron, nonmetallic elements are consumed at much greater rates than metallic ones.

Of the major mineral-producing countries, only a few—notably the United States and Russia—are also major processors and consumers. The other major processing and consuming centers, such as Japan and western European countries, lack strategic mineral deposits. Most key minerals will be exhausted within 100 years, and some will be depleted within a few years at current rates of consumption, assuming there are no new reserves. The United States is running short of domestic sources of many strategic minerals. Its dependence on imports has grown steadily; when measured in terms of percentage imported, U.S. dependency increased from 50% in 1960 to more than 82% in 2010. Minerals the United States expects to need in the future tend to be unevenly distributed around the world. Many of them, such as manganese, nickel, bauxite, copper, and tin, are concentrated in Russia and Canada and in developing countries.

Mineral extraction has a varied impact on the environment, depending on mining procedures, amount and form of nearby water, and the size of the operation. The environmental impact of mining also depends on the stage; exploration activities usually have less of an impact than mining and processing mineral resources. Minimizing the environmental impacts of mineral extraction poses challenges, because the demand for minerals continues to grow, leading to the mining of ever-poorer grades of ore. For example, in 1900, the average grade of copper ore mined was 4%; by 2010, ores containing as little as 0.4% copper were mined, which means that more rock has to be excavated, crushed, and processed to extract copper. The immense copper mining pits in Montana, Utah, and Arizona are no longer in use because foreign sources, mostly in the developing countries, have lower costs. As long as the demand for minerals increases, increasingly lower-quality minerals will have to be used and, even with good engineering, environmental degradation will extend far beyond excavation and surface plant areas.

ENERGY RESOURCES

Energy forms the lifeblood of modern economies and constitutes the biggest single item in international trade. Oil alone accounts for about one-quarter of the volume (but not value) of world trade. The U.S. economy consumes vast amounts of energy, overwhelmingly consisting of fossil fuels. With roughly 5% of the world's people, the United States consumes 25% of its fossil fuels. Transportation and industry consume most of the petroleum, whereas most coal goes toward electrical power generation. Several of the world's leading industrial powers—notably Japan, many western European countries, and the United States—consume much more energy than they produce, making them heavily reliant on imported oil, largely from the Middle East.

In contrast, less developed countries consume about 30% of the world's energy but contain about 80% of the population. Most developing countries consume meager portions of energy, well below levels required for even moderate levels of economic development. Commercial energy consumption in more developed countries has been at consistently high levels, whereas in developing countries, it has been at low but increasing levels.

Most commercial energy derives from nonrenewable resources. Most renewable energy sources, particularly wood and charcoal, are used directly by producers, mainly poor people in the developing countries. Although there is increasing interest in renewable energy development, commercial energy is the core of energy use at the present time.

In the next few decades, energy consumption is expected to rise significantly, especially because of the growing industrialization of less developed countries. Most of the future energy production to meet increasing demand will come from fossil energy resources—oil, natural gas, and coal. How long can fossil fuel reserves last, given our increasing energy requirements? Estimates of energy reserves have increased substantially in the past 20 years, and therefore, little short-term concern exists over supplies. Consequently, energy prices are relatively low. If energy consumption remained more or less at current levels, which is unlikely, proven reserves would supply world petroleum needs for 40 years, natural gas needs for 60 years, and coal needs for at least 300 years. Although the size of the world's total fossil fuel resources is unknown, they are finite, and production will eventually peak and then decline.

PETROLEUM

As an accident of geology, the world's fossil fuels are highly unevenly distributed around the globe. Two-thirds of the world's oil resources are located in the Middle East (Figure 6.7). Other large reserves are found in Northern Africa and Latin America—primarily Mexico and Venezuela—and in Russia and Nigeria. Offshore drilling, such as in the North Sea, forms another supply. Natural gas, often a substitute for oil, is also unevenly distributed, with nearly 40% in Russia and Central Asia and 34% in the Middle East.

The unevenness of the world's supply and demand for petroleum creates a distinct pattern of trade flows of petroleum, the most heavily traded commodity (by volume) in the world. Primarily, these flows represent exports from the vast reserves of the Middle East to Europe, East Asia, and North America, although the United States also imports considerable quantities from South America and Nigeria. Only a handful of countries produce much more commercial energy than they consume. If we take petroleum consumption and production as an example, the main energy surplus countries include Saudi Arabia, Iraq, Mexico, Iran, Venezuela, Indonesia, Algeria, Kuwait, Libya, Qatar, Nigeria, and the United Arab Emirates. Saudi Arabia is by far the largest exporter of petroleum and has the largest proven reserves. Nearly one-half of African countries are energy paupers.

Until the 1970s, most Westerners thought that the world's supply of commercial energy would always be sufficient to generate rising affluence. However, the **Organization of the Petroleum Exporting Countries (OPEC),** comprising the countries with the world's greatest oil supply, greatly reduced its output in 1973. This reduction occurred largely in response to American support for Israel during its 1973 war with its Arab neighbors. Suddenly, higher prices brought energy demands in the industrial countries to a virtual standstill, generating inflation,

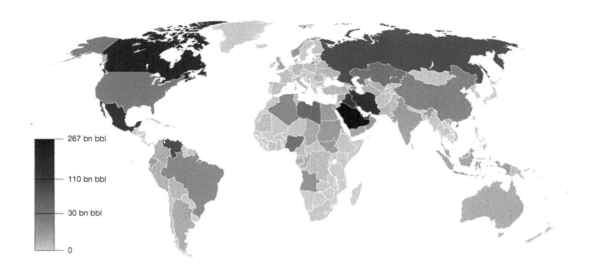

Figure 6.7: Global petroleum reserves reveal the unevenness of the distribution of the world's energy supplies. Roughly two-thirds of the world's petroleum deposits are located in the Middle East, making its supplies critical to the economies of Europe, North America, and Japan. Bbl = barrel unit.

unemployment, and accelerating deindustrialization. Thousands of factories shut down, and more than three million workers were laid off. During the 1980s and 1990s, oil prices decreased from $30 per barrel in 1981 to $14 per barrel in 2000. OPEC saw its share of world oil output drop steadily as non-OPEC countries expanded production. Oil-consuming countries, including less developed countries strapped by heavy energy debts, welcomed lower prices. In contrast, less developed oil-exporting countries, such as Mexico, Venezuela, and Nigeria, suffered because they depended on oil revenues. In 2008, however, world oil prices rose to $140 per barrel, only to fall again in 2009 and 2010 as the world's financial crisis and recession reduced global demand. Such oscillations suggest that petroleum prices will continue to rise and fall periodically, with huge impacts on other sectors of the world's economy.

Imported oil to the United States as a proportion of total demand increased from 11% in the late 1960s to 50% in the 1970s, although it is declining as domestic sources of petroleum have increased recently. The United States did become more energy-efficient. Manufacturing reduced its share of total U.S. energy consumption from 40% to 36%, and the burgeoning service economy consumed relatively little energy. In terms of conservation efforts, however, the United States lags behind Japan and Europe, where energy is more expensive. Gasoline taxes in Europe, for example, help to fund more energy-efficient public transportation. As the United States expanded drilling in Alaska and offshore, and increased its productivity of petroleum, it has instead become a net oil exporters.

NATURAL GAS

The political volatility of the world's oil supply has increased the attractiveness of natural gas, the fossil fuel experiencing the fastest growth in consumption. Natural gas production is increasing rapidly, and so too are estimates of proven gas reserves. Estimates of global gas reserves have increased during the past decade, primarily due to major finds in Russia, particularly in Siberia, and large discoveries in China, South Africa, and Australia. The distribution of natural gas differs from that of oil. It is more abundant than oil in the former Soviet Union, western Europe, and North America, and less abundant than oil in the Middle East, Latin America, and Africa. Russia exports vast amounts of natural gas, which gives it significant geopolitical influence in Europe. Cheap natural gas is also the major reason for the decline in the coal industry.

COAL

Coal is the most abundant fossil fuel and tends to be consumed in the country in which it is produced. Use of this resource, however, has been hampered by the inconvenience of storing and shipping and the environmental consequences of large-scale coal burning.

China is the world's largest consumer of coal, which is a major reason why it is the world's leading producer of gases, such as carbon dioxide, associated with climate change. With the exception of Russia, the United States has the largest proven coal reserves. Coal constitutes 67% of the country's fossil fuel resources, but only a small fraction of its energy consumption. However, the use of coal presents several problems. First, coal burning creates more pollution than other fossil fuels. Low-grade coal has large amounts of sulfur, which, when released into the air from the burning of coal, combines with moisture to form acid rain. Second, underground coal mining is costly and dangerous, and open-pit mining leaves scars difficult to rehabilitate. Third, coal is not a good fuel for mobile energy units, such as trains and automobiles.

Competition from cheap natural gas has led to a long, steady decline in the U.S. coal industry, which today employs only about 76,000 people.

6.7 TOURISM

An important part of contemporary globalization is tourism, which is usually defined as travel to a place outside one's usual residential environment involving a stay for at least one night. Tourism includes personal and business travelers, as well as conventions, and it stretches over a wide range of time and place. In most countries, domestic tourism exceeds international tourism by a substantial margin. The tourism industry consists of a variety of economic sectors, including international and local transportation, hotels and motels, eating and drinking establishments, entertainment, and retail trade. To sustain these activities requires enormous investments in transportation, housing, communications, water, and other infrastructure.

In seventeenth-century Europe, the elite visited spas for medical purposes; later, the supposed health benefits of seaside resorts attracted a broader clientele. The idea that traveling might be for pleasure and also for broadening the traveler's horizons first emerged in Europe during the eighteenth century. It grew as European nations developed their overseas empires and colonial officials and as travelers published accounts of exotic locales and peoples.

The growth of tourism reflects the rise in disposable income among a considerable segment of the world's population. Mass tourism began to form in the nineteenth century, with the introduction of the annual vacation (as opposed to holidays for religious observance), which was a product of the Industrial Revolution. By the late nineteenth century the vacation proper was established, and domestic seaside resorts in Europe, especially Britain, were becoming playgrounds for the working class. Railroads allowed easy access to beach resorts, for example.

Tourism grew rapidly after World War II, when it became an option for the working class. The automobile and wide-body jet airplane, which greatly reduced transport costs, contributed to this trend. Fundamentally, tourism reflects the demand for leisure that accompanies economic development, a desire to "get away from it all" and pursue education or novelty. Beginning about 1960, changes in employment patterns that allowed for more leisure time, additional discretionary income for many people, decreasing travel costs, and increasing numbers of retirees have helped to diversify tourism and make it a year-round industry. Recently, the Internet has allowed people to arrange all aspects of their vacation with ease and certainty in trip planning.

Today, tourism is arguably the world's largest industry in terms of employment. It forms a critical part of the economy of many countries, including, for example, Italy, France, and

Spain, as well as smaller island states in the Caribbean, Hawaii, and Florida, and cities, such as New Orleans, Las Vegas, Los Angeles, and New York. For most of the world's least-developed countries, tourism generates the largest percentage of foreign exchange. Kenya has an especially well-developed tourist industry, and other African countries are seeking to follow a similar path. Costa Rica has a well-developed ecotourism industry, and tourism in parts of Asia, such as Thailand and India, has been growing rapidly.

Table 6.2 indicates the distribution of the world's 1.18 billion international tourist arrivals in 2016 (12% of the world's population). More than one-half of the world's tourists, or 607 million people, visited Europe. France, Italy, and Spain boast the largest tourist industries in the world, attracting large numbers of tourists from colder northern climates. To a lesser extent, East Asia and North America are also significant.

There is a vast diversity in the types of tourism and tourist destinations, ranging from low-impact ecotourism to highly urbanized cores, from simple backpacking to luxury cruises, from safaris and writers' camps to honeymoon retreats, from individual exploration to package tours, from health resorts to Asian sex tourism, from dude ranches to night clubs, from museums and fine art institutes to tribal cultural events, and from small ski resorts to tropical playgrounds (Figure 6.8). Tourism is often highly seasonal, fluctuating greatly over the course of a year, with corresponding changes in prices for hotels and travel.

The volume of tourists visiting a given destination reflects, among other things, the information available to potential clients (which the Internet has greatly expanded); their disposable incomes and

TABLE 6.2: INTERNATIONAL TOURIST ARRIVALS, 2016

REGION	VISITORS (MILLIONS)	
Europe	609	
Asia-Pacific	277	
Americas	191	
Africa	53	
Middle East	54	
World	1,185	100

Source: World Tourism Organization.

Figure 6.8: Tourism represents a significant facet of globalization. In combining the economic and cultural dimensions, tourism has enormous impacts on host areas, including the injection of funds, generation of jobs, demands on the infrastructure, environmental repercussions, and effects on local ways of life.

willingness to travel; currency exchange rates; transportation and lodging supply and costs; the relative cultural familiarity or degree of exotic appeal the destination may have; concerns over crime; and unpredictable events, such as terrorist attacks. Political restrictions also exist; for example, until 2016, U.S. citizens could not legally visit Cuba, which attracts many visitors from Europe and Canada.

Culturally, the tourist industry can be seen as one more example of the dominance of the more developed over the less developed world. Critics argue that the activities of tourists both reflect and reinforce, perhaps even legitimize, existing patterns of inequality and the dominance of some groups over others. Increasingly, one reason for people to visit peoples and places in the developing world is to experience cultural difference. Yet—not surprisingly—those peoples and places are rarely authentic; in many cases they have been constructed specifically to satisfy the tourists' gaze. In fact, for many tourists, the places and peoples they visit are economic commodities, and—as with many other commodities—the advertising used to present them to prospective consumers relies more on fancy than on factual information.

Many of the favored tourist destinations today are in the less developed world—Mexico, various Caribbean and Pacific islands, and some Asian and African countries. For example, because of low labor costs these areas can usually offer competitive rates, and some may represent—at least for Europeans or North Americans—a new and perhaps "exotic" cultural experience. Certainly tourism is growing in the less developed world, especially in coastal areas, in response to increasing demand from the more developed world.

6.8 CULTURAL GLOBALIZATION

Another dimension of globalization involves the international transmission of culture. Recall that capitalism is not simply an "economic" system; it is also deeply political and ideological. This idea reflects one of the themes raised in Chapter 1: Human geographies must be understood in part by acknowledging the belief systems that people use to make sense of themselves and their worlds.

As American-style capitalism has spread throughout the world, it has had profound impacts on cultures everywhere. Among the mechanisms that allow this culture to spread spatially are various aspects of the mass media and consumer culture—newspapers, magazines, the Internet, music, television, films, videos, fast-food franchises, and fashions. Thus, Hollywood movies and television shows are popular the world over; billions of people dress like American teenagers and eat at McDonald's; music from the United States can be heard all over the planet. American sports, such as baseball and basketball, have become popular in Latin America and Asia. The globalization of culture—largely a one-way flow of American culture to the rest of the planet—is, thus, largely manifested through consumption and the commercialization of culture. It is true that Americans also import some aspects of foreign culture, including foreign movies and music (mostly from Europe), but cultural imports pale in comparison with cultural exports. As the world's largest economic, military, and political power, the United States is simultaneously envied, imitated, and despised. Admiration for American culture is typically strongest among the young, who often associate Western culture with status, fun, sex, and hope; the elderly tend to be more traditional, so that globalization creates a generation gap in terms of outlook and preferences. Some countries have imposed domestic content rules on television to limit the amount of foreign, notably American, shows shown within their borders.

It is also worth emphasizing that globalization has allowed elements of many foreign cultures to percolate into the United States (Figure 6.9). For example, foods that were once regarded as strange and foreign are often now seen as commonplace, including Chinese and Mexican food, Korean *kim chi*, Spanish tapas, and the global sushi craze. Some Americans watch foreign movies in theaters or on television (e.g., Monty Python or Bollywood films). Foreign music includes European rock, African jazz, and Latin American salsa. Karaoke and

Figure 6.9: The popularity of sushi reflects not only the globalization of Japanese cuisine but the penetration of a foreign food once regarded as bizarre into the United States, testimony to the power of the global economy to shape diets and everyday life.

anime, Japanese inventions, are popular, as are many foreign novelists. European fashions (e.g., skinny jeans) often catch on in the United States. Sports once regarded as foreign, such as soccer, have become widespread. Foreign television stations such as Al Jazeera have established footholds in the United States. In large cities, such as New York, one may witness celebrations and parades of Jamaicans, Trinidadians, Pakistanis, and Nigerians.

The dominance of Western culture worldwide has led many critics to fear that globalization is homogenizing cultures and places. There is considerable merit to this concern. Visitors to cities such as Los Angeles, Seoul, Frankfurt, Abu Dhabi, and Singapore may have difficulty telling them apart (Figure 6.10). In this reading, globalization produces a monoculture, a single way of being and thinking that is

Figure 6.10: In many respects, globalization does impose cookie-cutter homogenization on local landscapes, including skyscrapers, freeways, and businessmen in suits. Without knowing the names, can you tell which city is Los Angeles, Berlin, Abu Dhabi, and Seoul?

Figure 6.10: (*Continued*)

stamped like a cookie-cutter around the planet. Cultural homogenization helps to drive the annihilation of many small languages around the world (Chapter 3) and endangers indigenous cultures from Australia to Finland. The cultural homogenization unleashed by globalization, which is limited and partial, is not a positive development; often, creativity is stimulated by encountering difference, and as cultural differences around the world are reduced, so too are the opportunities for learning from people who view the world very differently from ourselves. Simultaneously, globalization has brought more people from different cultures into contact with one another than ever before, presenting new opportunities for cross-cultural learning, as well as new challenges.

Thinking about cultural globalization obliges us to acknowledge that there are no uncontested processes at play in the contemporary world. Globalization, in the sense of an ever-increasing connectedness of places and peoples, is a fact, but it is not the only important fact. Indeed, even as the world steadily becomes more globalized, many ethnic groups are reasserting their identities—at least partly in reaction against the declining importance of national political and cultural identities. For example, a pan-national Islamic religious identity has been promoted by Muslims in parts of Africa and the Middle East. In some African states, tribal identities have reasserted themselves, sometimes violently (e.g., the division between Sudan and South Sudan in 2011). Tribal peoples in Myanmar (formerly Burma) have waged a war against the national government. Additionally, indigenous peoples in Bolivia, Peru, and Brazil have exerted their rights to control their own lands.

6.9 MYTHS ABOUT GLOBALIZATION

Because it has received considerable media attention and lies at the core of many debates about economic trends and policies, globalization has often been plagued by erroneous or simplistic misconceptions. Six myths are especially widespread:

- Myth 1: Globalization is only an economic process.
- Myth 2: Globalization only erases cultural differences.
- Myth 3: Globalization began only in the past few decades.
- Myth 4: Globalization spells doom for the nation-state.
- Myth 5: Globalization is an unstoppable force.
- Myth 6: Globalization is always beneficial.

A common stereotype pertaining to globalization is that it is purely economic in nature. Many people think of globalization entirely in terms of international trade and foreign investment, especially in the context of transnational corporations. Yet, such a view is overly narrow and ignores the multiple ways in which globalization also operates as a political, cultural, and ideological force. For example, immigration is clearly a topic pertinent to globalization, with many so-called noneconomic dimensions associated with it. Equally, one can point to the globalization of education, crime, disease, or terrorism. Some of the aspects of globalization that are resisted most vehemently in parts of the world are its cultural dimensions, including, for example, the globalization of fast food, dress, or cinema, all of which are bound up with people's worldviews and daily lives.

A second myth equates globalization with cultural homogenization, as if the world economy stamped a monoculture throughout the world. For much of the world, globalization is synonymous with Americanization. Although there can be no denying that cultural homogenization often takes place in the wake of globalization, often at the expense of old, deeply held traditions, it is equally true that globalization generally means different things in different places—that is, it is geographically specific. National policies mediate global trends in different ways. The unique histories of individual places impart local flavor to global trends, such as, for example, when multinational corporations, such as McDonald's, must tailor their menus and advertising to local preferences (Figure 6.11). Local regions, thus, do not merely receive changes imparted to them by the global economy, but in turn, shape that global economy. The global and local are intimately intertwined, and geographers often use the term **glocalization** to describe the adaptation of worldwide trends to local contexts.

A third frequent misconception about globalization is that it began, or reached its most prominent stage, only in the late twentieth century. Clearly there is little doubt that the world is deeply globalized, increasingly so every day. However, the birth of capitalism on a global basis in the sixteenth century clearly marks an earlier epoch of globalization, as did colonialism in the following centuries. The Industrial Revolution unleashed waves of time-space compression that ushered in round after round of globalization. In terms of the relative magnitude of foreign investment, the late nineteenth century was at least as globalized as the present, if not more so. Moreover, globalization had even earlier roots: There was a world-system in the fourteenth century stretching throughout much of the Old World (Chapter 5), and some world-systems theorists have speculated on even earlier systems.

A fourth issue that is problematic in the study of globalization concerns its relations to the nation-state. Some analysts argue that true globalization could not have occurred prior to the emergence of the modern nation-state in the eighteenth and nineteenth centuries: It is, after all, difficult to be international if there is nothing national. However, this view of globalization is too narrow and ignores the extensive evidence of premodern globalization. A related issue is the question of whether globalization entails the end of the nation-state. Certainly some aspects of globalization have eroded the sovereignty of states in some matters. The globalization of financial capital, for example, has made national monetary controls increasingly ineffective, and international organizations, such as the European Union, the United Nations, the World Bank, and the International Monetary Fund have assumed some functions of the nation-state.

The bluntest manifestation of this view is that globalization is boundary-transcending and that localization is boundary-heightening. However, it is simplistic to assume that globalization

Figure 6.11: McDonald's, the quintessential food conglomerate associated with cultural homogenization, has tailored its menu to reflect different cultural preferences around the world. This forms an example of "glocalization," in which global trends intersect with the unique particularities of places. These examples include chicken burgers in India (no beef allowed!), lobster sandwiches in New England, grilled kofta (spiced meats) in the Arab world, kosher McDonald's in Israel, and McDonald's gazpacho soup in Spain.

leads inevitably to the end of states as they currently exist, to replace them with some seamless integrated market that will embrace the entire planet. Globalization is always interpreted through national policies—for example, those concerning labor, foreign investment, or the environment—one reason why it has spatially uneven influence across the world. Capitalism involves both markets and states, and the political geography of globalization is the interstate system, the existence of which is necessary for capital to play states and localities against one another.

A fifth stereotype about globalization is that it consists of some unstoppable force independent of human intervention. In this reading, globalization is inevitable—countries can do little to stop it and must accommodate to its needs and requirements. Such a view denies the historical origins of globalization and the fact that people create it. In fact, globalization has experienced reversals, such as during the trade wars of the 1930s, which led to a sharp decline in trade between the United States and Europe. Moreover, globalization is resisted, sometimes successfully and sometimes not, often by those who believe that it presents a secular, amoral threat to established local traditions. These critics, often social segments with values outside the market, view the market as a mechanism for reducing everyone to a consumer, annihilating all forms of identity except those that have to do with a commodity. Thus, the more globalization has disrupted local value systems around the world, the greater the backlash has been against it.

Finally, a sixth frequent misconception about globalization holds that it is always beneficial. This claim is often advocated by economists who focus on the notion of comparative advantage, discussed above. It is true that, generally, the most globalized societies are among the world's wealthiest, that is, the United States, the countries of the EU, and Japan, or most rapidly growing, such as the Asian newly industrializing countries. Conversely, some of the least globalized countries are also among the poorest, such as Bhutan or North Korea. In this reading, globalization is associated with lower consumer prices, technology transfer, and improved efficiency. However, the history of capitalism is characterized by uneven development, and globalization is no exception. Indeed, it represents capitalism at a global scale that creates poverty as well as wealth. Evidence for this argument includes local producers displaced by multinational firms, the exploitative labor conditions found in many sweatshops in the developing world, IMF austerity programs, and international economic crises. Among those who bear the costs but do not enjoy the benefits, globalization understandably breeds envy and resentment.

6.10 ANTI-GLOBALIZATION

Globalization—that diverse, complex set of processes that transcend national borders—is not a "natural" or inevitable phenomenon, but a historical product, and therefore changeable. Because globalization does not favor everyone, it often breeds resentment among those who lose more than they gain from it. For this reason, it may be said that globalization inevitably breeds its own opposition.

Resistance to globalization is as old as globalization itself. From the beginning, European colonial empires were met with heated opposition, often violent and generally unsuccessful. Examples include the Incan uprising against the Spanish in the early sixteenth century, Zulu attacks on Dutch and British pioneers in southern Africa, and the Sepoy Rebellion of India in the mid-nineteenth century (Chapter 5). A long series of anticolonial and guerilla struggles persisted well into the 1970s in Vietnam and much of Africa. For many contemporary opponents of globalization, current engagements are part of long history of opposition that reaches back centuries.

Antiglobalization movements take a variety of forms. For many, it is relatively peaceful in nature, including boycotts, demonstrations, and working through **nongovernmental organizations (NGOs)**. For others, opposition takes on a decidedly more active, even violent form. Benjamin Barber's famous book *Jihad vs. McWorld* noted this diversity of forms, using "jihad," the Arabic expression for holy war, as a metaphor for the vast umbrella of groups opposed to contemporary globalization, and "McWorld" as a metaphor for the American-led, information-intensive penetration of various societies by crass commercialism, including fast food, entertainment and media (especially cinema), and fashion, all of which are the most visible face of Western dominance.

ANTI-AMERICANISM AND ANTIGLOBALIZATION

Because of the dominant role of the United States in the contemporary world system, globalization in the minds of many people is synonymous with Americanization. Much of the world has a love/hate relationship with the United States, often adoring its popular culture but abhorring the foreign policies of the American government. In part, the United States has earned enmity for its long support of unsavory dictatorships in many regions, especially during the Cold War. Because globalization is so closely affiliated with Americanization for many people, opposition to globalization is often manifested in anti-Americanism (Figure 6.12). Opposition to the United States, for example, may take the form of attacks on Ronald McDonald, the clown

Figure 6.12: Because globalization and Americanization are synonymous in the minds of many people around the world, protests against globalization often take the form of anti-American demonstrations, often fueled by resentment against American foreign policy.

statue that serves as mascot for that chain, as a symbolism of American commercialism.

Some people in western Europe have expressed a different form of antipathy to globalization. Many residents of France and neighboring countries, for example, are disgusted with what they perceive to be the crassest aspects of American culture, its obsession with money, commodities, and status, its neglect of tradition and leisure. (Vacation times in Europe tend to be longer than those in the United States.) For example, when French farmer José Bové drove his truck into a McDonald's restaurant to demonstrate his hostility to fast food, which he viewed as a threat to French cuisine, he became a national hero. Other Europeans strenuously object to imports of American genetically modified foods. Many Europeans dislike the intensity of American individualism, its denial of the social origins and obligations of people, and what they perceive as the correspondingly lack of empathy of the poor and misfortunate. The United States, in this view, is overly tolerant of inequality and social injustice. Moreover, in Europe, an increasingly secular continent, there is widespread dislike of the prominent role that organized religion plays in American public life and the profound religiosity of the American people, manifested, for example, in the rise of the "religious right" and attempts to limit the teaching of evolution in schools. Such phenomena are seen as symptomatic of a generalized anti-intellectualism in the United States, which is stereotyped (accurately or not) as a culture in which ideas and the life of the mind are held in low regard. Finally, many Europeans view the United States as a cowboy culture, given its widespread ownership of firearms and rates of violent crime that greatly exceed those of all industrialized countries. This celebration of violence extends to the American use of the death penalty, which is absent in all other economically developed states.

GRASSROOTS ANTIGLOBALIZATION

Hostility toward globalization is often strongest among those socioeconomic groups that have benefited the least. This includes many working-class people in Europe and the United States. Recent political events, including Britain's decision to leave the EU and the election of Donald Trump as U.S. president, testify to these concerns.

BREXIT

Although Britain joined the EU in 1973, and segments of British society benefited considerably (especially in the London metropolitan area), many older, less skilled workers in northern England were much less enthusiastic. In 2016, after a public debate, the British electorate voted on a referendum to leave the EU. Proponents of the so-called Brexit played on fears of immigrants (only those from other EU member states can enter legally without special permission), blaming them for job losses, crime, and social problems. Others argued that Britain could "go it alone." Support for Brexit was greatest among older Britons, especially those living in northern England, some of whom support the United Kingdom Independence Party (UKIP); the young generally did not support the decision, and Scotland voted overwhelmingly against leaving. Some thought it might accelerate the move toward Scottish independence.

Leaving the EU is a long, complicated process with many dimensions, including renegotiating Britain's trade relations with the EU, immigration, customs procedures, international student exchanges, and a host of other issues. Britain gave formal notice by invoking Article 50 in March 2017. Final departure is scheduled for March 2019. The EU has a vested interest in making the process burdensome and expensive for Britain as a disincentive for other members to leave the union.

The long-term consequences of Brexit are unclear. Some fear that the London region's status as Europe's premier banking center may be threatened, and some firms are already considering relocating to cities such as Paris, Frankfurt, Amsterdam, and Milan. If the United Kingdom loses the preferential trade status it enjoys with the EU, it will have to abide by trade rules set by the World Trade Organization, and the cost of imports may rise. British farmers will lose EU agricultural subsidies. The legal status of three million immigrants from EU states, as well as 1.5 million Britons living in the other EU countries, will be called into question. Finally, the border between Northern Ireland and the Republic of Ireland, an EU member, will become more rigid and fixed.

THE ELECTION OF DONALD TRUMP

In November, 2016 billionaire real estate developer and reality television host Donald J. Trump was elected president of the United States after running on a "Make America Great Again" campaign platform that was strongly anti-immigrant and opposed to foreign trade deals. He ran on a platform of bringing back industrial jobs, building a wall on the border with Mexico, combating fears of Muslim terrorism, and reviving the coal industry, which has faced long-term declines in employment. Republican Trump received strong, enthusiastic support from poorly educated blue-collar workers in many parts of the country, including several deindustrialized states that had long tended to vote Democratic (e.g., Pennsylvania, Michigan, Wisconsin, and Ohio). Many such voters were overtly hostile to immigrants (especially nonwhite ones) as well as loosely defined group of "elites" whom they perceived to run the government for their own purposes. (On the left, Bernie Sanders made a similar argument).

Since assuming office, Trump has enacted several steps that run contrary to the long-standing Republican Party embrace of globalization. He withdrew the United States from the Trans-Pacific Partnership and the Paris Accord to reduce greenhouse gas emissions that cause global warming. He has called for renegotiating NAFTA, but he has not proposed concrete steps. He also criticized Canadian lumber imports and threatened tariffs on steel. He has criticized China and argued it manipulated its currency. He also called Germany's foreign trade surpluses "very bad" for the United States. The Trump administration also sought to impose strict travel bans on visitors from several predominantly Muslim countries, although these have been challenged in the courts. The status of several million undocumented immigrants who have lived in the country for decades has become legally ambiguous. Moreover, Trump called into question the long partnership between the United States and western Europe, making statements that indicated low American commitment to the alliance and pushing allies to pay a higher share of the defense costs. More broadly, many critics argue that his policies represent a decline of American leadership from the existing international order.

Both Brexit and the pro-Trump movement illustrate the upsurge in ethnonationalism that has shaken many Western countries, where many people are disgusted with the mainstream political parties that have long held power. High rates of unemployment, wage stagnation, rising wage inequality, and the influx of refugees from Africa and the Middle East contributed to mass discontent. In France, the National Front, led by Marine Le Pen, articulates many of the same anti-immigrant political positions as Trump, and called for a French exit from the EU, or "Frexit." Similar movements can be found in the Netherlands (e.g., the Freedom Party of Geert Wilders), Greece (e.g., Golden Dawn), Sweden, and Germany. Conservative political parties now control or are influential in many governments in eastern Europe, such as Austria, Poland, and Hungary. However, their agendas interfere with well-established corporate supply chains and transnational modes of doing business, and thus contradict the interests of many transnational corporations and banks.

GLOBALIZATION AND RELIGIOUS FUNDAMENTALISM

Given that globalization is often viewed as a secularizing force, it is not surprising that some of the most heated violent opposition has emanated from religious groups. Indeed, in the wake of the end of the Cold War, religious fundamentalism has erupted around the world, often coupled with antiglobalization sentiments. As a long tradition of social science has noted, modernity generates numerous changes in identity and behavior associated with the rise of markets, individualism, and commodity-based norms. Thus, there are those who hold the belief that Western, modern forms of life reduce identity to that of a buyer or seller of commodities, obliterating many time-honored, noncapitalist forms of life. Especially for people who experience severe disorientation through rural-to-urban migration, and those experiencing the loss of systems of meaning that provided ontological security (honor, family, ancestors, god, and so forth), modernity can be viewed as a sinister, morally offensive force.

The upsurge of globalization that began in the twentieth century and continues in the twenty-first has given birth to several religious fundamentalist movements. Although these differ according to the specific religious beliefs involved, all of them to one extent or another involve the militant assertion of religious beliefs, often violently. In India, for example, Hindu fundamentalists include the Bharatiya Janata Party, which views the country as a Hindu nation with little tolerance for other faiths, notably Islam. In the 1990s, it instigated the destruction of the Muslim Ayodhya mosque, and has since promoted the revival of *suttee*, or widow burning, and celebrated India's acquisition of nuclear weapons.

In the world of Judaism, the most rapid growth has been among ultra-orthodox Hasidim, who support extremely restrictive policies concerning the Palestinians, including settlements in the West Bank. In 1996 a fundamentalist Jew assassinated the prime minister, Yitzhak

Rabin, whom he believed was making concessions to the Palestinians. The growth of ultra-Orthodox Jews has become a serious point of contention within Israel, as they are exempt from the military draft and dependent on state subsidies.

The upsurge in religious fundamentalism also includes Islam, the world's second-largest religion. Although the Western media often portrays the Muslim world in negative terms, the vast majority of Muslims are peaceful and law abiding. Nonetheless, a tiny minority, fueled by indifferent, corrupt governments in the Arab world, blames their culture's relative powerlessness in the world, and particularly against Israel, on an ostensible departure from the teachings of the holy Koran. Radical Islamists toppled the Shah of Iran in 1979 and installed the Ayatollah Khomeini, turning Iran into a medieval theocracy. Others include the Muslim Brotherhood, which assassinated Anwar Sadat in 1982. In Afghanistan, the Taliban drove out the Soviets in 1989, reorganized the country along strictly fundamentalist lines, hosted al-Qaeda in 2001, and has fought the United States since the American invasion later that year. Among the Palestinians, whose nationalist movement has long been fairly secular, as exemplified by the Palestine Liberation Organization, fundamentalists have led to the

Figure 6.13: The September 11, 2001, attacks against the World Trade Center exemplify both the virulence of anti-Americanism among some Muslims and the way in which global processes can be telescoped into local arenas.

rise of groups such as Hamas, with associated suicide attacks against Israel. In 2011, popular revolts toppled the governments of Egypt, Tunisia, and Libya in the so-called "Arab Spring." Although these were led by secular forces, in the power vacuum that emerged afterward religious groups such as the Muslim Brotherhood began to exert their influence in elections.

The organization called Al-Qaeda (Arabic for "the base") exemplifies the most pernicious aspects of this trend. Led by the infamous Osama bin Laden, member of a Yemeni family that grew rich in Saudi Arabia, al-Qaeda has become the most visible face of violent Muslim opposition to globalization and the United States, as spectacularly exemplified by the attacks on the World Trade Center (Figure 6.13) and Pentagon on September 11, 2001. Al-Qaeda operated from a base in Afghanistan with its allies, the Taliban, until the U.S. invasion toppled the government of that country. It also organized attacks against a U.S. Navy ship in Yemen and orchestrated bombings in London and Madrid. Affiliates of Al-Qaeda have been active in Somalia, Yemen, Nigeria, Mali, and Indonesia. To understand such movements, it is imperative to understand the social origins that drive individuals and groups to seek such goals: the poverty, frustration, humiliation, and sense of powerlessness that globalization often generates.

■ KEY TERMS

Capital markets: markets that exchange financial resources, including currency exchange, stocks, bonds, and loans among banks and governments

Comparative advantage: the economic theory that some places produce different goods and services than do other places, leading to specialization of production

European Union: the economic union of 25 European countries that has essentially eliminated barriers to trade, investment, and migration among them, forming one large economy

Foreign direct investment: foreign funds invested in land, buildings, machinery, and equipment

Globalization: the set of processes through which different regions and countries are interconnected, including international flows of people, goods, capital, ideas, and disease

Glocalization: the way in which global processes play out in different ways in different regions

International Monetary Fund: an institution founded during the Bretton-Woods Conference that plays a major role in regulating international loans and debt

Mineral: a naturally occurring inorganic material in Earth's surface

Nongovernmental organizations (NGOs): nonprofit agencies that often address problems and issues that the market cannot or will not, including poverty, human rights, and environmental issues

Nonrenewable resources: resources with a fixed stock that cannot be replenished, such as soils, minerals, and petroleum

North American Free Trade Agreement (NAFTA): the agreement that removed trade barriers among the United States, Canada, and Mexico

Organization of Petroleum Exporting Countries (OPEC): a cartel of most, but not all, of the world's major oil-producing states

Regional economic integration: the process of tying together different national economies by removing obstacles to international trade, investment, or migration. This process takes a variety of forms and levels of intensity

Renewable resources: resources whose stock can be replenished, including water, wood, and animal products

Transnational corporations: firms that operate in more than one country

Unequal exchange: the process by which countries trade goods and services of highly uneven value, leading some to prosper and others to stagnate

Washington Consensus: also known as neoliberalism, a loosely connected set of international policies that centers on deregulation, privatization, and free trade

World Bank: one of two organizations set up at the Bretton-Woods Conference, it is the largest lender in the world, financing long-term development projects in many countries

World Trade Organization: an organization set up in 1995 dedicated to the reduction of trade barriers and resolution of trade disputes worldwide

■ STUDY QUESTIONS

1. What forces drive international trade? Why has it grown so much?

2. What is foreign direct investment, and who is responsible for generating it?

3. Does the presence of a transnational corporation in a country generate more benefits than costs?

4. Why do some countries enter into regional trade agreements such as NAFTA?

5. What are the costs and benefits of tourism for host countries?

6. Why is it simplistic to think of globalization as something that only started within the last century?

7. Is saying that globalization changes the role of the nation-state the same as saying the nation-state will soon disappear?

8. Is globalization causing the world's cultures to become homogeneous, or does it adapt to the specifics of individual cultures?

9. Why does globalization produce much resentment and hostility among some groups?

BIBLIOGRAPHY

Amin, A. (2002). Spatialities of globalisation. Environment and Planning A, 34, *385–399.*

Appadurai, A. (1996). Modernity at Large: Cultural Dimensions of Globalization. *Minneapolis: University of Minnesota Press.*

Barber, B. (1995). Jihad vs. McWorld: How Globalism and Tribalism Are Reshaping the World. *New York: Ballantine.*

Beaverstock, J., Smith, R. & Taylor, P. (2000). "World city network" A new metageography for the future? Annals of the Association of American Geographers, 90, *123–134.*

Blaut, J. (1993). The Colonizer's Model of the World: Geographical Diffusionism and Eurocentric History. *New York: Guilford Press.*

Chandler, A., & Mazlish, B. (Eds.). (2005). Leviathans: Multinational Corporations and the New Global History. *Cambridge: Cambridge University Press.*

Chase-Dunn, C. (1989). Global Formation: Structures in the World-Economy. *Oxford: Basil Blackwell.*

Chase-Dunn, C., & Hall, T. (1997). Rise and Demise: Comparing World-Systems. *Boulder: Westview Press.*

Cox, K. (Ed.). (1997). Spaces of Globalization: Reasserting the Power of the Local. *New York: Guilford Press.*

Dicken, P. (2015). Global Shift: The Internationalization of Economic Activity *(7th ed.). New York: Guilford Press.*

Featherstone, M., ed. (1990). Global Culture: Nationalism, Globalization and Modernity. *London: Sage.*

Friedman, T. (2005). The World Is Flat: A Brief History of the 21st Century. *Picador.*

Herod, A. (2009). Geographies of Globalization: A Critical Introduction. *Wiley-Blackwell.*

Herod, A., Ó Tuathail, G., & Roberts, S. (Eds.). (1998). An Unruly World? Globalization, Governance and Geography. *London: Routledge.*

Hirst, P., & Thompson, G. (1999). Globalisation in Question *(2nd ed.). Cambridge: Polity.*

King, A. (1997). Culture, Globalization and the World-System. *Minneapolis: University of Minneapolis Press.*

Sassen, S. (1991). The Global City: New York, London, Tokyo. *Princeton, NJ: Princeton University Press.*

Scott, A. (1997). The Limits of Globalization. *London: Routledge.*

Sheppard, E. (2002). The spaces and times of globalization: Place, scale, networks, and positionality. Economic Geography, 78, 307–330.

Stiglitz, J. (2002). Globalization and its Discontents. *New York: Norton.*

Storper, M. (1997). The Regional World: Territorial Development in a Global Economy. *New York: Guilford Press.*

Waters, M. (1995). Globalization. *London: Routledge.*

IMAGE CREDITS

- Figure 6.3: Copyright © Sitris (CC BY-SA 3.0) at http://commons.wikimedia.org/wiki/File:IDE_in_stock_2009.jpg.
- Figure 6.4: Copyright © E Pluribus Anthony (CC BY-SA 3.0) at http://commons.wikimedia.org/wiki/File:WTOmap_currentstatus.png.
- Figure 6.6: Copyright © Júlio Reis (CC BY-SA 3.0) at http://commons.wikimedia.org/wiki/File:EU_accession_map.svg.
- Figure 6.7: HêRø, http://commons.wikimedia.org/wiki/File:Oil_Reserves_Updated.png. Copyright in the Public Domain.
- Figure 6.8: Copyright © CTHOE (CC BY-SA 3.0) at http://commons.wikimedia.org/wiki/File:Albufeira-Strand.JPG.
- Figure 6.9: Copyright © Andrew Sullivan (CC BY-SA 4.0) at https://commons.wikimedia.org/wiki/File:Sweet_Potato_Sushi.jpg.
- Figure 6.10a: Copyright © Thomas Pintaric (CC BY-SA 3.0) at https://commons.wikimedia.org/wiki/File:DowntownLosAngeles.jpg.
- Figure 6.10b: Copyright © A.Savin (CC BY-SA 3.0) at https://commons.wikimedia.org/wiki/File:Siegessaeule_Aussicht_10-13_img4_Tiergarten.jpg.
- Figure 6.10c: Copyright © Challiyan (CC BY-SA 3.0) at https://commons.wikimedia.org/wiki/File:Dubai_night_birds_eye_view.jpg.
- Figure 6.10d: Copyright © Ian Muttoo (CC BY-SA 2.0) at https://commons.wikimedia.org/wiki/File:Korea-Seoul-Sejongno-01.jpg.
- Figure 6.11a: Copyright © Yusuke Kawasaki (CC by 2.0) at http://commons.wikimedia.org/wiki/File:Chicken_Maharaja_Mac_Combo_%283155972375%29.jpg.
- Figure: 6.11b: Copyright © Alper Çuğun (CC by 2.0) at http://commons.wikimedia.org/wiki/File:McArabia_Tagine-2009.jpg.
- Figure 6.11c: Copyright © Geogast (CC BY-SA 3.0) at http://commons.wikimedia.org/wiki/File:Kosher_McDonald%27s,_Abasto_Shopping,_Buenos_Aires.jpg.
- Figure 6.12: Copyright © Loavesofbread (CC BY-SA 4.0) at https://commons.wikimedia.org/wiki/File:Shaw_Day_2_Photo_18.jpg.
- Figure 6.13: National Park Service, "National Park Service 9-11 Statue of Liberty and WTC fire," https://commons.wikimedia.org/wiki/File:National_Park_Service_9-11_Statue_of_Liberty_and_WTC_fire.jpg. Copyright in the Public Domain.

Chapter Seven

GEOGRAPHIES OF DEVELOPMENT

MAIN POINTS

1. "Development" seems like a simple term, but is actually complex and loaded with multiple meanings. The world's less developed or underdeveloped countries contain roughly three-quarters of the planet's population and the vast bulk of the increase in population.

2. Measures of economic development include national output per person (gross domestic product per capita), a country's labor force composition (in agriculture, manufacturing, and services), its production of consumer goods, and education and health levels.

3. As the world urbanizes, cities in developing countries are rapidly expanding. These cities are often surrounded by slum areas filled with decrepit housing and impoverished people.

4. Poverty in the developing world stems from many problems such as low incomes, high unemployment, insufficient investment and trade, foreign debt, unequal land ownership, and government corruption.

5. Several development strategies exist to alleviate poverty, including enhanced investment, trade, and foreign aid.

6. Some parts of the developing world are industrializing. The world's most successful region over the past five decades has been East Asia. This region includes Japan, the "tiger" economies of South Korea, Taiwan, and Singapore, parts of Southeast Asia, and above all, China.

When you watch the news on TV, consider the contrasts in the life opportunities that people face around the world, such as starving babies in Africa, a cholera outbreak in Haiti, and desperate farmers in India. Between these stories may be advertisements for luxury cars, the latest season's fashions, and a newer, more powerful smartphone. An observer cannot help but be struck by the contrast between the poverty in the real world and the plenty touted in the ads. How can two such different sides of life coexist on the planet? Are they related?

One of the most enduring and striking characteristics of the modern world is the division between rich and poor countries. By the end of the nineteenth century, this division was achieved through an international system in which a wealthy minority of Western countries and Japan, which had industrialized used primary sector products produced by the impoverished majority of the countries they colonized or dominated (Chapter 5). More recently, this colonial global division of labor has given way to a new one: The wealthy minority increasingly engages in office work while parts of the developing world find hands-on manufacturing jobs on the global assembly line as well as in agriculture and raw-material production.

The creation of today's world, with a rich core and a poor periphery, is the outcome of a systemic process by which the world's political economy reproduces uneven spatial development. As we have seen (Chapter 5), capitalism is capable of creating enormous wealth in some regions and, simultaneously, poverty in others. The landscapes of capitalism are typified by uneven spatial development, which was generated on a planetary scale during the era of colonialism and continues today. In short, the birth of the modern capitalist world system and the division between rich and poor countries are two sides of the same process at the heart of global political economy.

The developed and **less developed countries** of the world are clearly separated on a map of the planet. A line drawn at 30° north latitude would put most of the developed countries to the north and the underdeveloped to the south, a division often known as the North-South split (Figure 7.1). Note that this dichotomy is not identical to the northern and southern hemispheres of the globe. Most of the world's landmass is in the northern hemisphere, including much of the developing world—most notably, India and China. Despite this categorization, Australia and New Zealand, which are in the southern hemisphere, nonetheless belong economically to the North. North and South are thus shorthand terms to describe the more developed and less developed or developing countries (or what during the Cold War used to be called the First and Third Worlds). They are social products, not referents to the world's physical geography.

The less developed countries of the world—which contain the vast majority of the planet's population—include diverse societies in Latin America, East Asia (except Japan and South Korea), Southeast Asia, South Asia, the predominantly Muslim world of Southwest Asia and North Africa, and sub-Saharan Africa.

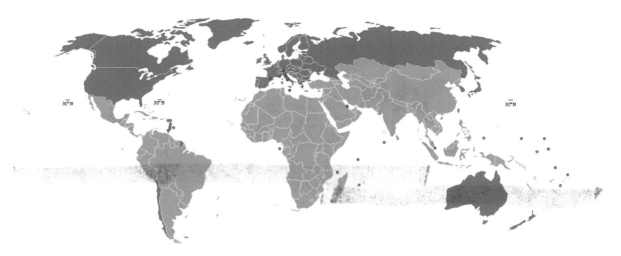

Figure 7.1: The division between the world's developed and developing countries has several names, such as the global North (blue) and South (red). "North" in this context does not refer to the northern hemisphere, in which many poor countries are also located. Rather, the United States, Canada, Europe, and Japan are positioned to the north of Latin America, Africa, and most of Asia. By this measure, Australia and New Zealand are part of the world's North.

In this chapter we will consider how this world of unequal development came about. We begin by analyzing the word and idea of **development**, noting how it embodies many different concepts and measures. We will then discuss the characteristics of less developed countries and some of the barriers to their development. This section also focuses on urbanization in the developing world. In the last section, we will consider some development strategies and examine some of the potentials and pitfalls of industrialization in the developing world.

7.1 MEASURING ECONOMIC DEVELOPMENT

Geographers and other social scientists measure economic development through a number of social, economic, and demographic indexes. The principal ones are per capita GDP, the distribution of the labor force by economic sector, ability to produce consumer goods, educational and literacy levels, the status of health of a population, and its level of urbanization.

PER CAPITA GDP

By far the most common measure of wealth and poverty internationally is the per capita **gross domestic product** (GDP)—that is, the sum total of the value of goods and services produced by a national economy divided by its population. As shown in Figure 7.2, per capita GDP is more than $30,000 in most developed nations. At the same time, the United Nations estimates that three billion people live on less than $2 a day, or $750 per year, and that one billion people live on $1 per day or less (Figure 7.3). Japan, North America, western Europe, Australia, and New Zealand have the highest per capita incomes in the world. The Middle East, Latin America, South Asia, East Asia, Southeast Asia, and sub-Saharan Africa have the lowest. However, GDP fails to measure nonmarket, noncommercial economic activity—for example, barter and subsistence production or household domestic labor. It is also vulnerable to fluctuations in exchange rates and the costs of living.

Per capita purchasing power is a more meaningful measure of actual income per person (Figure 7.4). The relative purchasing power in developed nations is more than $20,000 per capita per year, whereas in the less developed countries of Africa it is much less than $1,000 per capita per year. Per capita purchasing power includes not only income, but the price of goods in a country. The United States is surpassed by Japan, Scandinavia, Switzerland, and Germany in per capita income. However, it surpasses almost all countries in per capita purchasing power because its goods and services, especially housing, are relatively inexpensive compared with those in other industrialized nations. In other respects, including poverty rates and income inequality, the United States lags behind much of Europe.

ECONOMIC STRUCTURE OF THE LABOR FORCE

The distribution of jobs among economic sectors in a country also reflects its economic development. Economists and economic geographers divide employment into three major categories:

1. The **primary economic sector** involves the extraction of materials from the earth, such as mining, lumbering, agriculture, and fishing (Chapter 10).

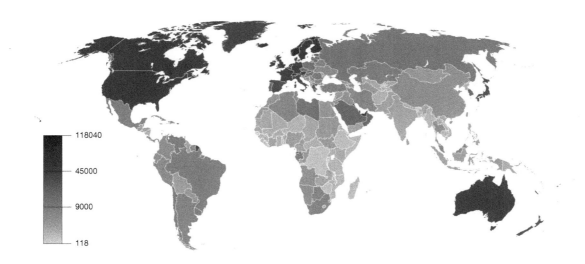

Figure 7.2: Map of per capita GDP, 2015. Per capita GDP is by far the world's most common measure of national wealth and poverty. However, this index ignores important factors such as cost of living, exchange rates, noncommodified labor and output, inequality, and many other ways of defining development.

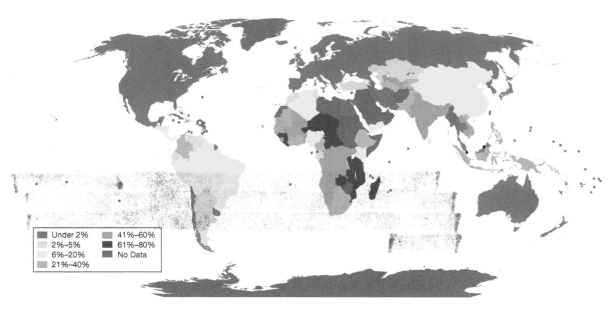

Figure 7.3: The percentage of people in each country living on $1.25 per day or less. Even with the low incomes of the developing world, it is difficult to survive on such a wage. This is, therefore, a map showing the poorest of the world's poor.

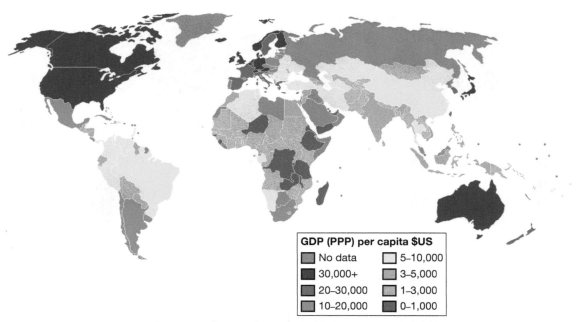

Figure 7.4: A map of per capita GDP that is adjusted for purchasing power parity, including cost of living, which simple per capita GDP maps do not. Because the cost of living varies widely around the world, this is a more meaningful measure of national wealth and poverty.

2. The **secondary economic sector** includes assembling raw materials and manufacturing, that is, the transformation of raw materials into finish products (Chapter 11).
3. The **tertiary economic sector** consists of the provision of services—producer services (finance and business services), wholesaling and retailing, personal services, health care, entertainment, transportation, and communications (Chapter 12).

In less developed countries, a large share of the labor force works in the primary sector, mostly as peasants and farmers who either own small plots of agricultural land or, if landless, must sell their labor to landowners. Because much agricultural labor in the developing world consists of people using essentially preindustrial methods (human and animal power) to grow food for their own survival (rather than for the market), productivity levels tend to be low and many people must work the farms. In the United States, only 2% of the labor force is engaged in agriculture (Chapter 10), whereas in India, China, and certain African nations, a much larger share of laborers works in this sector.

PRODUCTION OF CONSUMER GOODS

The quantity and quality of consumer goods that are purchased and distributed in a society form another measure of its level of economic development. Easy availability of consumer goods means not only that a country's economic resources have fulfilled the basic human needs of shelter, clothing, and food, but also that more resources remain to provide nonessential household goods and services. Automobiles, textiles and clothing, home electronics, jewelry, watches, refrigerators, washing machines, and personal computers are some of the major consumer goods produced worldwide on varying scales. In industrialized countries, more than one television, telephone, or automobile exists for every two people. In developing nations, relatively few of these products exist per thousand people. For instance, the average ratio of persons to television sets in developing countries is 150 to 1, and the ratio of population to automobiles is 400 to 1. The amount of consumer goods per capita provides a good indication of a country's level of economic development.

EDUCATION AND LITERACY OF A POPULATION

Social scientists also measure economic development by the extent and quality of education in a country, including the proportion of children who attend school. The **literacy rate** of a country is the proportion of people who can read and write (Figure 7.5). The number of students per teacher provides another measure of access to education. Small classes allow more student–teacher interaction and greater facilitation of learning. Richer societies in the developed world generally have low student–teacher ratios, whereas poor countries have high ones. Moreover, many developing countries have low teachers' salaries, decrepit school buildings, and insufficient funds for textbooks or equipment such as computers. Despite these shortages, many such countries have ample funds for their militaries, which often amount to far more than what they spend on development. This state of affairs indicates that insufficient investment in education is a policy choice, not a "natural" limitation. We will discuss the influence of politics on development later in this chapter.

Often there are also vast gender differences in literacy rates within developing countries. The highly developed world has nearly identical literacy rates for men and women. In many impoverished societies, however, desperately poor rural families can often afford school fees for only one child (and may need other children to work the fields). The educated child is very likely to be male, given the entrenched sexism that much of the world exhibits. Worldwide,

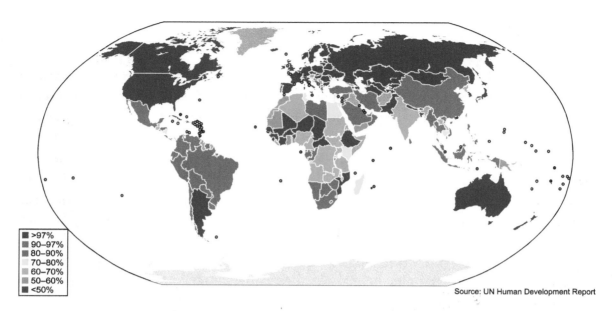

Figure 7.5: Mass literacy is the first step toward economic development. Although near-universal literacy is found in the economically developed world, in many of sub-Saharan African nations a large share or the majority of adults cannot read or write, particularly women.

only 75 girls attend school for every 100 boys. In areas especially biased against women, their literacy rates tend to be much lower than men's. In many countries, the literacy rate of women is less than 25%, whereas the literacy rate of men is between 25% and 75%. The Middle East, sub-Saharan Africa, and South Asia show the greatest disparities. For example, in Nigeria, 68% of men are literate but only 43% of women are; in Afghanistan, 47% of men but only 13% of women can read or write. The regions that have low percentages of women attending secondary school invariably have poor social and economic conditions for women in other respects, including job opportunities. Indeed, raising women's literacy rates has been shown to lower fertility rates and to empower women economically and politically. It therefore constitutes one of the most important forms of economic development.

In a society in which many people can read and write, a proliferation of newspapers, magazines, and (in some areas) online forums improves and fosters communication and exchange. This exchange lead to further development by informing people of opportunities (e.g., jobs, crop prices), allowing them to obtain more skilled and better-paying jobs, circulating best practices, and so forth.

HEALTH OF A POPULATION

Measures of health and welfare, in general, are much higher in developed nations than in less developed countries. One measure of health and welfare is diet, typically measured as caloric consumption per capita (Figure 7.6). In developed nations, the population consumes approximately one-third more than the minimum daily requirement and therefore maintains a higher level of health. Some areas of each country, however, lack sufficient calories and food supplies. Even in the United States significant pockets of hunger and malnutrition exist. Conversely, overly abundant, cheap food and inadequate exercise have generated an obesity epidemic in the United States, and to some extent in many other countries, leading to widespread diabetes, heart disease, and strokes. Obesity has become a worldwide epidemic. Today here are more obese people than malnourished ones. It is even possible to be obese and malnourished at the same time because of calorie-rich, nutrient-poor diets.

People in developed nations also have better access to doctors, hospitals, and health care providers. For relatively developed nations, there is one doctor per 1,000 people. In developing countries, however, each

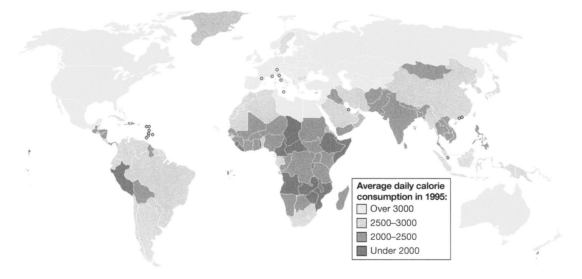

Figure 7.6: Average daily caloric consumption indicates economic development or lack thereof. In the developed world, food is cheap and obesity is a growing problem. In contrast, in much of Africa and South Asia, low-caloric consumption leads to malnutrition and even starvation.

person shares a doctor with many thousands of others. Africa by far has the worst access to health care, followed by Southeast Asia and East Asia. In some African countries there are more than 15,000 people per physician, effectively meaning that most people *never* see one. Tanzania and Malawi, for example, have 50,000 people per physician. Everywhere, wealthier societies have better access to health care, although there are huge discrepancies within them as well, such as in the United States, with its pockets of poverty and a large pool of uninsured people.

Infants and children are the most vulnerable members of any society, In part, this fact stems from their immunological systems being less well-developed than those of adults. In part, it is because they lack effective political power to shape public policy. In developed nations, on average, fewer than 10 babies in 1,000 die within the first 100 days. In many less developed nations, more than 100 babies die per 1,000 live births. Most of these deaths stem from poor prenatal care, malnutrition, and infectious diseases. When there are economic downturns or disruptions of the food supply brought on by droughts or war, they are generally the first to die. The geography of infant mortality (Figure 7.7)—the proportion of babies who die before their first birthday—is thus perhaps one of the best measure of economic development,

Figure 7.7: Infants are the most vulnerable members of any society. The geography of infant mortality rates reflects the uneven access to food and health care found throughout the world.

or the lack of it. In much of Africa, more than 10% of infants do not live through their first year. In the developed world, by contrast, infant mortality rates are very low. Notably, Cuba's **infant mortality rate**—6.0 deaths per 1,000 live births—is lower than that of the United States whose rate is 7.0, a discrepancy that reflects the former's investment in the health care and the latter's widespread unavailability of health insurance amid its poor.

Acquired immune deficiency syndrome (AIDS) has emerged as a significant health threat worldwide. Since 1980, more than 25 million people have died of this disease. An additional 33 to 45 million people currently live with the HIV virus. The epicenter of the AIDS epidemic is sub-Saharan Africa (Figure 7.8), where in some countries 40% of the adult population is infected with the HIV virus. The sub-Saharan region accounts for more than 60% of the people living with HIV worldwide, or some 25 million men, women, and children. AIDS is tightening its grip outside the United States and Europe. In India, researchers estimate that 50 million people could be HIV positive by the year 2020. Half the prostitutes in Bombay are already infected. Doctors report that the disease is spreading along major truck routes

Figure 7.8: HIV, the virus that leads to AIDS, is most prevalent in southern Africa, where up to 40% of the population may be infected. In some countries, the disease is leading to population decline and millions of orphans. It robs many economies of young adults in their prime working years.

and into rural areas, as migrant workers bring the virus home. In some African countries, politicians have denied that the HIV virus causes AIDS. In China, too, AIDS is spreading rapidly. The social consequences are catastrophic, as millions of children have lost their parents to AIDS and may show symptoms of the disease in the future. Because the disease has a long lead time in which infected people do not show symptoms, and because sexual behavior is very difficult to change, many health experts fear that AIDS could depopulation parts of the world. However, recently the widespread introduction of new drugs has allowed people with HIV to live long, relatively healthy lives. Still, for many people this path remains unaffordable without generous foreign subsidies.

The reliability and quality of the food supply, access to clean drinking water, public health measures, ability to control infectious diseases, and access to health services all shape how long we live. People in economically advanced countries live a relatively long time—often more than 75 years, on average—whereas those in poorer countries live considerably fewer years (Figure 7.9). In parts of Africa, few adults can realistically expect to live past their 50th birthday. Thus, the geography of life expectancy is another measure of the health and welfare of a population.

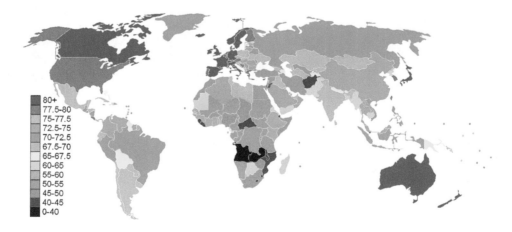

Figure 7.9: Map of average life expectancy at birth. In economically developed countries, people can expect to live into their 70s, or even higher. In parts of Africa, in contrast, life expectancies can be as low as 32, largely due to malnutrition, high infant mortality rates, and AIDS.

7.2 URBANIZATION IN DEVELOPING COUNTRIES

In the industrialized West, intensive urbanization occurred as a result of the Industrial Revolution (Chapter 5). Although parts of the developing world are urbanized, in general less developed countries lag behind the economically advanced countries in this regard. However, cities throughout the developing world are growing much more quickly than those in Europe, North America, Japan, or Australia. Today, about 56% of the world's people, an all-time high, live in cities (Figure 7.10). For the first time in history, the majority of people on the planet reside in urban areas.

The proportion of each country's population who live in cities varies widely around the globe (Figure 7.11). In parts of Latin America, urbanization rates resemble those of North America or Australia, where 75% or more of the people live in cities. In Asia and Africa,

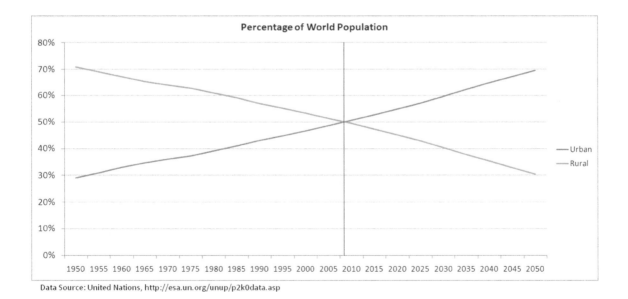

Data Source: United Nations, http://esa.un.org/unup/p2k0data.asp

Figure 7.10: Percent of the world population residing in settlements of 5,000 people or more. In the early twenty-first century the world passed a threshold, in which the majority of the planet lives in an urban area.

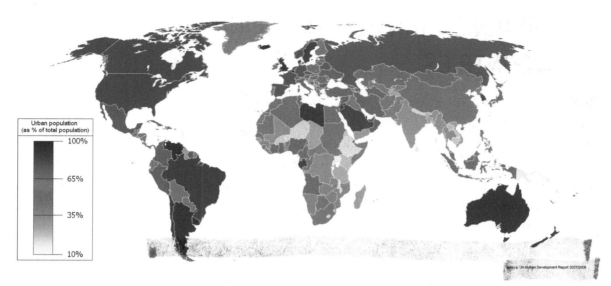

Figure 7.11: This map shows the proportion of people living in cities. Whereas cities were fueled by the Industrial Revolution in the economically developed world, in developing countries urban growth is propelled largely by rural-to-urban migration. Now, for the first time, more than half of the world's people live in urban areas.

however, the proportions are generally lower. About 50% of China's people live in cities, and in wide swaths of Africa less than 20% do so.

Urbanization in the developing world differs significantly from that in the developed one. This difference exists because the historical contexts of less developed countries differ from those of their former colonial powers, and because their mode of incorporation into the global division of labor was very different. Some countries in the developing world, such as the Arab world and China, had well-established traditions of urbanization before the Europeans came. In China, for example, Guangzhou was one of the world's largest cities in the fourteenth century, and during the height of the Arab Empire from the seventh to the thirteenth centuries, Damascus, Alexandria, and Baghdad were major cosmopolitan centers of trade and scholarship. In others, colonial powers such as Britain played a major role, as they constructed cities such as Calcutta in India, Rangoon in Myanmar (formerly Burma), and Singapore. Similarly, the Dutch started Batavia, Indonesia, which later became Jakarta (Chapter 5). Colonialism had powerful impacts on many cities, including the construction of wide roads, European-style buildings, rail lines, and water infrastructure.

Today, the vast bulk of the world's urban growth is in the developing world. The world's largest urban areas, for example, are in Latin America and Asia, not in Europe or North America (Table 7.1). In 2012, the greater Tokyo metropolitan region, with 36 million, was the world's most populous urban area, while New York, Mexico City, Mumbai, Guangzhou, and São Paolo vied for second place, with roughly 20 million inhabitants each. Many of the other urban areas, with populations greater than five million, are located in China and India (Figure 7.12).

TABLE 7.1: WORLD'S LARGEST URBAN AREAS, 1975–2025 (MILLIONS)

RANK	AGGLOMERATION, COUNTRY	POP. 1975	POP. 2000	POP. 2015	POP. 2025 (EST.)
1	Tokyo, Japan	26,615	34,450	35,676	36,371
2	New York, USA	15,880	17,846	19,040	19,974
3	Ciudad de México, Mexico	10,690	18,022	19,028	20,189
4	Mumbai (Bombay), India	7,082	16,086	18,978	21,946
5	São Paulo, Brazil	9,614	17,099	18,845	20,544
6	Delhi, India	4,426	12,441	15,926	18,669
7	Shanghai, China	7,326	13,243	14,987	17,214
8	Kolkata (Calcutta), India	7,888	13,058	14,787	17,039
9	Dhaka, Bangladesh	2,221	10,285	13,485	17,015
10	Buenos Aires, Argentina	8,745	11,847	12,795	13,432
11	Los Angeles, USA	8,926	11,814	12,500	13,160
12	Karachi, Pakistan	3,989	10,019	12,130	14,855
13	Cairo, Egypt	6,450	10,534	11,893	13,465
14	Rio de Janeiro, Brazil	7,557	10,803	11,748	12,775
15	Osaka-Kobe, Japan	9,844	11,165	11,294	11,365
16	Beijing, China	6,034	9,782	11,106	12,842
17	Manila, Philippines	4,999	9,958	11,100	12,786
18	Moscow, Russia	7,623	10,016	10,452	10,524
19	Istanbul, Turkey	3,600	8,744	10,061	11,177
20	Paris, France	8,558	9,692	9,904	10,007
21	Seoul, South Korea	6,808	9,917	9,796	9,740
22	Lagos, Nigeria	1,890	7,233	9,466	12,403
23	Jakarta, Indonesia	4,813	8,390	9,125	10,792
24	Chicago, USA	7,160	8,333	8,990	9,516
25	Guangzhou, China	2,673	7,388	8,829	10,414

308 HUMAN GEOGRAPHY: A SERIOUS INTRODUCTION

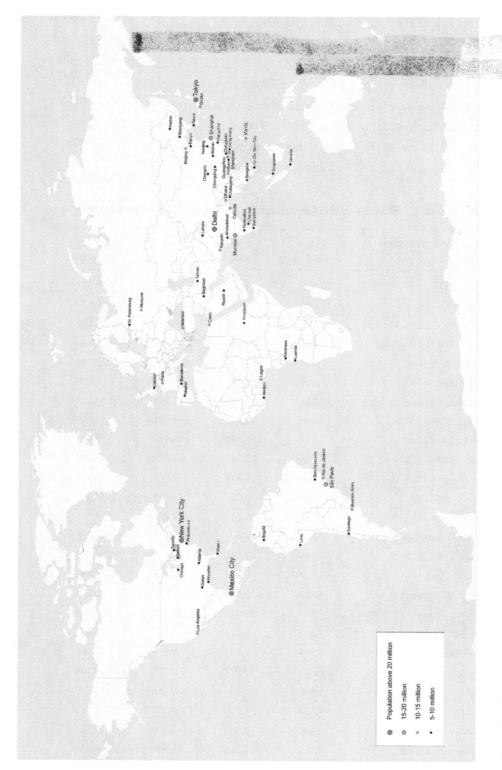

Figure 7.12: The largest metropolitan areas of the world are mostly found in developing countries, where urban areas of 20 million people or more are often found. Given high population growth and rural-to-urban migration, most urban growth in the future will continue to be in the developing world.

Moreover, cities in the developing world are growing much more rapidly than their counterparts in the developed countries. Although natural growth rates in cities in the developing world are a little higher than those in the West, urbanization in less developed countries is primarily attributable to the massive influx of rural-to-urban migrants. Many of them are displaced by agricultural mechanization, unequal land distribution, unemployment, lack of public investment, low crop prices, war, and high population growth in rural areas. To some extent, this pattern held true of cities in the West as well, as it developed during the Industrial Revolution. However, because few developing countries have generated the industrial job growth that the West did during the nineteenth and twentieth centuries, the labor markets and employment conditions in the developing world are fundamentally different. In part this difference results from the roughly five centuries that the developing world spent as European colonies (Chapter 5). Thus, the urban growth and rural crisis occurring in less developed countries are two sides of one coin. Because many migrants move to the cities with perceptions—often erroneous or imperfect—that they will have greater opportunities there, they often find themselves in desperate circumstances.

INFORMAL EMPLOYMENT

The lack of sufficient employment opportunities in the labor markets of less developed countries generally leads to high unemployment and underemployment rates (discussed later in this chapter). Many migrants find work in the **informal economy,** which includes a wide array of jobs that are not regulated by the state or taxed. Most positions in the informal economy are oppressive, low-paying, and offer little financial security. This sector includes a wide variety of occupations, including very low-end retail trade, pulling rickshaws, prostitution, smuggling, day laborers, and begging (Figure 7.13).

The labor market is often segmented on the basis of gender, with women performing the least attractive labor as measured by the quality of the work environment and remuneration. Child labor is also a common circumstance, including working in the home and in the informal sector. Children are especially valued by some employers because they can be forced to work long hours for little pay in often unsafe and unhealthy environments, such as mining and brickmaking.

In many poor countries, the informal sector constitutes the bulk of employment opportunities. In Manila, for example, 30,000 people earn their living recycling garbage on the Payatas garbage dump, as adults and children work together, stepping over rotting debris to scavenge bits of plastic, metal cans, or pieces of destroyed buildings (Figure 7.14). Others find marginal incomes selling trinkets and food on the streets; engaging in prostitution; selling illegal drugs; participating in the black market, especially illicit currency exchange; or, as casual day laborers,

Figure 7.13: The dynamics of the global economy, national policies, and local social circumstances all contribute to the geography of beggars, the world's poorest of the poor.

working "off the record" in construction. In contrast, formal sector, export-oriented jobs (such as those in the garment industry or electronic assembly plants) that tie cities in developing countries to the global economy still offer higher wages than most of those in the informal economy, however exploitative they appear to those of us in the West. But these tend to be the exception, not the norm. Jobs in the formal sector tend to be in foreign-owned corporations that invest there looking for cheap labor, resources (e.g., petroleum or mineral mining), or in government.

Much of the available employment in the formal sector is low-paying, and sometimes very unsafe, but the workers usually regard such work as temporary. For migrants, the attraction to cities is not the squatter settlement or immediately available temporary employment, but the perception that the city offers better schooling, better medical services, a better water supply, and most importantly, the prospect of permanent

Figure 7.14ab: Recycling plastic, paper, metal cans, or parts of old buildings is part of the informal economy, which makes up the bulk of employment opportunities in the developing world. Child labor is common in such circumstances.

employment. As squatter settlements grow, enormous pressure is put on urban governments to provide the needed services. Inevitably, however, extended urban services attract even more migrants and the cycle begins again.

RESIDENTIAL PATTERNS IN THE DEVELOPING WORLD

Residential patterns in cities of developing countries differ from those in the West. Whereas in the developed countries the poor tend to be a minority, often consigned to the city center, in the developing world, the poor may be a majority of the city's inhabitants, depending on the overall level of economic development of that nation. Frequently, the relatively wealthy command the city centers, whereas the poor live on the urban periphery. In many Latin American cities—for example, Buenos Aires and Mexico City—housing near the old colonial plaza in the center is relatively expensive.

Many cities in the less developed world are economically and socially vibrant. Tall buildings that are home to major corporations and impressive architecture indicate that these cities are command centers of regional and national economies. Many of the cities are also increasingly involved in the global economy. Unfortunately, however, parts of these cities often suffer from problems of overcrowding, crime, poverty, disease, limited provision of services, traffic congestion, environmental damage, and ethnic conflicts. In all these situations the poor and underprivileged suffer most.

GROWTH OF SLUMS

Cities that expand so rapidly that their growth is not controlled—termed as undergoing "premature urbanization"—always experience problems. The slums of English industrial cities in the midnineteenth century were just as bad as those in many cities in the less developed world today. People moved to them for essentially the same reason: No matter how difficult life there might be, for many people the opportunities in the cities outweighed the problems.

More than two billion people today live in a "planet of slums" (Figure 7.15). Densities in such places are often far higher than in any Western city. In Africa, about 70% of the urban population lives in slums. The highest incidence is in Sierra Leone, where slum dwellers make up 96% of the urban population. The figure for the Central African Republic is 92%, and for Nigeria, 79%. In dilapidated houses, often self-made from cheap materials such as cinder blocks or sheets of corrugated tin, many urbanites inhabit squalid neighborhoods that go by a variety of names: slums, Brazilian *favelas*, Indonesian *kampong*, Turkish *gacekondu*,

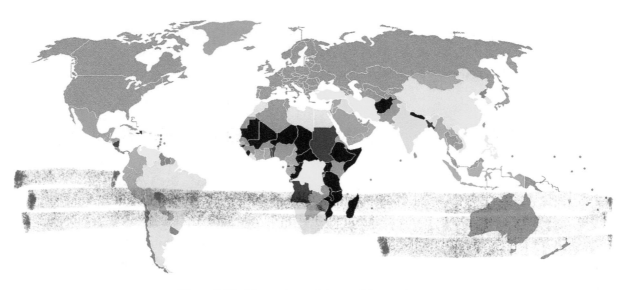

Figure 7.15: A large share of the world's population lives in slums, large communities of informal housing that is not professionally made because the formal housing sector is small and not profitable.

South African townships, and West African *bidonvilles* (or "tin can cities") (Figure 7.16). Countless numbers of people are forced into conditions that lack adequate housing, electricity, roads, transportation systems, schools, clean water, or sewers. High unemployment, lack of access to health care, inadequate educational facilities, and poor road systems all conspire to produce vast pools of suffering. Often densities in such places are far higher than in any Western city. When migrants seize buildings or land that belongs to landowners, the government may deem such "squatter settlements," as slums are often called, to be illegal and tear them down with bulldozers. (The term "squatting" refers to the legality of landownership and not to the issue of rent payment.) These communities can become more than just cesspools of misery—they can turn into breeding grounds of resentment and political activism against corrupt and uncaring governments.

One consequence of rapid and unplanned squatter expansion is often a chaotic, uncoordinated governance system. In Bangkok, the city core is governed by a single authority, but the larger metropolitan area is divided into about 2,000 small areas, each with its own local government. It is easy to realize just how difficult it is to build transportation lines and other components of the urban infrastructure under these circumstances.

a

b

Figure 7.16 [a-d]: Although they go by different names around the world, shantytowns or slum districts are fueled by similar mechanisms—for example, rural-to-urban migration. A serious lack of adequate housing forces people to construct their dwellings out of available materials such as cardboard or corrugated tin.

Figure 7.16 [a-d]: (*continued*)

HEALTH AND ENVIRONMENTAL CONCERNS

There are two principal health concerns in slums: first, conventional environmental health problems, such as limited land for housing and lack of services; and second, problems related to rapid industrialization, such as waste disposal and pollution. Cities in the less developed world must address both sets of problems, which remain unsolved even by many cities in the more developed world.

Those living in slums face serious health problems, many caused by unsafe water and inadequate sanitation. For example, Bangui in the Central African Republic has a current population of about 700,000 but a sewage system built for a population of 26,000 that has never been extended or improved. Also, Cairo treats less than half of its sewage; the remainder ends up in the Nile or in local lakes. Acute respiratory infections and diarrhea are major causes of death in many urban slums, especially among young children. Malaria, typhoid, and cholera also remain major problems in some cities. Air pollution is often at a dangerous level because of the proximity of industrial activities and the often minimal control of emissions (Figure 7.17).

Figure 7.17: Rapid growth, congestion, and poor environmental safeguards often lead to severe air pollution in cities in the developing world, which in turn triggers a variety of respiratory problems. Chinese cities are among the most polluted in the world, with enormous health repercussions. Some Indian cities are even worse.

Slums in less developed world cities are especially vulnerable to some of the possible consequences of global environmental change, including both sea-level rise and increased incidence of extreme weather events such as hurricanes, earthquakes, and tsunamis. Often located in areas prone to flooding, lacking infrastructure and protection, unplanned, and poorly if at all managed, these areas suffer most during disasters. For example, cities in Haiti, such as the capital, Port-au-Prince, suffered tremendously following the earthquake there in 2010, which killed 250,000 people. Such calamities are worsened by the widespread use of cheap building materials (e.g., unfortified concrete), which do not hold up well in earthquakes.

7.3 PROBLEMS OF DEVELOPING COUNTRIES

The developing world encompasses a vast array of societies with enormous cultural and economic differences. Nonetheless, several characteristics tend to exist to one degree or another in all these nations. Obviously, the more economically developed a country is, the less likely it is to exhibit these qualities.

RAPID POPULATION GROWTH

We must be careful to avoid the simplistic Malthusian error (Chapter 8) of ascribing all of the world's problems to population growth. Nonetheless, in many countries, the rate of population growth exceeds by far the productivity gains in agriculture and other sectors, sorely diminishing the average standard of living. Many scholars argue that population growth must decline if development is to succeed.

In many less developed countries, especially those in Africa, rapid population growth rates tax the available food supplies, generating food insecurity and even starvation. Poor countries tend to have the highest fertility, and thus natural growth, rates (Chapter 8). Although we cannot attribute every problem such as inadequate farming land, malnutrition, or famine to population growth, we nonetheless cannot ignore the steadily rising demand for resources that population growth entails. Rapid population growth also reduces the ability of households to save, inhibiting the accumulation of domestic investment capital and helping to perpetuate a cycle of poverty (Figure 7.18). With rapid population growth, more investment is required to maintain a level of real growth per person (i.e., constant GDP per capita). For example, in 2010–2011, Chad had a GDP growth rate of 2.5%, but a population growth rate of 2.9%, meaning that real average incomes declined slightly. If public or private investment fails to

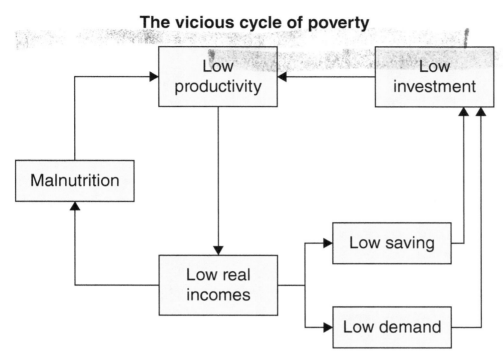

Figure 7.18: The cycle of poverty takes many forms, including low productivity, low incomes, and low investments. In the less developed world in particular, escaping this cycle is very difficult.

keep pace with population growth, each worker—having fewer tools and equipment with which to produce goods and services—will be less productive. This declining productivity results in reduced per capita incomes and economic stagnation.

Rapid population growth in agriculturally dependent countries means that the land must be further subdivided and used more heavily than ever. Smaller plots from subdivision inevitably lead to overgrazing, overplanting, and the pressing need to increase food production from a limited amount of space for a growing population. Many less developed countries are rapidly urbanizing. Rapid population growth generally entails large flows of rural residents to urban areas and greater urban problems. As we have discussed above, availability of housing, congestion, pollution, crime, and lack of medical attention are all seriously worsened by the rapid urban population growth.

A rapid increase in population—especially the number and proportion of young dependents—creates serious problems in terms of food supply, public education, and health and social services. It also intensifies the employment problem. However, a high rate of population growth was once a characteristic of present-day developed countries, and it did not prevent their development. Thus, it is difficult to argue that population growth necessarily leads to underdevelopment or that population control necessarily aids development. Indeed, population growth is only one of several factors that contribute to poverty. These factors include the legacy of colonialism, the dynamics of the world system, and governments' indifference to investing in their populations and infrastructure.

UNEMPLOYMENT AND UNDEREMPLOYMENT

Unemployment and underemployment are major problems in less developed countries. Unfortunately, their economies rarely generate enough jobs for all, for a variety of reasons. *Unemployment* is a condition in which people who want to work cannot find jobs. *Underemployment* means that working people are not able to work as much as they would like, usually 8 hours per day. It also means that they cannot fully utilize their skills and talents. Reliable statistics on unemployment and underemployment in less developed countries are difficult to obtain, but unemployment in many developing countries often exceeds 20%. For example, in 2015 the unemployment rate was 28% in Bosnia, 59% in Djibouti, and 60% in Tajikistan.

Unemployment and underemployment are not the sole reasons for the problems of less developed countries. Most less developed countries lack sufficient investment to generate jobs. (We discuss this topic in detail later in the chapter.) Because domestic pools of investment capital are limited, poor countries rely on foreign direct investment (FDI) (see Chapter 6) supplied by transnational corporations. However, such opportunities can be fickle and often come with social, economic, and environmental costs. Thus, unemployment is much more complex than simply the willingness or unwillingness of people to work. It involves a complex, globalized institutional framework that includes the allocation of public and private resources, the generation (or not) of jobs, the creation and maintenance of occupations and skills, and the channeling (or not) of investment capital.

LOW LABOR PRODUCTIVITY

Are the problems of less developed countries a result of low labor productivity? It is true that a day's toil in a typical developing country, especially in agriculture, produces little commercial value compared with a day's work in a typical developed country. Human productivity in a developing country may be as little as 10% of that in a developed country.

One reason for low levels of productivity is the small scale of farming operations that often exist in less developed countries, which are often subsistence-based. Another is that capital investment rates for new machinery and equipment are often low, interest rates (the cost of borrowing to invest) are relatively high, and most capital is generated by foreign-owned firms, whose major incentive is exports and profits, not local job generation. Most developing countries lack the buildings, machines, engines, electrical power networks, and factories that enable people and resources to produce efficiently. In addition, less developed countries are less able to invest in **human capital**. Investments in human capital—such as education, health, and other social services—make unskilled workers into productive workers. As we have discussed above, schools are often inadequate, literacy rates may be comparatively low, and technical and management skills are in short supply. Many highly skilled professionals, such as engineers and doctors, emigrate to economically advanced countries such as the United States, where they are more highly paid. This immigration has contributed to a so-called **brain drain**, whereby less developed countries lose talented people to the developed world.

Although almost all less developed countries have low rates of labor productivity, this factor does not cause their lack of development. What has prevented labor productivity from improving in developing countries? To answer this question, we must examine other characteristic problems of developing countries.

LACK OF CAPITAL AND INVESTMENT

Most less developed countries suffer from a lack of capital in the forms of machinery, equipment, factories, public utilities, and infrastructure in general. The more capital there is, the more tools are available for each worker and the higher the rate of productivity. Thus a close relationship exists between output per worker and per capita investment and, in turn, between investment and income. If a nation hopes to increase its output, it must find ways to increase per capita capital investment.

In most cases, increasing the amount of arable (farmable) land is no longer a possibility. Most cultivable land is already in use. Therefore, capital accumulation for less developed countries must come from savings and investment. If a less developed country can generate export revenues by selling goods and services on the international market, and invest some of its earnings, the production of consumer goods and capital goods (goods that make other goods) will provide additional resources. But barriers to saving and investing are often high in less developed countries, including high interest rates, political unrest, corruption, and inefficient regulations. Domestic output is often so low that the country must use all of the output to support the country's many urgent needs.

Many less developed countries suffer from **capital flight**—wealthy individuals and firms in these countries have invested and deposited their monies for safekeeping in overseas ventures and banks in developed countries. They have done so because of fear of confiscation of their monies by politically unstable governments, unfavorable exchange rates brought on by inflation, high levels of taxation, and the possibility of business and bank failures. Inflows of foreign aid and bank loans to less developed countries are almost completely offset by capital flight to banks in Europe and North America. In 2010, for example, Mexicans held abroad approximately $130 billion in assets, an amount roughly equal to their country's international debt.

Finally, investment obstacles in less developed countries have impeded capital accumulation. The two main problems with investment in less developed countries are (1) lack of investment opportunities comparable with those available in developed countries, and (2) lack of incentives to invest locally. Because less developed countries usually have low levels of domestic spending per person, their markets are weak compared with those of advanced nations. Without domestic industries, consumers must turn to imports to satisfy their needs, especially for high-value-added products such as manufactured goods. In addition, a lack of infrastructure to provide transportation, management, energy production, and community services impedes improvement of the environment for investment activity.

INADEQUATE AND INSUFFICIENT TECHNOLOGY

Typically, less developed countries lack the technologies necessary to increase productivity and accumulate wealth. Some less developed countries acquire new production techniques through technology transfer that may accompany investment by transnational corporations, as happened in the newly industrializing countries of East Asia (which we discuss later in this chapter). Similarly, Organization of the Petroleum Exporting Countries (OPEC) nations benefited from foreign technology in oil exploration, drilling, and refining. Unfortunately, for less developed countries to put this technology to use, they must have a minimal level of capital goods and human capital for the installation, repair, and maintenance of machinery. Needed is a flow of technologically superior, highly reliable capital goods from the developed countries to the developing nations so that they can improve their productivity.

In developed countries, technology has been developed primarily to save labor and to increase output, resulting in capital-intensive production techniques. However, in less developed countries, labor-saving technology tends to displace workers, eliminating critically needed jobs. Thus less developed countries need technology that is labor-intensive, that is, uses more workers per unit of output than is used in the developed world. In agriculture, much of the midlatitude technology of the developed countries is unsuited for low-latitude agricultural systems with tropical or subtropical climates and soils (Figure 7.19). For example, imported wheat harvesting combines were not optimal in Brazil, where they displaced workers and turned fecund tropical areas into virtual deserts by exposing soils to rainfall, leading to a leaching of nutrients and depletion of soil quality. In short, to understand the impacts of technology, we must examine them in the social and economic context in which they occur.

UNEQUAL LAND DISTRIBUTION

In most developing countries, in which a large part of the population lives in rural areas, land is a critical resource essential to survival. Unfortunately, a small minority of wealthy landowners often controls the vast bulk of farmable land. In Brazil, for example, 2% of the population owns 60% of the arable land; in Colombia, 4% owns 56%; in El

Figure 7.19: The mechanization of agriculture in the developed world occurred under conditions of labor scarcity. In the developing world, it occurs under conditions of labor surplus, meaning many rural workers are rendered redundant. The same technology can therefore have different impacts depending on local economic and environmental circumstances.

Salvador, 1% owns 41%; in Guatemala, 1% owns 36% of the land; and in Paraguay, 1% owns an astounding 80% of all land suitable for farming. These numbers reflect the historical legacy of the Spanish land grant system (*encomiendas*) imposed over centuries of colonialism. Wealthy rural landowners often control vast estates and plantations designed primarily for export crops, not domestic consumption. The majority of the rural population, consequently, is landless and must sell their labor. Often, such rural landless farmers earn below subsistence wages, and live in serf-like conditions (Figure 7.20). Farmers who are forced to sell their labor for very low wages are prone to cyclical economic changes such as global food prices, and are apt to live in squalid housing, suffer malnutrition, and offer few opportunities for their children.

Shortages of land are accentuated by agricultural mechanization and high population growth. The result is frequent political turmoil, including demands for land redistribution. In much of Latin America, social movements of peasants and indigenous people to regain control of the land and violence over land possession are frequent. Sometimes large landowners hire private armies to confront landless farmers, setting the stage for prolonged violence. Further, rural areas with high natural

Figure 7.20: Without land, rural workers have nothing to sell but their labor, trapping them in poverty and destitution. When people own land, they are productive and often self-sufficient. Struggles over land ownership are therefore an important part of economic development efforts.

growth rates and widespread poverty become the source of waves of rural-to-urban migration, which as we have discussed above often swamps cities with desperately poor people seeking jobs.

Throughout much of the developing world, **land reform**, which is vital to increased agricultural output, is lacking because the government is too inept or too corrupt to redistribute the land owned by a few wealthy families. For some nations, such as the Philippines, land reform is the single most pressing problem deterring economic development. In contrast, strong action taken by the South Korean government after the Korean War allowed for increased productivity and the development of an industrial and commercial middle class that made South Korea a success story of East Asia.

POOR TERMS OF TRADE

A country's **terms of trade** refer to the relative values of its exports and imports (Chapter 6). When its terms of trade are good, a country sells, on average, relatively high-valued commodities—for example, manufactured goods—and imports relatively low-valued ones—for example, agricultural products. This situation generates foreign revenues that allow it to pay off debt or to reinvest revenues in infrastructure, human capital, or new technologies. Unfortunately, many less developed countries have economies distorted by centuries of colonialism in which they became mere suppliers of minerals and foodstuffs. These countries still lack strong indigenous manufacturing, so they must export low-valued primary-sector goods and import expensive high-valued, manufactured ones.

Thus, these less developed countries rely heavily on the export of goods such as petroleum, copper, tin and iron ore, timber, coffee, and fruits and must import items such as automobiles and trucks, pharmaceuticals, office and telecommunications equipment, and machine tools. This situation makes it difficult to generate foreign revenues and intensifies a country's foreign debt problems, as we will see in the next section. These poor terms of trade perpetuate a country's complex cycle of poverty and inhibit the prospects for long-term development.

FOREIGN DEBT

Much of the developing world is deeply in debt to foreign governments and banks. In 2016, total world debt amounted to more than $54 trillion. In absolute terms, the largest debtors are in the developed world (Figure 7.21), led by the United States, which owes the world more than $15.8 trillion (Table 7.2). However, when we examine this issue in terms of relative debt, or debt as a percent of GDP, quite a different picture emerges. Measured in terms

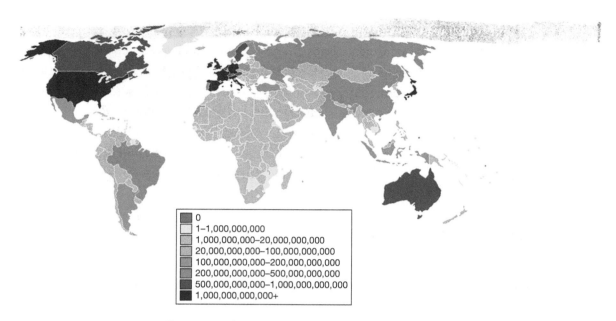

Figure 7.21: Absolute national debt, 2011. In absolute terms, the United States, whose national debt is more than $18 trillion, is by far the largest in the world.

TABLE 7.2: WORLD'S MAJOR DEBTOR COUNTRIES, 2017 ($ TRILLIONS)

Country	
United States	18.6
United Kingdom	7.8
France	5.3
Germany	5.1
Netherlands	4.3
Luxembourg	4.0
Japan	3.5
Italy	2.3
Ireland	2.2
Spain	2.0
Canada	1.7
Switzerland	1.7
Australia	1.6
China	1.5

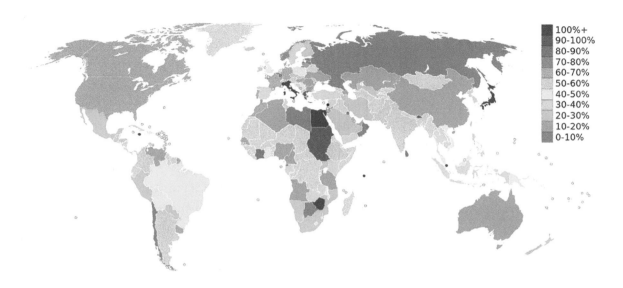

Figure 7.22: Relative national debt, 2012. Debt as a percent of GDP reflects a country's ability to pay back that debt. By this measure, the United States is not as badly off as many European states or Japan. However, in most of the developed world, debt levels are as high as 80% to 90% of GDP, and sometimes even higher.

of national income, developing countries owe the most (Figure 7.22). Some countries such as Egypt, Sudan, or Zimbabwe have foreign debts in excess of 100% of their GDPs. Having borrowed hundreds of billions of dollars, they find themselves unable to repay either the interest or the principal. Debt repayments—annual payments of interest and the debt principal—often consume a large share of their export revenues, which might otherwise be used for development. Indeed, the average developing country spends 50% of its export revenues on debt repayment.

The origins of the debt crisis lay in the 1970s and 1980s, when many less developed countries took out large loans with the expectation of paying them off when their economies improved in the future. Western banks had a huge influx of dollars earned by OPEC countries from the sale of petroleum and were eager for borrowers. Developing countries were happy to take advantage of this unaccustomed access to cheap loans with few strings attached. The borrowing enabled them to maintain domestic growth. Even oil-exporting nations such as Mexico, Libya, and Nigeria borrowed heavily based on their expectations of earning revenues from future oil sales. When oil prices fell significantly in the

1980s and 1990s, these nations found themselves saddled with debts that they could not afford. In addition, some governments spent heavily on their militaries rather than improving their economic infrastructure, and others supported overly ambitious, unrealistic development schemes.

The mounting debt caused concerns about the stability of the international monetary system. In 1982, Mexico was unable to meet interest and capital payments on its debts and essentially defaulted; Brazil and Argentina also appeared ready to default. A collapse of the financial system was forestalled by a series of emergency measures designed to prevent large debtor countries from defaulting on their loans. These measures involved banks, the International Monetary Fund (IMF), the World Bank, and the governments of lending countries in massive bail-out exercises that accompanied debt reschedulings (redefining the payment terms of debt, such as granting a temporary moratorium or stretching payments over a longer period of time).

The debt restructuring policies imposed by the IMF typically tie debt relief to **structural adjustment policies**, which typically include reductions in government subsidies to the poor (such as that for kerosene, cooking oil, or mass transit), a rise in interest rates to attract foreign investors, and devaluations of currencies (which drive up the costs of imports). Such policies lowered the quality of life for hundreds of millions, if not billions, of people, and have made the IMF a much hated institution in many countries (Figure 7.23). For example, a poor mother in a less developed country with a sick infant relies on imported pharmaceuticals when her country lacks a domestic industry. IMF-imposed currency devaluations drive up the cost of these necessities, often making them out of reach for the poor, who need them the most. For these reasons, scholars and politicians involved in international development often call for a debt moratorium (a temporary period in which debt repayments are halted) to allow less developed countries a respite from what are often crushing debt burdens.

As the debt crisis grew, some less developed countries required foreign banks to rewrite their loans and cancel or decrease a portion of the principal and interest. The result was a loss of confidence in the future ability of many less developed countries to repay. Higher oil prices, declining prices for raw material exports, higher world interest rates, and a decline in public and private lending to less developed countries because of loss of confidence all contributed to the debt crisis that continues today.

RESTRICTIVE GENDER ROLES

Deeply entrenched social stratification systems work against people, especially women, in many developing countries. Gender refers to the socially reproduced differences—including both privileges and obligations—between women and men, and permeates every society's allocation

Figure 7.23: Because it has enforced austerity measures on many countries through its "structural adjustment" policies, the International Monetary Fund (IMF) has become a widely disliked organization throughout the world, leading even to riots in some cases.

of resources and people's life chances (Chapter 13). In most of the world, but especially in developing countries, gender roles work to the advantage of men and the disadvantage of women. Although a few women in the developing world have risen to positions of power at various times, becoming, for example, the leaders of Chile, Argentina, Brazil, Panama, the Philippines, and Liberia, these are the exceptions. Most women in developing nations have lower incomes, less status and power, and fewer opportunities than men do.

The economic, political, and social status of women varies around the world. Women account for most of the world's 1.7 billion people living in extreme poverty. In most countries, women do most of the field work as well as household chores such as cooking, carrying water, and raising children. In some countries with more mechanized forms of agriculture, such as most of Latin America, men assume the job of farming. In these cases, women work primarily in the paid labor market, including both the formal sector (e.g., assembly plants) and the informal one.

In Latin America, labor-force participation of women in the economy is increasing but mostly outside of the agricultural sector.

Sub-Saharan Africa, India, and Southeast Asia depend on female farm and market income as well as wage labor. In Muslim areas, most women are not economically active outside the home because of stringent religious prohibitions. At a world scale, women generally garner only 60% to 70% of what men earn, often for the same work. This ratio is remarkably widespread. It includes the United States, in which women make up the bulk of the poor and occupy less than 3% of the highest levels of management and ownership.

CORRUPT AND OPPRESSIVE GOVERNMENTS

Often, less developed countries' governments are controlled by bureaucrats whose primary interest is catering to the wealthy elite and foreign investors, creating governments that are at best indifferent and often outright hostile to the needs of their own populations. The public policies of many less developed countries are ineffectual or counterproductive, and typically ignore the rural areas in favor of cities. Patronage, not a merit system, often decides the allocation of government jobs. For example, in Africa, government jobs often go disproportionately to members of the same tribe as the president. The military is often the most well-funded and well-organized institution and may be a source of political instability, as during military takeovers of the government in coups d'état. For example, since 1990, the military has overthrown the governments of Afghanistan, Algeria, the Central African Republic, the Democratic Republic of the Congo, Cote d'Ivoire, Ethiopia, Fiji, Guinea, Haiti, Liberia, Madagascar, Mali, Niger, Pakistan, Sierra Leone, Thailand, and Turkey. Corruption often is endemic and may become a way of life, generating inefficiency and inequality (Figure 7.24).

Widespread and well-protected civil liberties are more common with economic development. In many poor countries, dictatorial governments curtail their citizens' civil liberties, censoring the media and imprisoning or executing dissidents. Censorship of the media and the Internet is common in countries such as China, Vietnam, Iran, and Cuba. Attacks on journalists, labor union leaders, student movements, religious groups, and ethnic minorities are often common in countries where governments rely on violence rather than political legitimacy to retain control. For this reason, the world's poorest countries tend to be the worst abusers of human rights. Understandably, many people feel alienated from their own governments under these conditions and may sympathize with various resistance movements, contributing to frequent political instability, which hampers economic development. For example, opposition to dictatorial governments, along with increasingly widespread usage of cell phones and the Internet, helped to fuel the "Arab Spring" of 2011, which toppled regimes in Tunisia, Egypt, and Libya and led to widespread unrest, even war, in Yemen, Bahrain, and Syria.

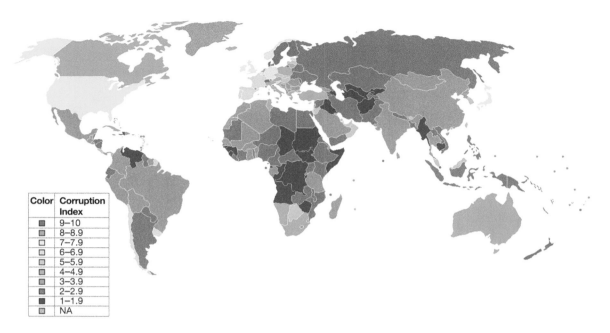

Figure 7.24: The geography of government corruption, as measured by this index, reveals that it is relatively low in economically developed countries but very high in poor ones. Corruption leads to inefficiencies and inequalities in the provision of public services, and takes a variety of forms. Many government bureaucrats enrich themselves through bribery, but corruption also extends into border control, customs, and many other offices. Governments in Afghanistan, Iraq, and in parts of sub-Saharan Africa are often held up as notorious examples of this phenomenon.

7.4 DEVELOPMENT STRATEGIES

It is imperative that economically developed countries come to the aid of less developed countries today. How can this occur peacefully? Economists generally cite three methods for developed countries to help less developed countries:

1. Expand trade with less developed countries
2. Invest private capital in less developed countries
3. Provide foreign aid to less developed countries

EXPANSION OF INTERNATIONAL TRADE

Economists suggest that more developed countries can help less developed countries by expanding trade with them. Indeed, reducing tariffs and trade barriers with less developed countries will improve their situation somewhat. Eliminating protectionism levied against developing countries' producers gives them access to the large, wealthy markets in the United States, Europe, and Japan, increasing their export volumes and revenues. With the North American Free Trade Agreement (NAFTA), the United States removed all trade barriers with Mexico, for example. Mexican manufacturing flourished as a result. However, free trade can have its costs. Trade liberalization also opened the doors for U.S. exports, especially heavily subsidized agricultural products, which have lowered crop prices so much that millions of farmers in the developing world have been forced into bankruptcy.

INTERNATIONAL PRIVATE INVESTMENT

Less developed countries are also a destination for investments from multinational corporations, private banks, and large corporations. Some of these take the form of FDI. For example, major U.S. automakers have now built numerous **maquiladora** plants in Brazil and Mexico (Figure 7.25) that assemble electronics products, automobiles, and clothing for the American market. Other types of capital flows are purely financial: Citicorp and Chase Manhattan Bank have made loans to governments of numerous less developed countries.

A positive international trade climate also demands reliable financial and marketing systems, a favorable tax rate, an adequately maintained infrastructure, and a reliable flow of sufficiently skilled labor. Often, however, less developed countries cannot guarantee that a politically stable environment will prevail. Many countries in today's world are torn by tribal conflicts, ethnic strife, religious struggles, and civil wars. For example, religious strife has repeatedly erupted in Nigeria between Christians and Muslims and secessionist movements have led to unrest in Morocco, Sudan, Pakistan, Indonesia, the Philippines, and Myanmar. These obstacles often thwart the major capital flows that might otherwise exist. African states in particular have not been able to tap private capital flows from large corporations and commercial banks because of problems with these conditions. For example, sustained unrest in countries such as Somalia, Democratic Republic of Congo, Liberia, and Sierra Leone has kept investors at bay.

Figure 7.25: The assembly plants known in Mexico as *maquiladoras* (or simply *maquilas*) are typically foreign-owned. They use imported parts to assemble clothes, automobiles, electronics products, and other goods, and are a major source of employment and foreign revenues.

FOREIGN AID

To reverse the vicious cycle of poverty, more developed countries need to provide foreign aid in the form of direct grants, gifts, and public loans to less developed countries. Capital is necessary for the less developed countries to finance companies, build infrastructure, generate jobs, increase productivity, and retrain their labor pool.

In absolute terms, the United States has been a major world player in foreign-aid programs. U.S. foreign aid averages $15 billion per year, for example, less than 1% of GDP, which is the smallest share among all developed countries (Figure 7.26). Unfortunately, developed country contributions are much too small to make a meaningful difference to most of the developing world. However, for some countries in Africa, aid can form as much as 15% of their national output.

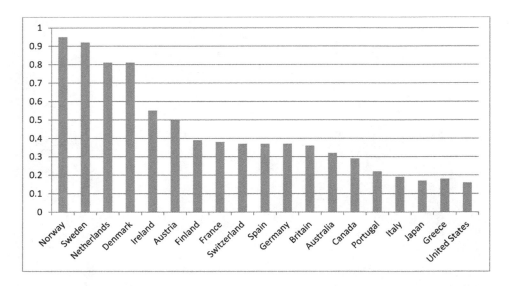

Figure 7.26: Foreign aid as a percent of GDP, 2016. Although many Americans mistakenly believe that foreign aid is a large share of the federal budget, in reality it is less than 1% of government expenditures and .19% of the U.S. GDP, the lowest of the industrialized world.

Unfortunately, most foreign aid has stipulations such as purchase requirements that force the less developed countries to patronize the developed country's products and services, including construction contracts and imports of agricultural goods. For example, almost 75% of U.S. foreign aid is military in nature, and the vast bulk of it flows to Israel and Turkey on the basis of political ties and geopolitical strategy as opposed to economic need. These nations are neither the most populous nor the neediest of the developing world, but they occupy strategic areas of the Middle East where vast oil deposits exist and where the United States struggles against Islamic fundamentalism.

■ KEY TERMS

Brain drain: the phenomenon by which skilled workers leave underdeveloped countries for developed ones, sometimes getting there education in the destination country and staying there

Capital flight: the movement of capital, or money, from developing to developed countries

Development: improvements in standards of living over time

Gross domestic product (GDP): the sum total of the value of a country's goods and services in a year

Human capital: the skills and education level that people have

Informal economy: the unregulated, and often untaxed part of the economy in which jobs are generally unskilled and low-paying

Land reform: political movements to redistribute rural land from large landowners to small ones

Less developed countries: also called developing countries or the global South, these include most of Latin America, Africa, and Asia except for Japan and South Korea

Literacy rate: the proportion of the population over age 15 that can read and write

Maquiladora: assembly plants in Mexico typically owned by foreign firms and designed to export goods

Primary economic sector: the part of the economy concerned with the extraction of raw materials, such as agriculture, mining, logging, and fishing

Secondary economic sector: the part of the economy that transforms raw materials into finished goods (i.e., manufacturing)

Structural adjustment: programs imposed on developing countries to improve their ability to pay back foreign loans, including currency devaluation, deregulation, and privatization

Terms of trade: the relative value of a country's exports and imports

Tertiary economic sector: the part of the economy that produces intangibles (i.e., services)

Underdeveloped: countries that have not experienced significant growth in productivity and standards of living

STUDY QUESTIONS

1. What are five meanings and measures of development?
2. What problems in the developing world come from colonialism?
3. Where is the world's AIDS crisis the most pronounced? Why?
4. Are all the problems in the developing world attributable to overpopulation? Why or why not?
5. What generates shantytowns in poor countries?
6. Why is land often unequally distributed in developing countries, and what is the impact?
7. What employment options do landless peasants have?
8. What drives rural-to-urban migration?
9. Explain the size and nature of the informal sector.
10. What are poor terms of trade and why do so many developing countries suffer from them?
11. Why are productivity levels typically lower in developing countries than developed ones?
12. How did the developing world's debt crisis begin?
13. How do inadequate health care services in the developing world affect their economic performance?
14. Why are girls less likely than boys to be literate in much of the world?
15. What are some major ecological problems that confront many developing economies?

BIBLIOGRAPHY

Cole, J. (2011). Development and Underdevelopment: A Profile of the Third World. *London: Routledge.*

De Silva, S. (2012). The Political Economy of Underdevelopment. *London: Routledge.*

Escobar, A. (2011). Encountering Development: The Making and Unmaking of the Third World. *Princeton, NJ: Princeton University Press.*

Forbes, D. (2011). The Geography of Underdevelopment: A Critical Survey. *London: Routledge.*

Lawson, V. (2011). Making Development Geography. *London: Routledge.*

Nagle, G. (2005). Development. *Philadelphia: Trans-Atlantic Publications.*

Peet, R. (2009). Unholy Trinity. The IMF, World Bank and WTO. *London: Zed Books.*

Potter, R., Binns, T., Elliott, J., & Smith, D. (2008). Geographies of Development: An Introduction to Development Studies. *London: Routledge.*

Toussaint, E. (2010). Debt, the IMF, and the World Bank: Sixty Questions, Sixty Answers. *New York: Monthly Review Press.*

World Bank. (2009). World Development Report 2009: Reshaping Economic Geography. *Washington, DC: World Bank.*

IMAGE CREDITS

- Figure 7.1: Kingj123, http://commons.wikimedia.org/wiki/File:North_South_Divide_4.png. Copyright in the Public Domain.
- Figure 7.2: Emilfaro, http://commons.wikimedia.org/wiki/File:GDP_per_capita.png. Copyright in the Public Domain.
- Figure 7.3: Copyright © Tony0106 (CC BY-SA 3.0) at http://commons.wikimedia.org/wiki/File:Percentage_population_living_on_less_than_$1.25_per_day_2009.svg.
- Figure 7.4: Copyright © Roke (CC BY-SA 3.0) at http://commons.wikimedia.org/wiki/File:Gdp_per_capita_ppp_world_map_2005.PNG.
- Figure 7.5: Ichwan Palongengi, http://commons.wikimedia.org/wiki/File:World_Literacy_Map_UNHD_07-08.svg. Copyright in the Public Domain.
- Figure 7.6: Copyright © Interchange88 (CC by 3.0) at http://commons.wikimedia.org/wiki/File:Food_consumption.gif.
- Figure 7.7: Copyright © Sbw01f (CC by 3.0) at http://commons.wikimedia.org/wiki/File:Infant_Mortality_Rate_World_map.png.
- Figure 7.8: PDH, http://commons.wikimedia.org/wiki/File:HIV_Epidem.png. Copyright in the Public Domain.

- Figure 7.9: Copyright © Digowan (CC by 3.0) at http://commons.wikimedia.org/wiki/File:Life_Expectancy_2011_CIA_World_Factbook.png.
- Figure 7.10: Taylorluker, http://commons.wikimedia.org/wiki/File:Percentage_of_World_Population_Urban_Rural.PNG. Reprinted with permission.
- Figure 7.11: Copyright © Sbw01f (CC BY-SA 3.0) at http://commons.wikimedia.org/wiki/File:Urban_population_in_2005_world_map.PNG.
- Figure 7.12: Freeworldmaps.net, "Map World's Largest Cities," http://www.freeworldmaps.net/cities/top50/top50-cities-world.png.
- Figure 7.13: Copyright © Peter van der Sluijs (CC BY-SA 3.0) at https://commons.wikimedia.org/wiki/File:Two_female_beggars_n._l_a_mother_and_a_daughter.jpg.
- Figure 7.14a: Copyright © Praveenp (CC BY-SA 3.0) at https://commons.wikimedia.org/wiki/File:E-waste.jpg.
- Figure 7.14b: Copyright © Jcaravanos (CC BY-SA 4.0) at https://commons.wikimedia.org/wiki/File:E-waste_workers.jpg.
- Figure 7.15: Fabienkhan and Korrigan, http://commons.wikimedia.org/wiki/File:Urban_population_living_in_slums.png. Copyright in the Public Domain.
- Figure 7.16a: Copyright © Emmanuel DYAN (CC by 2.0) at https://www.flickr.com/photos/emmanueldyan/5933893026/.
- Figure 7.16b: Copyright © Stefan Fussan (CC BY-SA 3.0) at https://commons.wikimedia.org/wiki/File:Kampong_Phlouk_01.jpg.
- Figure 7.16c: Copyright © Matt-80 (CC BY-SA 2.0) at http://commons.wikimedia.org/wiki/File:Soweto_township.jpg.
- Figure 7.16d: Copyright © Peter (CC by 2.0) at https://commons.wikimedia.org/wiki/File:Favela_in_Sao_Paulo.jpg.
- Figure 7.17: Copyright © Bobak (CC BY-SA 2.5) at https://commons.wikimedia.org/wiki/File:Beijing_smog_comparison_August_2005.png.
- Figure 7.19: Copyright © José Reynaldo da Fonseca (CC BY-SA 3.0) at http://commons.wikimedia.org/wiki/File:Arando_150706_REFON.jpg.
- Figure 7.20: Copyright © Jonathan McIntosh (CC by 2.5) at https://commons.wikimedia.org/wiki/File:Jakarta_farmers_protest12.jpg.
- Figure 7.21: Copyright © Roke (CC BY-SA 3.0) at http://commons.wikimedia.org/wiki/File:Debt.PNG.
- Figure 7.22: Copyright © Master Uegly (CC BY-SA 3.0) at http://commons.wikimedia.org/wiki/File:Public_debt_percent_gdp_world_map_(2007).svg.
- Figure 7.23: Copyright © Elvert Barnes (CC BY-SA 2.0) at http://commons.wikimedia.org/wiki/File:IMF-WB.March.WDC.20oct07.jpg.
- Figure 7.24: Copyright © MrNett1974 (CC BY-SA 3.0) at http://commons.wikimedia.org/wiki/File:World_Map_Index_of_perception_of_corruption.png.
- Figure 7.25: guldhammer, http://commons.wikimedia.org/wiki/File:Maquiladora.JPG. Copyright in the Public Domain.

Chapter Eight

POPULATION AND RESOURCE CONSUMPTION

MAIN POINTS

1. The world's population is unevenly distributed and reflects a complex series of environmental, historical, economic, and cultural forces. The most populous regions are East Asia, South Asia, and Europe.

2. Population change reflects natural growth, or the difference between birth and death rates, and net migration, the difference between in-migration and out-migration.

3. Thomas Malthus was a hugely influential early theorist of population change who argued that, because populations grew much faster than food supplies, the world was doomed to starvation and poverty. However, he did not foresee the impacts of industrialization on agriculture and the fact that birth rates decline as societies industrialize.

4. Neo-Malthusians argue that although Malthus's argument suffered from flaws, in the long run he was right. They coupled his view to a concern for the ecological impacts of population growth and advocated birth control.

5. There are several alternatives to Neo-Malthusian thought. One school, Marxism, argues that poverty is not generated by overpopulation but comes from the dynamics of global capitalism, including the uneven distribution of wealth and life opportunities.

6. The demographic transition theory examines birth, death, and natural growth rates as societies industrialize and how and why they change over time, explaining why poor societies growth rapidly and wealthy societies tend to grow slowly.

7. The population structure of a society is the distribution of people by age and sex. This can be summarized handily using population pyramids, whose shape reflects changing birth and death rates.

8. Migration, or at the international scale, immigration, is also a major force behind population change. There are many causes of migration, including "push" causes that drive people from their home countries (e.g., poverty, disasters) and "pull" causes that attract them, especially job opportunities.

Human geographers study, above all, human beings; this issue involves where they are located and how populations in size and composition. These questions underlie many other issues of interest to geographers, such as level of economic development, migration, the size and nature of the labor force, the consumption of food and resources, and the environmental impacts of rising population levels. As the world's population continues to grow, reaching 7.4ven billion in 2017, we face the critical question of whether we humans may be straining the limits of our environmental and economic resources. Does population growth inhibit sustainable development? Does it lead to poverty, unemployment, and political instability?

This chapter examines the causes and consequences of population change in both developed and developing countries. It analyzes the world's population distribution, the fundamental characteristics such as age–sex structures, and growth trends. It also reviews competing theories that explain why populations grow and what the implications of growth might be, including the hugely influential works of Thomas Malthus and neo-Malthusians as well as alternatives such as Marxism, technological optimism, and the demographic transition.

8.1 THE GROWTH AND DISTRIBUTION OF THE WORLD'S POPULATION

Although today a widespread belief exists that there are "too many" people in the world, in fact the presence of large numbers of human beings is a relatively recent phenomenon (Figure 8.1). For most of human existence, including the long millennia when our ancestors lived as hunters and gatherers, world population was relatively small. The average growth rate hovered around zero; because births equaled deaths, there was virtually no increase in the number of people in the world. Following the development of agriculture in the Neolithic Revolution (Chapter 2), societies became able to support more people because farming generates more calories per unit area than does hunting and gathering. However, the really big gains in population did not occur until the onset of the Industrial Revolution of the late eighteenth and nineteenth centuries, when the industrialization of agriculture freed millions from lives of rural toil and allowed for

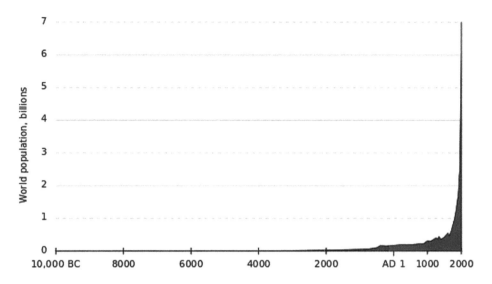

Figure 8.1: World population growth was stable throughout the long millennia of the Paleolithic era. Although both birth and death rates were high, the difference between them—natural population growth—was slight. Dramatic increases in the number of the planet's people occurred only after the Industrial Revolution began in the late eighteenth century.

large, dense urban settlements. Thus, exponential population growth is a recent phenomenon, a product of modernity and global capitalism.

The Earth's population is very unevenly distributed across the planet's surface. Most people are concentrated in a few parts of the world (Figure 8.2), particularly along coastal areas and the floodplains of major river systems. Four major areas of dense settlement are: (1) East Asia (China, Korea, and Japan), (2) South Asia (India, Pakistan, and Bangladesh), (3) Europe, and (4) the Eastern United States and Canada. In addition, clusters exist in Southeast Asia, Africa, Latin America, and along the U.S. Pacific Coast. Figure 8.3 is a cartogram, a map that deliberately distorts areas in proportion to a variable, in this case, population size, revealing the large masses of humanity in Asia. In 2013, Asia contained 3.8 billion people or 56% of the world's population. Europe (including Russia) had 736 million residents, about 12%; Africa had 1.0 billion, or 14%; Latin America had 600 million, or 8.5%; the United States and Canada had 340 million, or 5%; and Oceania had 35 million, less than 1%. The populations

Figure 8.2: The uneven distribution of the planet's people reflects various factors. Environmental conditions affecting agriculture play a role: climate, length of growing seasons, soil quality, water availability, and availability of farmable land. Colonialism reshuffled vast numbers of people through voluntary and involuntary migration, as well as effecting genocide against some indigenous peoples. Cultural preferences regarding family size, as well as economic conditions such as job opportunities also matter a great deal.

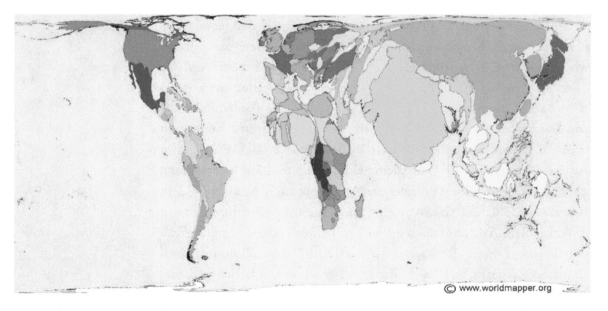

Figure 8.3: A cartogram is a map distorted in proportion to a given variable, in this case, national population size. This map demonstrates that more than half of the world's population lives in Asia, including more than one billion each in China and India. The Western Hemisphere, in contrast, is relatively lightly populated.

of the developing world—Africa, Asia (excluding Japan), and Latin America—accounted for three out of every four humans.

National population figures show even more variability. Ten out of the world's nearly 200 countries account for two-thirds of the world's people (Table 8.1). Five countries—China, India, the United States, Indonesia, and Brazil—contain half of the world's population. The world's most populous country, China, has 1.4 billion people or almost 20% of the planet, almost as large as the Americas and Europe combined. India is second largest, with 1.3 billion, but is growing more quickly, and at the current rate will surpass China in population in roughly 10 years. The United States is the world's third most populous, with 325 million, followed by Indonesia (263 million), Brazil (207 million), Pakistan (197 million), Nigeria (191 million), Bangladesh (162 million), Russia (146 million), and Japan (126 million). Other countries with significant populations include Mexico (123 million), Ethiopia (104 million), the Philippines (104 million), Egypt and Vietnam (93 million each), Germany (83 million), the Democratic Republic of the Congo (82 million), Iran and Turkey (80 million each), Thailand (68 million), France and the United Kingdom (67 million

TABLE 8.1: THE WORLD'S TEN MOST POPULOUS COUNTRIES, 2013.

COUNTRY	POPULATION, 2017 (MILLIONS)	% ANNUAL GROWTH RATE	ESTIMATED POPULATION, 2050 (MILLIONS)
China	1,409	0.6	1,314
India	1,339	1.4	1,652
United States	325	0.6	400
Indonesia	263	1.4	366
Brazil	209	1.0	260
Pakistan	197	2.0	295
Nigeria	191	2.4	440
Bangladesh	164	2.1	202
Russia	143	-0.4	119
Japan	127	-0.1	101

Source: Population Reference Bureau World Population Data Sheet.

each), and Italy (60 million). Only two of the 10 most populous nations are economically developed (the United States and Japan).

POPULATION DENSITY

Because countries vary so greatly in area, national population totals tell us nothing about crowding. This ratio of people to land is called population density—the average number of people per unit area. Several countries with the largest populations have relatively low population densities. For example, although they have significant and dense metropolitan areas, the United States and Canada form one of the more sparsely populated areas of the world.

Excluding countries with a very small area (such as Singapore), Bangladesh is the world's most crowded nation, with more than 158 million people in an area the size of Iowa. Three of the 10 most densely populated countries—the Netherlands, Japan, and Belgium—are economically developed, whereas another three—South Korea, Taiwan, and Israel—are newly industrializing countries (NICs). The rest are less developed nations, reminding us that no clear relationship exists between population density and economic development.

Contrary to popular opinion, not all crowded countries are poor, and not all poor countries are crowded. In fact, the impoverished Sahel states of Africa have very low population densities, which partly reflect their harsh desert climates and limited ability to grow crops. But how do many people in the Netherlands or Singapore live well on so little land? Part of the explanation lies in their historical development and position within colonial and contemporary world economies; the Netherlands was an important colonial power and Singapore today is a vital, highly globalized nation. This history has led them to have productive, industrialized economies. Their governments pursue policies that encourage economic growth, and their people have shown a remarkable ability to adapt to change.

Average national population densities conceal much variation within countries. Egypt has 206 people per square kilometer, but 96% of the population lives on irrigated, cultivated land along the Nile

River valley where densities are extremely high. In China, the vast majority of people live in the eastern third of the country, near the Pacific Coast, where most of the large cities are concentrated (Figure 8.4). Similarly, the United States includes both densely and sparsely populated areas (Figure 8.5). Large areas to the west of the Mississippi River have very low population densities, whereas the Northeast is densely settled.

What explains the uneven distribution of the world's people? Along with other factors, the physical environment plays an important role. Most of the world's people tend to be concentrated along the edges of continents, in river valleys, at low elevations, and in humid midlatitude and subtropical climates. Lands low in rainfall, such as the Sahara Desert, tend to have sparse populations because they are inhospitable to agriculture. Few people live in very cold regions where growing seasons are short, such as northern Canada, arctic Russia, and northern Scandinavia, Many mountainous areas—whether because of climate, thin and stony soils, or steep slopes—also have small populations.

We must use caution when attributing population distribution to the natural environment. To hold that environmental factors, including climate and resources, control population distribution is simplistic and often factually incorrect. Japan, for example, has a small, mountainous landmass with one of the world's greatest population densities, but became a developed country through the use of its human capital and linkages to the world economy. Certainly climatic extremes, such as insufficient rainfall, present difficulties for human habitation and cultivation. However, humans adapt, and they can use technology to compensate for difficult climates; air conditioning, heating, water storage, and irrigation all allow residents to live in harsh environments.

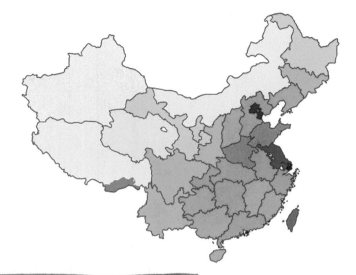

Figure 8.4: Uneven population distribution occurs not only among countries, but also within them. Most of China's 1.38 billion people live in the eastern third of the country (more in the dark red areas above), near the Pacific Coast, which has a moister climate and more arable land than do the western parts of the country (light pink).

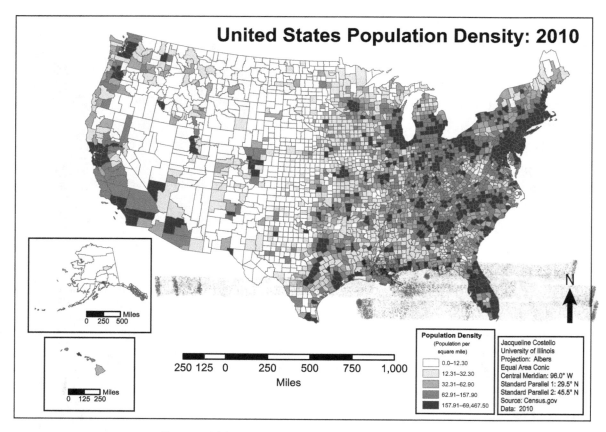

Figure 8.5: The U.S. population is also highly unevenly distributed geographically. The bulk of the country lives east of the Mississippi River, including the numerous large cities of the Northeast and industrial Midwest. A secondary concentration appears along the Pacific Coast. Much of the western interior of the country, as well as Alaska, has a sparse population.

If physical environments alone do not explain the world's population distribution, what other factors contribute? To understand present population distribution, we must look to the past. Geographies are never created instantaneously, and the location of the world's people is the accumulation of forces operating at the global, national, and local scales for centuries or longer. For example, the high population densities of Europe or the northeastern United States reflect the accumulated impacts of the Industrial Revolution and its associated waves of urbanization. China's large population reflects long periods of fertile agriculture, irrigation, and a cohesive social system stretched over vast areas. The distribution of people in the developing world largely reflects centuries of colonialism. As Chapter 5 indicates, development in colonies focused on coastal areas surrounding port cities such as Kolkata in India, Lima, Peru, or Jakarta, Indonesia, which were the centers of maritime trade in the colonial world economy.

Economic, political, and social factors also influence where people live. The geography of economic activity—labor markets, job opportunities, housing costs and availability, and infrastructure—generates large, dense populations in many countries. Population sizes, age structures, and growth rates reflect the economic, political, and cultural contexts in which people live, including factors such as birth and death rates, life expectancy, and migration. Therefore, populations cannot be understood separately from the social contexts of different areas. Policy decisions, such as tax policies or zoning and planning ordinances, are eventually reflected on the population map. Population sizes and distributions also stem from demographic components of fertility, mortality, and migration. Social disasters as war or famine may alter population distribution on any scale.

8.2 POPULATION CHANGE

The size and geography of the world's population constantly change. Total global population continues to increase, albeit at a decreasing rate. Each year an additional 110 million people inhabit the world, which means the planet adds 300,000 people daily, about 3.5 per second. Russia and many other European countries, as well as Japan, are losing population as their deaths exceed births. Most world population growth occurs in the developing countries, home to more than three-fourths of humankind. How will the vast population increase affect efforts to improve living standards? Will the planet's poorest countries become a permanent underclass in the world economy? Or will the imbalance between population and resources lead to waves of immigration and other spillovers such as political unrest to the developed countries?

Demographers, social scientists who study populations, measure the rate of natural increase for a country or a region as the difference between the **birth rate** and the **death rate** (or **mortality**). (Although some people use *fertility* as a synonym for the birth rate, technically the **fertility rate** is the average number of children that a woman will bear in her lifetime, whereas the birth rate is an average per 1,000 people in a given year.) The simplest and most common measure of fertility is the *crude birth rate* (CBR), the total number of live births in a given period (usually 1 year) for every 1,000 people already living.

Births and deaths represent two of the three basic population change processes; the third is migration, which we will discuss in more detail later. Every population combines these three processes to generate its pattern of growth. We can express the relationship among them using the following equation:

$$\text{Population change} = \text{Births} - \text{Deaths} + \text{In-migration} - \text{Out-migration},$$
$$\text{or,}$$
$$\Delta P = BR - DR + I - O,$$

where ΔP represents the rate of population change, BR is the crude birth rate, DR is the crude death rate, I is the total in-migration rate (**immigration**, if internationally), and O is the total out-migration (or **emigration**, if internationally) rate. The **natural growth rate** (NGR)—the most important component of population change in most societies—is defined as the difference between the birth and death rates:

$$NGR = BR - DR,$$

whereas the **net migration rate** (NMR) is the difference between in-migration and out-migration rates:

$$NMR = I - O.$$

Thus,

$$\Delta P = NGR + NMR,$$

where NGR is the natural growth rate, and NMR is the net migration rate.

For the world as a whole, net migration is obviously zero. However, any scale smaller than the globe sees the effects of both natural growth and net migration. Natural growth, also called natural increase, accounts for the greatest population growth in most societies, especially in the short run. However, in the long run, migration contributes far more than just the number of migrants because many will have children, who add to the population base.

One way of understanding population growth rates is **doubling time,** the number of years that it takes a population to double in size, given a particular rate of growth. In general, the doubling time for a population can be determined by using the rule of 70, which means dividing 70 by the average annual rate of growth. For example, at 1.2%, the average rate of world population increase, the doubling time is 58 years. In the year 2070 the world may have 14 billion people *if* current rates of growth continue unchanged. At an annual increase of 0.6%

per year, the doubling time for the U.S. population is 116 years (meaning, if the current rate continues unchanged, the United States will have 600 million people in 2128). As growth rates increase, doubling times decrease accordingly.

8.3 MALTHUSIAN THEORY

One of the first social scientists to tackle the matter of population growth and its consequences was the British Reverend Thomas Robert Malthus. Malthus's ideas, contrived in the early days of the Industrial Revolution in the late eighteenth century, had an enormous impact on the subsequent understanding of this topic. He offered his most concise explanation in his 1798 book, *Essay on the Principle of Population Growth*. Concerned with the growing poverty in British cities, Malthus focused on their high rates of population growth, which are common to early industrializing societies. Thus, Malthus originated the theory of overpopulation. His pessimistic worldview earned economics the label of the "dismal science" and stood in sharp contrast to the utopian socialism emanating from France in the aftermath of the French Revolution of 1789.

Malthus believed that human populations grow exponentially (or in the wording of his times, geometrically). A geometric series of numbers increases at an increasing rate over time. For example, in the sequence 1, 2, 4, 8, 16, 32, and so on, the number doubles each time, and the increase rises from 1 to 2 to 4 to 8 and so forth. Exponential population growth is widely observed in animal species, including bacteria and rodents, in the absence of constraints such as predators or a limited food supply. Note the assumption regarding fertility embedded in Malthus's analysis; he portrayed fertility as a biological inevitability, not a social construction. This argument was in keeping with the large size of British families at the time and the excess of fertility over mortality. In short, in Malthus's view, humans, like animals, always reproduced at the biological maximum. Malthus's argument included a strong moral dimension; it was not just anyone who reproduced rapidly, he observed, but most particularly the poor.

Second, Malthus maintained that food supplies, or resources more generally, grew at a much slower rate than did population. Specifically, he held that the food supply grew linearly (or arithmetically, in his terminology). An arithmetic sequence of numbers, in contrast to an exponential (geometric) one, grows at a constant rate over time. For example, in the sequence 1, 2, 3, 4, 5, and so on, the difference from one number to the next is always the same. Malthus's view that agricultural outputs (crops) increased linearly as a function of inputs (land, labor) reflected the preindustrial farming systems that characterized his world. In such circumstances, an increase in outputs is accomplished only with a proportional increase in inputs such as labor. However, Malthus held that proportional increases in agricultural output were the most optimistic scenario.

He argued that in the face of limited inputs of land and capital, agricultural output was likely to suffer from **diminishing marginal returns**, or declining output per additional unit of input. For example, as farmers moved into areas that were only marginally hospitable for crops, perhaps because they are too dry, wet, cold, or steep, they would need increases in inputs that are proportionately much larger than the increases in output. Diminishing returns, he held, would actually lead increases in agricultural output to become less pronounced over time.

When one plots the exponential growth of population against the linear growth of food supplies (Figure 8.6), it is clear that sooner or later, the former must exceed the latter. Thus, in the Malthusian reading, populations always and inevitably outstrip their resource bases, leading to suffering and misery. Malthus blamed much of the world's problems on rapid population growth, and subsequent generations of theorists influenced by his thoughts have invoked overpopulation to explain everything from famine to crime rates to deviant social behavior. Malthus himself entered into a famous debate with his friend David Ricardo over whether the British government should subsidize food for the poor, Malthus maintaining that such subsidies only encouraged the poor to have more children and thus exacerbated poverty in the long run. Indeed, he was contemptuous of the poor, blaming them for their poverty, and even advocating mass death to keep their numbers in check:

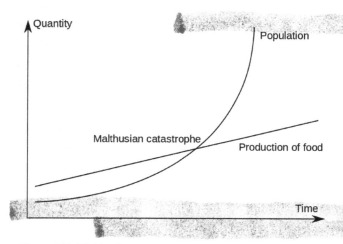

Figure 8.6: Malthus's theory that populations rose geometrically, or exponentially, and food supplies rose arithmetically, or linearly, led to the inescapable conclusion that ultimately people would exhaust their available resources, leading to catastrophe, famine, and declining standards of living.

Instead of recommending cleanliness to the poor, we should encourage contrary habits. In our towns we should make the streets narrower, crowd more people into the houses, and court the return of the plague .... The necessary mortality must come.

Malthusianism thus attributes to rapid population growth a variety of social ills, including poverty, hunger, crime, and disease.

Malthus refined his argument to include checks to population growth. Given that natural population growth is the difference between fertility and mortality, "preventative checks" are factors that reduce the fertility rate. Contraceptives are an obvious example, although Malthus objected to their use on religious grounds, advocating instead moral restraint or abstinence. Other preventative checks include delayed marriage and prolonged lactation, which inhibits pregnancy. Should preventative checks fail, as he predicted they would, population growth would ultimately be curbed by "positive checks" that increased the mortality rate, particularly disease, famine, and war.

Malthus's ideas became widely popular in the late nineteenth century. They became part of the prevailing social Darwinism of the time, which erroneously applied the theory of "survival of the fittest" to economic contexts. In this view, the rich succeeded and the poor failed based only in their individual abilities. Malthusianism was often used, for example, to legitimate the inequalities of the Industrial Revolution as necessary and inevitable. Its explanation of poverty—blaming the poor—appealed to politicians who preferred to blame the weak rather than invest in real solutions to this problem. Malthusianism also entered into popular consciousness as an explanation of problems such as poverty, crime, and environmental degradation.

However, many observers recognized that Malthus's predictions of widespread famine were wrong. The nineteenth century saw the food supply improve, prices decline, and famine and malnutrition virtually disappear from Europe (except for the Irish potato famine of the 1840s). By the early twentieth century, Malthusianism was out of favor. Critics noted that Malthus made three major errors. First, he did not foresee, and probably could not have foreseen, the impacts of the Industrial Revolution on agriculture. The mechanization of food production meant that food supplies increased exponentially instead of linearly. Nor did he anticipate that food supplies could be increased not only by increasing the supply of land but also by improving fertilizers, crop strains, and so forth. Indeed, the world's supply of food has consistently outpaced population growth, meaning that productivity growth in agriculture has been higher than the rate of increase in the number of people. This observation implies that hunger results not simply from overpopulation but from factors including colonialism, the dynamics of the world economy, unequal resource distribution, and corrupt and indifferent governments.

Second, Malthus did not foresee the impacts of the opening up of midlatitude grasslands in much of the world, particularly North America, Argentina, and Australia, which increased the world's wheat supplies during the formation of a global market in agricultural goods. To this day, industrialized agriculture in such ecosystems feeds a large share of the planet (Chapter 10). Third, and perhaps most important, Malthus's analysis of fertility as biologically predetermined and unchangeable was deeply flawed. During the Industrial Revolution, fertility rates actually declined and family sizes decreased.

8.4 NEO-MALTHUSIANISM

In the 1960s, when the world experienced average population growth rates in excess of 2.6% annually, Malthusianism underwent a revival in the form of neo-Malthusianism. Neo-Malthusians acknowledged the errors that Malthus made but maintained that although he may have been wrong in the short run, much of his argument was correct in the long run. Thus, influential authors such as Paul Ehrlich and Lester Brown sounded the alarm about population increases. In keeping with the growing environmental movement of the times, neo-Malthusians also added an ecological twist to Malthus's original argument.

The most famous expression of neo-Malthusian thought was the Club of Rome, an international organization of policymakers, business

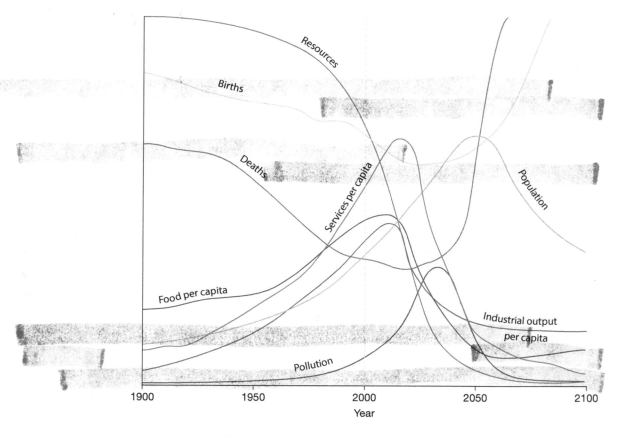

Figure 8.7: The Club of Rome undertook some of the first systematic computer simulations of the future of the world based on estimates of population growth, resource consumption, and pollution levels.

executives, scholars, and others concerned with the fate of the planet. The Club of Rome funded a famous study of the planet's future, published as *The Limits to Growth* (1972), which modeled the world's population growth, economic expansion and resource consumption, and energy and environmental impacts (Figure 8.7). It concluded that the rapid population and economic growth rates of the post–World War II boom could not be sustained indefinitely, and that ultimately there would be profound worldwide economic, environmental, and demographic crises. Much of this argument was framed in terms of the exhaustion of nonrenewable resources and ecological catastrophe. Unlike Malthus, neo-Malthusians advocated sharply curtailing population growth through the use of birth control. They had an important impact on international programs promoting contraceptives and family planning, such as the Peace Corps and Agency for International Development.

Although neo-Malthusianism retains a credibility that the original Malthusian doctrine does not, it exhibited a simplistic understanding of resource production. (For example, when oil becomes scarce, the price rises, and corporations find more oil.) In addition, family planning programs in the developing world have often disappointed because they ignored the reasons why people in impoverished countries have large families. These shortcomings indicate that neo-Malthusianism must be contrasted with other models of population growth that originate from different premises and often arrive at different conclusions.

8.5 OTHER VIEWS OF POPULATION GROWTH

In addition to Malthusian and Neo-Malthusian views of population, which remain dominant, there are several other interpretations that come to different conclusions. These include the Marxist view, technological optimists, and later, the demographic transition theory.

MARXISM

One of the earliest and most powerful critics of Malthusian theory was Karl Marx, who objected to the assumptions and conclusions of Malthus. Marx believed that population growth must be considered in relation to the prevailing form of social organization in a given society, including its class relations, ownership of property, and political institutions. For example, Marxists insist that it is no accident that high population growth rates occur in poor, largely agricultural societies in which families require children for farm labor. For Marx, Malthusianism represented a bourgeois viewpoint (that of the ruling class) primarily aimed at maintaining existing social inequalities. Malthus saw population growth as the primary cause of poverty, whereas Marx saw the inequalities of capitalism as the primary cause. For Marx the problem was not a population problem at all, but a resource-distribution problem caused by capitalism.

Malthus was concerned with overpopulation; he believed that a society could be said to be overpopulated when food and other necessities of human life were in such short supply that life-threatening circumstances arose. For Marx, by contrast, the key concept was surplus population; unemployed workers represented a reserve army of labor that capitalists used to keep wages low and profits high. Surplus population is, therefore, an inevitable consequence of capitalism.

TECHNOLOGICAL OPTIMISM

In direct contrast to neo-Malthusians, a loosely related school of thought may be called technological optimists. In this view, capitalism has a long history of rising productivity levels, especially in agriculture, where gains in output have steadily kept pace with population growth. Indeed, since the Industrial Revolution began, optimists asked, agriculture has experienced soaring productivity due to chemical fertilizers and farm machinery (Chapter 11). Why, then, should we expect that these gains should suddenly stop?

The most dramatic example of technological optimism was the **Green Revolution**, an effort initiated by the famed American agronomist Norman Borlaug in the 1960s. Borlaug and colleagues working in Mexico developed new, genetically engineered types of rice that could withstand drought, resist pests, and generate more protein than traditional forms. These new forms of rice were implanted throughout numerous Asian countries, including India, China, and the Philippines, with dramatic results. Rice yields rose steadily, and India, once a food importer, is now a food exporter. Other countries such as Mexico followed suit. These gains went far in helping to reduce malnutrition in many societies. Critics of the Green Revolution point out that these efforts require farmers to buy seeds from U.S. agribusiness corporations, and to use energy- and machinery-intensive forms of cultivation, which are unaffordable to small farmers.

Another example of technological optimism is aquaculture, the commercial growing of fish in lakes and oceans. Because the oceans have been so severely overfished and their stocks depleted, aquaculture forms a large share—perhaps half—of the world's fish supply today. Fish is a relatively cheap source of protein for many people around the world, and as wild stocks plummet, aquaculture will continue to grow in importance.

8.6 THE DEMOGRAPHIC TRANSITION

Developed by several American demographers in the 1950s, **demographic transition theory** stands as an important alternative to Malthusian and neo-Malthusian notions of population growth. Essentially, this is a model of a society's fertility, mortality, and natural population growth rates over time. Because this approach is explicitly based on the historical experience of western Europe and North America as they went through the Industrial Revolution, time represents industrialization and all of its economic, social, and geographic characteristics. The demographic transition examines how and why birth, death, and growth rates change as a society moves from an impoverished, rural, and traditional context into a progressively wealthier, more urbanized, and more modern one.

This approach can be demonstrated with a graph of birth, death, and natural growth rates over time that divides societies into four major stages (Figure 8.8). We will address each stage in detail.

STAGE 1: PREINDUSTRIAL SOCIETY

The first stage corresponds to a traditional, rural, preindustrial society and economy. Fertility rates are high. Families are large, and most women are pregnant much of the time. Thus, impoverished countries such as Rwanda have exceptionally large families with an average of 8.5 children per mother. In contrast, in wealthy countries in Europe and North America, as well as some of the newly industrializing countries in Asia, women have on average less than two children apiece.

Preindustrial societies have very high fertility rates for many reasons. In agrarian economies, children provide vital farm labor, helping to plant and sow crops, tending farm animals, performing chores, carrying water and messages, and helping with younger siblings (Figure 8.9). Econometric studies reveal that even children as young as four can generate more income than they consume. Even in North America, summer vacations were originally necessary so that school kids could return to the farms to help their families, an excuse that no longer holds today.

Families in this context are typically extended, with several generations living together. Children bear primary responsibility for care of their elderly parents, in the absence of institutionalized social programs such as Social Security. Finally, in such societies with high infant mortality rates, having many children ensures that some will survive. In short, there are very good reasons why poor societies have high crude birth rates. In contrast to Malthusianism, which explains this observation by appealing to human genetics, the demographic transition

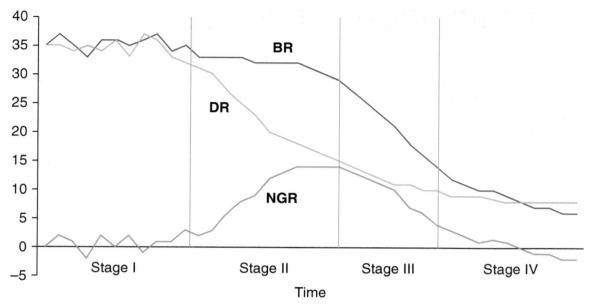

Figure 8.8: The demographic transition is a model of a society's birth rates (BR), death rates (DR), and natural growth rate (NGR) as it shifts from a traditional, agrarian culture to an urbanized, postindustrial one. Death rates decline first as improved nutrition reduces infant mortality and susceptibility to diseases. Birth rates decline as more women work in paid jobs outside the home and cultivate careers. Natural growth rates start out very low, rise significantly as death rates drop but birth rates do not, and then fall again as average family sizes decline.

portrays high total fertility rates as a rational strategic response to poverty.

Thus, a map of crude birth rates around the world (Figure 8.10) reveals that the poorest societies have the highest rates in the world, particularly in Africa and most of the Middle East. In contrast, for reasons we shall soon see, crude birth rates in North America, Europe and Russia, Japan, Australia, and New Zealand are relatively low. The world's lowest birth rates are found in Spain and Italy. In societies with high birth rates and thus many children, the age distribution of populations tends to be young. Thus, the proportion of the population aged less than 19 (the median age in many developing countries) is largely a reflection of fertility rates.

However, in preindustrial societies, mortality rates are also quite high, which means that average life expectancy is relatively low. The

Figure 8.9abcd: Child labor is the norm in the developing world. In rural, agrarian societies, children play an important economic role helping to cultivate crops, tend farm animals, and assist in a wide variety of chores. Even in some urban environments, families often need child labor to supplement their income.

Figure 8.10: The geography of crude birth rates is largely a map of wealth and poverty. In poor countries, large family size is a rational economic response to the need for child labor and to high infant mortality rates. In wealthier countries, higher incomes and women's careers raise the cost of having a child, leading to lower fertility.

primary causes of death in poor, rural contexts are the result of inadequate diets, particularly lack of protein, and malnutrition, which weaken the immune system, as well as unsanitary drinking water and bacterial diseases. The most common diseases in this context are diarrheal ones that lead to severe dehydration, including cholera, as well as others such as malaria, dengue fever, yellow fever, schistosomiasis, bilharzia, tuberculosis, plague, and measles, although historically smallpox was also important. Table 8.2 lists the most dangerous infectious diseases in the world in 2016. Because disease and malnutrition are ever-present threats to people in poor societies, a significant proportion of babies do not live to see their first birthday. Deaths due to infectious diseases are the first stage in what is frequently called the **epidemiological transition**, a model of how the causes of death change as societies industrialize over time. As we shall see, the major causes of death in wealthy, urbanized countries are very different.

Acquired immune deficiency syndrome (AIDS) was first identified in 1981, and the human immunodeficiency virus (HIV) was recognized as its cause in 1984. Today, in the more developed areas of the world, drugs prevent people with HIV from developing AIDS. However, the

TABLE 8.2: WORLD'S MOST DANGEROUS INFECTIOUS DISEASES, 2016 (MILLIONS OF VICTIMS ANNUALLY)

Respiratory infections	4.0
Diarrheal diseases	1.5
Tuberculosis	1.5
Malaria	1.3
AIDS	1.1

Source: World Health Organization.

worst-affected regions are in less developed areas—notably sub-Saharan Africa—where drugs are largely unavailable, and it is not unusual for a person to die just 6 months after infection with HIV. AIDS is the one of the greatest threats to human health in the world, not because of the number of people dying from AIDS but because of the speed with which the virus spreads through a population. Consider, for example, that in 1990 the affected population in South Africa was 1%, but by 2010 it was about 22%. In 2016, some 37 million people globally were infected with HIV. More than 20 million have died of AIDS since the early 1980s. The fact that AIDS has now spread to all parts of the world indicates that the disease is appropriately described as a **pandemic**. Although it is especially prevalent in sub-Saharan Africa and among the poorest members of individual populations, AIDS honors no social or geographical boundaries. Because AIDS is most often transmitted sexually, the most vulnerable population is the 15–49 age group—the very group that ought to be most highly productive, and that is most likely to have dependents, both young and old.

The world geography of death rates (Figure 8.11) thus closely reflects the wealth or poverty of societies (including their historical development and role in the global economy). The amount of wealth manifests in a variety of issues that shape national mortality rates. These include the amount, quality, and consistency of adequate food; access to health care; the public health infrastructure; care for expectant mothers, babies, and young children; smoking rates; and several other factors. Countries with the highest death rates—and thus lowest life expectancies—are found primarily in sub-Saharan Africa, although Afghanistan, Pakistan, Russia, and Central Asian states also

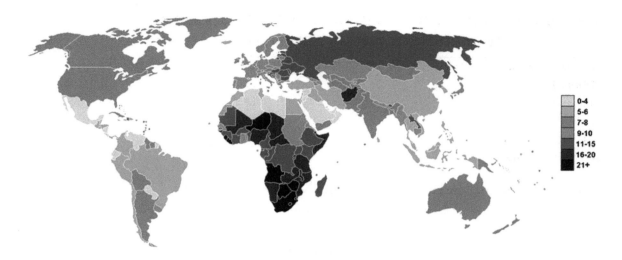

Figure 8.11: Like birth rates, death rates are highest in poor countries, a reflection of low standards of living, poverty, malnutrition and occasional famine, contaminated drinking water, inadequate health care, and frequent wars.

have relatively high mortality. Conversely, the developed world, as well as Latin America, China, and India, has relatively low death rates.

Life expectancy throughout most of human history has been relatively low, often only in the 20s, although once people survived infancy their chances of living to old age improved considerably. The geography of life expectancy around the world (Figure 8.12) closely reflects that of crude death rates but is also shaped by differences in standards of living. Living for a long time is a luxury enjoyed by the populations of economically developed societies, whereas people in most of Africa, the Middle East, and Russia tend to die before they reach age 70.

Because both fertility and mortality rates are high, the *difference* between them—natural population growth—is relatively low, often fluctuating around zero. Thus, although families are large and parents have many children, growth rates are curtailed by malnutrition, diseases, and infant mortality. For this reason, for thousands of years, human growth rates worldwide have been very slow, occasionally even negative (Figure 8.1), and new arrivals to a community were welcomed. Indeed, prior to Malthus, rapid population growth was celebrated as a way to increase the local labor force, diversify the division of labor, and raise standards of living. Although relatively few societies in the world live in the circumstances described here—that is, few people today live

Figure 8.12: How long you live largely reflects where you live, that is, your geography. Life expectancies vary widely around the world, from less than 50 in some African states to more than 80 in wealthy countries such as Sweden and Japan. The geography of life expectancy reflects the historical, economic, and political forces that shaped a society's status.

isolated from the world economy and its demographic aftermath—Stage I may be held to describe certain tribes in parts of central Africa, Brazil, or Papua New Guinea.

STAGE 2: EARLY INDUSTRIAL SOCIETY

The second stage of the demographic transition pertains to societies in the earliest phases of industrialization, when manufacturing jobs are growing in urban areas. Such conditions pertain, for example, to the Britain of Malthus's day, the United States in the early nineteenth century, or selected countries in the developing world today, such as Mexico or the Philippines (although these countries have experienced fertility declines). In this context, economic changes in the labor market as well as in consumption, particularly diet and public health care, have important demographic consequences.

Early industrial societies retain some facets of the preindustrial world, particularly high fertility rates. Because most people still live in rural areas, children remain an important source of farm labor. The major difference is the decline in mortality rates, which leads to longer life expectancies. Why do mortality rates decline as societies industrialize? One might think the reason would be access to better medical care, particularly hospitals and vaccinations from diseases. However, the historical evidence does not sustain this view. Early hospitals were often filthy, and patients may have been more likely to die in them than if they stayed home! Moreover, the introduction of vaccinations often came *after*, not before, declines in deaths due to many diseases occurred. For example, the vaccines for measles, scarlet fever, typhoid, diphtheria, tuberculosis, and pneumonia were invented in the 1920s, 1930s, and 1940s. This all occurred after the United States had seen the majority of the declines in the death rate from each disease in the first two decades of the twentieth century. Indeed, rather than private health care providers, public health measures including clean drinking water and sewers played a significant role in lowering diseases. Better housing was also important. Finally, the industrialization of agriculture brought cheaper food, which led to better diets. This improvement in diet strengthened people's immune systems and raised life expectancies, including lowering infant mortality rates (percentage of babies who die before their first birthday).

Death rates for different demographic groups do not drop evenly as an economy develops over time. Infant mortality rates tend to drop earliest and most quickly. Many premodern or preindustrial societies had an infant mortality rate above 200 infant deaths per 1,000 live births—20% or more of all babies died before reaching their first birthday. Nowhere in the world today are rates that severe, but the highest rates are found in sub-Saharan Africa. In much of this region, poverty, disease, inadequate diets, contaminated drinking water, and insufficient health care services conspire to kill 10% of all infants before age one.

Because the drop in the death rate disproportionately affects the very young, it acts much like an increase in the birth rate. The death rate initially drops as communicable diseases are brought under control. The very young, who were most susceptible to disease, survive to grow to adulthood. Life expectancies likewise increase as the elderly, another vulnerable group, are spared. The control of communicable disease also reduces the overall illness level in society, thus promoting increased labor productivity. Workers miss fewer days of work, are healthier when they do work, and are able to work productively for more years than when death rates are high. Eventually, as death rates drop, the timing of death shifts increasingly to the years beyond retirement when the economic impact on the labor force is minimal. Although death rates have declined throughout the world, mortality is usually, but not always, lowest in the developed world (especially in northwestern Europe and in Japan) and highest in the underdeveloped world (especially in sub-Saharan Africa). Variations in fertility tend to be more pronounced, with much higher levels in the developing world.

In early industrial societies, because the death rate has dropped but the birth rate has not, the population grows explosively. This situation characterized the world Malthus observed at the end of the eighteenth century and is evident in a wide number of countries in the developing world today such as much of Latin America or China. In short, countries in the early stages of industrialization have rapid increases in population not because they have more children than before, but because fewer people die, and they live longer.

STAGE 3: LATE INDUSTRIAL SOCIETY

Societies experiencing rapid industrialization and widespread urbanization exhibit a markedly different pattern of birth, death, and growth rates than those earlier in the transition. Death rates remain relatively low, but fertility rates also decline. It is important to note that decreases in crude birth rates almost always occur *after* decreases in the death rate; societies are much more amenable to death control than birth control.

Why do crude birth rates fall and families get smaller as societies become wealthier? People face changing incentives as their worlds shift from primarily rural to primarily urban, with a corresponding increase in the size and complexity of labor markets. For many people, the decision to have or not to have children is the most important question they will ever face, with profound consequences not only for their personal well-being but also for society at large. As we have seen, urbanization and industrialization lead to children becoming more of an expense and less of a benefit to the household economy. In addition, when large numbers of women enter the paid labor force—engage in commodified labor outside of the home, rather than unpaid workers inside of it—the constraints to child rearing become formidable. Women, typically have primary responsibility for child care, and working outside of the home while taking care of young children in urbanized societies poses many obstacles. Many new mothers who have the option choose to drop out of the labor market, if only temporarily, to take care of their children. As a result, they do not earn an income while staying at home, relying on their partner for support. Even mothers who continue working with young children, including in countries with subsidized day care such as in much of northern Europe, find that children are an obstacle to career progress and leisure time. Economically, this process generates an **opportunity cost** to having children, that is, the price one pays in terms of foregone benefits; the more children a couple have, or the longer a mother refrains from working outside of the home, the more income the family forgoes. As women's incomes rise either over time, the opportunity cost of children rises accordingly, leading to lower fertility rates.

Although neo-Malthusian family planning (such as subsidized contraception or family planning clinics) tends to ignore the social context of fertility, this model links labor markets and fertility behavior. Getting women into paid work outside of the home is the surest form of birth control. As fertility rates decline, so too does the natural growth rate. For this reason, relatively prosperous societies tend to have smaller families. The decline in family size, coupled with a broad tendency for young families to establish their own households, often distant from their parents, has led to a corresponding shift from extended to **nuclear families** in the process. In wealthy societies, it is rare for several generations to live with one another.

Historically, fertility levels fell first in western Europe, followed by North America, Japan, and then the remainder of Europe. In most of those areas of the world, birth rates are below the level of generational replacement. The United States is the only major exception, with rates just barely above the level of replacement. Elsewhere in the world, however, crude birth rates remain at much higher levels. They are dropping quickly in China and Southeast Asia. There has been a modest decline in South Asia, the Middle East, much of Latin America, and parts of sub-Saharan Africa, where fertility rates tend to be very high (Figure 8.10). In addition to these patterns of natural increase, migration also affects many areas of the world, as we will see.

Increasing life expectancy in a region or country indicated social progress. Between 1980 and 2011, the world's average life expectancy at birth increased from 61 to 68 years. In developed regions, average life expectancy is 74 years for males and 81 years for females; in developing regions, 65 years for males and 68 years for females; in least-developed countries, the values are 53 and 56, respectively.

STAGE 4: POSTINDUSTRIAL SOCIETY

The fourth and final stage of the demographic transition, postindustrial society, depicts wealthy, highly urbanized worlds with their own configurations of birth, death, and growth rates. In this context, indicative of Europe, Japan, and North America today, death rates are very low, and life expectancies correspondingly high. Do death rates ever reach zero? No, for that would mean life expectancies become infinite! Even the declines in the death rates will not exhibit much improvement, and they face diminishing marginal returns; the easy causes of death have been largely eliminated, and overcoming the remaining ones will be much harder.

As societies industrialize and become progressively wealthier, the causes of death change from infectious diseases to lifestyle-related ones, particularly those associated with smoking and obesity, as well as car accidents, suicides, and homicides (Figure 8.13). Medically, these causes are manifested in heart disease, cancers of all forms, and strokes, the first, second, and

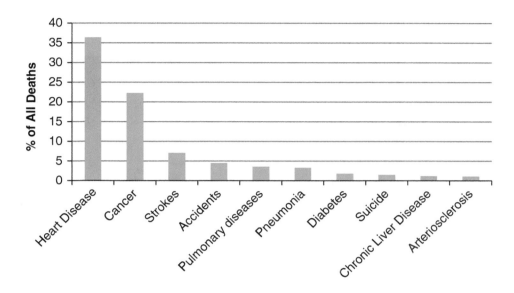

Figure 8.13: The causes of death are reflected in several lethal medical conditions, notably heart disease, cancers of various types, and strokes, the three leading causes in the United States.

third leading causes of death, respectively (Figure 8.14); many leading causes of death are related to smoking. Other major causes include the opioid epidemic, gunshot wounds, and car accidents. These causes of death contrast with the reasons why people die in economically underdeveloped countries, although in some developing countries obesity, heart attacks, and diabetes are on the rise.

Birth rates continue to fall in these societies, as families grow smaller. Many couples elect to go childless or have only one child. Around the world, national income and population growth rates are inversely related. When crude birth rates drop to the level of crude death rates, a society reaches zero population growth (ZPG). When birth rates drop below death rates, as they have in virtually all of Europe, the society experiences negative population growth. Japan, the demographically oldest society in the world, will see its population decline by 30% or more in the next 50 years. Such societies have large numbers of the elderly, a high median age, and a relatively small number of children, all of which have dramatic implications for public services (e.g., few schools, rising expenditures on health care). In such contexts, governments may try to increase the birth rate with ample

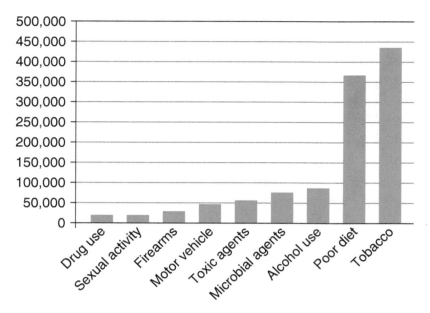

Figure 8.14: The most common causes of death in the United States—tobacco use and poor diet—are related to behavior and life style. In contrast, infectious diseases cause most deaths in the developing world.

rewards for childbirth (e.g., subsidized child care and long paid parental leaves) and publicity campaigns. In societies with extremely low or negative population growth, the major cause of demographic growth is often immigration.

Globally, uneven economic development—the legacy of colonialism, different national policies, and national cultures—generates uneven patterns of natural population growth. The geography of natural growth rates shows the difference between birth and death rates (Figure 8.15). The most rapid rates of increase are found throughout the poorer parts of the developing world (i.e., in Africa as well as parts of the Arab and Muslim worlds). The economically developed regions and countries, including the United States, Canada, Japan, Europe, Australia, and New Zealand, in contrast, have low rates of population growth, often hovering around zero or even lower.

These patterns have significant implications for the nature and future of the world's people. Although the world's average natural growth rate has been declining (Figure 8.16), it still adds approximately 110 million people per year, roughly the population of Mexico. Declining fertility

Figure 8.15: The world's geography of natural growth rates, or birth rates minus death rates, reflects how different regions fall into various phases of the demographic transition. Growth rates tend to be highest in the poorest countries, where death rates have fallen but birth rates have not. Conversely, natural growth in wealthy countries is often very low, or even negative, as in eastern Europe and Russia.

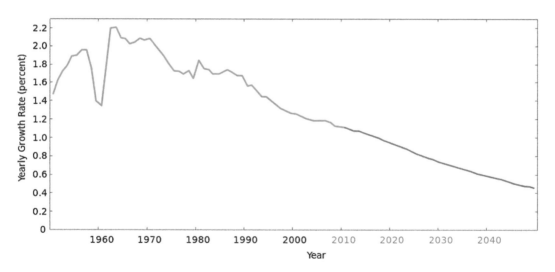

Figure 8.16: The world's average natural growth rate reached its peak in the 1960s, when the Club of Rome issued its famous warning. It has been dropping ever since, and today it is about 1%. The change reflects improved quality of life for many women.

levels will probably lead to slower rates of demographic growth throughout the twenty-first century. However, because there are so many people of child-bearing age in the developing world, the total population of the planet is projected to rise to roughly 10 billion by the year 2100. The vast bulk of these additions will be in developing nations. However, these projections are based on different assumptions about the future of fertility (Figure 8.17). Assuming that fertility rates remain high, for example, leads to much higher expectations about future world population levels than does assuming that they fall either gradually or sharply. Should fertility rates decline more rapidly than expected, the increase in the world's people may not be as dramatic as some believe.

However, the population explosion in the developing world will have enormous impacts. Rapid population growth will accelerate, among other things, agricultural overcultivation and soil depletion,

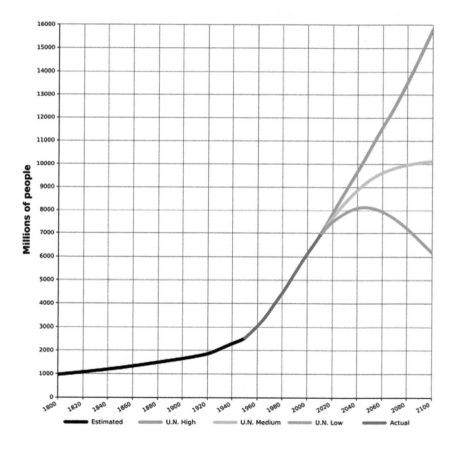

Figure 8.17: Estimates about the future size of the world's people hinge heavily on assumptions about how fertility rates will change. These different assumptions lead to predictions that the world will have between 8 and 13 billion people by the year 2050.

overfishing, poaching, deforestation, depletion of water resources, loss of biodiversity, and rural-to-urban migration, which strains cities' abilities to provide jobs and services. Further, by generating an infinite pool of poor people, it keeps wages low in both developing and developed nations, as globalization pits the labor forces of countries against one another. Thus, the dynamics of population growth affect environmental degradation and economic development.

CONTRASTING THE DEMOGRAPHIC TRANSITION AND MALTHUSIANISM

Malthusian theory and the demographic transition offer markedly different assumptions, analyses, and conclusions. Whereas Malthusianism tends to take fertility for granted—arguing that people are prisoners of biological imperatives to reproduce uncontrollably—the demographic transition reveals that fertility relates to social context (i.e., families have or don't have many children depending on the costs and benefits that children offer). Moreover, whereas Malthusian scenarios depict the population as growing uncontrollably to the point of resource exhaustion, the demographic transition predicts steadily declining levels of world population growth as crude birth rates converge upon death rates. The evidence supports the latter view. After accelerating for two centuries, the overall rate of world population growth is slowing down. In the 1960s, it reached a peak of 2.6%, declining to 1.7% a year in the 1990s, and dropping further to 1.2% in 2011. However, the absolute size of the world's population will continue to increase.

CRITICISMS OF DEMOGRAPHIC TRANSITION THEORY

Although the demographic transition has wide appeal because it links fertility and mortality to changing socioeconomic circumstances, it has also been criticized on several grounds. Some critics point out that it is a model derived from the experience of the West. Non-Western societies may not be bound to repeat the exact sequence of fertility and mortality stages that occurred in Europe, Japan, and North America. Critics have pointed out that the developing world differs from the West in many ways, in no small part because of the long history of colonialism. Further, demographic changes in the developing world have been much more rapid than in the West. For example, whereas it took decades, or even centuries, for mortality rates in Europe to decline to their modern levels, in some developing countries the mortality rate has plunged in only one or two generations. Because mortality rates do not vary geographically as much as fertility rates, most of the differences in natural growth around the world reflect differences in fertility. These caveats caution us to examine the historical context of theories and explanations. We must be wary of blindly importing models developed in one social and historical context into radically different ones.

8.7 POPULATION STRUCTURE

Except for total size, the most important demographic feature of a population is its age-sex structure. The age-sex structure affects the needs of a population as well as the supply of labor; therefore, it has significant policy implications. A rapidly growing population implies a high proportion of young people under the working age. A youthful population also puts a burden on the education system. When this cohort reaches working age, it requires a rapid increase in available jobs. By contrast, countries with a large proportion of older people must develop retirement systems and medical facilities to serve them. Therefore, as a population ages, its most pressing needs change from schools to jobs to medical care.

The age-sex structure of a country is typically summarized or described through the use of **population pyramids**. They depict 5-year age groups with the base representing the youngest group and the top delineating the oldest. Population pyramids show the distribution of males and females of different age groups as percentages of the total population. The shape of a pyramid reflects long-term trends in fertility and mortality, as well as short-term effects of baby booms, migrations, wars, and epidemics. It also shows the potential for future population growth or decline.

Two basic, representative types of pyramid exist (Figure 8.18). One is the squat, triangular profile. It has a broad base and sides that taper to a narrow tip. It is characteristic of developing countries with high crude birth rates, a young average age, and relatively few elderly. Natural growth rates in such societies tend to be high. In contrast, the pyramid for economically developed countries such as the United States describes a slowly growing population. Its shape reflects a history of declining fertility and mortality rates, augmented by substantial immigration. With lower fertility, fewer people have entered the base of the pyramid; with lower mortality, a greater percentage of the births have survived until old age. In short, the structure of the population pyramid closely reflects a country's stage of the demographic transition.

Like all developed countries, the U.S. population has been aging, meaning that the proportion of older persons has been growing. The relatively small numbers of the elderly in the pyramid reflect the baby dearth of the Great Depression in the 1930s, when total births

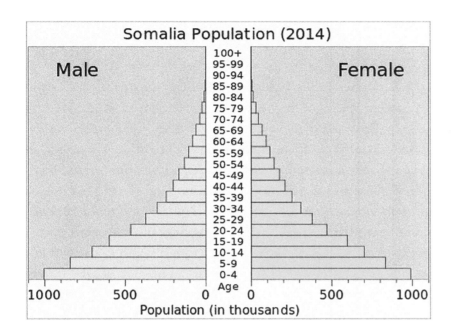

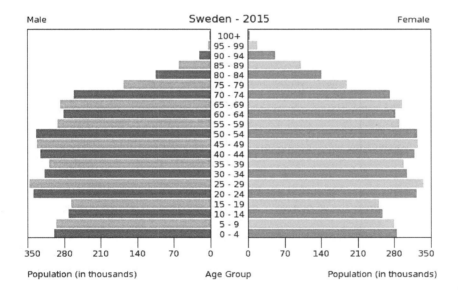

Figure 8.18ab: The population pyramids for Somalia and Sweden reflect two very different societies and demographic structures. Somalia's age composition indicates an underdeveloped society with high birth rates and hence many young people. Sweden's, in contrast, has a low birth rate and a relatively large share of middle-aged and elderly people.

dropped from 3 million to 2.5 million annually. The bulge of the middle aged at the waist of the pyramid is a consequence of the **baby boom** that followed World War II. By 1976, the fertility rate had fallen to 1.7 children per mother, a level below replacement. When members of the baby-boom generation had children in the 1980s and 1990s they drove the total fertility rate up to 2.0, almost at replacement level and the highest in the developed world. Thus, the U.S. population continues to grow from natural increase as well as from immigration. Because different parts of the United States have very different socioeconomic conditions, cultures, and migration patterns, various places in the country have very different population pyramids.

Discussions of the number of males and females in a population refer to the **sex ratio**. Sex ratio data for individual countries frequently are estimated. Globally, there are about 101 males for each 100 females. Sex ratios vary with age. Under normal circumstances, in most parts of the world about 104 boys are born for each 100 females. The major exceptions are countries where a strong cultural preference for sons leads to reductions in the number of girls through abortion and female infanticide. This practice is most prevalent in India, Taiwan, South Korea, and China. Worldwide, the surplus of males at birth decreases over time as male mortality rates are generally higher and male life expectancy generally lower. In most countries, the number of males in the population of a country will be overtaken by the number of females by middle age. The elderly population usually is predominantly female.

A few developed countries have very low rates of population growth—in some cases zero population growth (ZPG) or negative population growth. They have low crude birth rates, low death rates, and, in some cases, net out-migration (more people leave than arrive). For example, France is experiencing negative population growth because of very low fertility. Although there is a steady stream of foreigners (especially Arabs and Africans) into the country, France tries to limit immigration. Population decline is an economic concern to many European countries, as well as Japan, the world's oldest nation demographically. Who will fill the future labor force? Is the solution the importation of workers from developing countries? In these respects, demographic changes profoundly influence immigration policies and the size and nature of the labor force.

8.8 MIGRATION

Migration is a movement involving a change of permanent residence either within a country or between countries. It is a complex phenomenon that raises many questions. Why do people move? What factors influence the extent of migration? What are the effects of migration? What are the main patterns of migration?

CAUSES OF MIGRATION

Most people move for economic reasons. They relocate to take better-paying jobs or to search for jobs in new areas. They also move to escape poverty or to improve opportunities for their children. Noneconomic reasons for migration include escape from adverse political conditions and fulfillment of personal dreams.

The causes of migration encompass push and pull factors. Push factors include high unemployment rates, low wages, poverty, shortages of land, famine, or war. In the late 1970s and early 1980s, communist purges in Vietnam, Cambodia/Kampuchea, and Laos pushed out approximately one million refugees from their countries, who resettled in the United States, Canada, Australia, China, Hong Kong, and elsewhere. Pull factors include job and educational opportunities, relatively high wages, the hope for agricultural land, or the "bright lights" of a large city. The jobs available in rich oil-exporting countries in the Middle East act as a pull factor for millions of immigrants seeking employment. In Kuwait and the United Arab Emirates, nearly 80% of the workforce is composed of foreigners, mostly drawn from South Asia and the Philippines. Geographic differences in economic opportunities, therefore, help to explain why young people often leave rural areas, the influx of Mexicans into the United States, and the immigration of non-Westerners into Europe. Indians and Bangladeshis have frequently migrated to Britain, Turks to Germany, and Arabs to France.

Migrations can be voluntary or involuntary and reflect the historically specific cultural, economic, and political circumstances in which migrants live. Most movements are voluntary, such as the westward migration of pioneer farmers in the United States and Canada. Involuntary movements may be forced or impelled. In forced migration, people have no choice; their transfer is compulsory. One example is the African slave trade, in which 20 million people were shipped to the New World. The deportation of British convicts to the United States in the eighteenth century and to Australia in the nineteenth century also exemplifies forced migration. In impelled migration, people choose to move under duress. In the nineteenth century, many Jewish victims of Russian persecution elected to move to the United States and the United Kingdom. Civil wars in Central America in the 1980s led hundreds of thousands of Salvadorans and Nicaraguans to emigrate to the United States. In Africa, multiple civil wars have displaced millions of refugees into neighboring nations.

Barriers to migration, including legal obstacles and immigration restrictions, imperfect information, lack of skills, inability to afford transportation, and the powerful economic and familial bonds that hold people in place prevent a free flow of labor among, and often within, countries. At best, only a small portion of the population in a low-wage region can gain access to higher pay in developed nations. Therefore, there will continue to be a discrepancy in per capita earnings between less developed countries and developed countries and between depressed regions and economically healthy regions within countries. Migration is thus intertwined with local and national labor markets that shape incomes and unemployment rates.

The availability of work and wages accounts for migration throughout the world from countries lacking in jobs and high wages to countries with jobs available at relatively higher wage rates. Major world labor flows occur (1) from Mexico and the Caribbean to the United States and Canada; (2) from North Africa and southern European nations to northwestern Europe; (3) from Asia to Saudi Arabia and the Persian GUlf; and (4) from Indonesia to Malaysia, Singapore, and Australia.

All countries regulate the flow of immigration. The United States limits legal immigration to approximately 600,000 people annually, although a total of roughly 1.1 million enter the United States legally or illegally every year (a volume that dropped markedly during the recession of 2008–2012). Altogether, about 33.1 million immigrants live in the United States, making up 11% of the population. Of this group, an estimated total of 5,000,000 people live in the country illegally, often under constant threat of being caught and deported. The United States spends billions of dollars annually to police its borders, in the attempt to keep Mexicans and other Latin Americans from entering the country illegally. The status of illegal immigrants is a significant political issue in the United States. Some argue that by paying income and sales taxes, undocumented workers generate more wealth than they consume. Others argue that such migrants consume public services and compete with unskilled American residents, many of whom are ethnic minorities, and drive down wages in that sector of the labor market.

CONSEQUENCES OF MIGRATION

Migration has numerous and important demographic, social, and economic effects. Many of these effects stem from the fact that migrants tend to be young adults and are often the more ambitious and well-educated members of the society from which they originated. Obviously, the movement of people from one region to another causes the population of the origin country to decrease and of the destination country to increase. If the migrants are young adults, their departure increases the average age, raises the death rate, and lowers the birth rate in the origin country. For the destination region, their arrival tends to lower the average age and the death rate but increase the fertility rate. If migrants to the destination are retirees, their effect is to increase the average age, raise the death rate, and lower the birth rate. Arizona and Florida, for example, have attracted a large number of retirees, resulting in higher-than-average death rates.

Social conflict is a fairly frequent social consequence of migration. It often follows the mass movement of people from poor countries to rich. Tensions existed in Boston and New York after the Irish arrived in the 1840s. Fleeing the potato famine in Ireland, they were the first Catholics to arrive to the United States in large numbers. Similar friction has come with more recent migrants, such as Cubans to Miami. Social unrest and instability also follow the movement of refugees from poor countries to other poor countries. Many immigrants face discrimination and are blamed for all the problems in their new country, especially during economic downturns. In much of Europe, for example, nationalists blame Turks, Arabs, Pakistanis, and other immigrants for unemployment. Generally, poor migrants have more difficulty adjusting to a new environment than do the relatively well educated.

The economic effects of migration vary. With few exceptions, migrants contribute enormously to the economic well-being of places to which they come. For example, immigrant laborers known as guest workers from Turkey and Yugoslavia were indispensable to the economy of Germany in the 1960s and 1970s. Without them, assembly lines would have closed down, and patients in hospitals and nursing homes would have been unwashed and unfed. Without Mexican migrants, fruits and vegetables in Texas and California would go unpicked and service in restaurants and hotels would be much more expensive. Migrants to the United States also pay income and sales taxes, but illegal ones do not reap the benefits of programs such as Social Security.

In the short run, the massive influx of people to a region can cause problems. The southern and western U.S. states have benefited from new business and industry but are hard pressed to provide the physical infrastructure and services required by population growth. In Mexico, migrants to Mexico City accelerate the competition for scarce food, clothing, and shelter. Despite massive relief aid, growing numbers of refugees in the developing world impoverish the economies of host countries.

Emigration can relieve problems of poverty by reducing the supply of labor. International migration reduced poverty somewhat in Jamaica and Puerto Rico in the 1950s and 1960s. However, emigration can also be costly. Some of the most skilled and educated residents of developing countries migrate to developed countries. Each year, the income transferred to the through this "brain drain" (Chapter 7) to the United States alone amounts to significant quantities of funds. However, these immigrants also send billions of dollars back home in the form of remittances, or payments, to family members who stayed behind. Indeed, remittances often constitute a major source of income for impoverished villages in the developing world.

PATTERNS OF MIGRATION

People who examine patterns of migration, often consider migration internationally (external migration) or within a country (internal migration). They also subdivide external migration into migration between continents or within a continent. Internal migration can occur between regions, from rural to urban settings, and from one urban area to another area. International migrations are much smaller than internal population movements, especially to and from cities.

The great overseas exodus of Europeans and African slaves to the Americas is a spectacular example of intercontinental migration. In the five centuries before the economic depression of the 1930s, these population movements contributed greatly to a redistribution of the world's population. It has been estimated that between 10 and 20 million slaves, mostly from Africa, were hauled by Europeans into the sparsely inhabited Americas. The Atlantic slave trade, however, was dwarfed by the voluntary intercontinental migration of Europeans. Mass emigration to the United States began slowly in the 1820s and peaked on the eve of World War I, when the annual flow reached 1.5 million (Figure 8.19). At first, migrants came from densely populated northwestern Europe. Later they came from poor and oppressed parts of southern and eastern Europe. Between 1840 and 1930, at least 50 million Europeans emigrated. Their main destination was North America, but the wave of migration spilled over into Australia and New Zealand, Latin America (especially Argentina), and southern Africa. These new lands were important for Europe's economic development. In addition to offering outlets for population pressure, they provided new sources of foods and raw materials, markets for manufactured goods, and openings for capital investment. Another large-scale intercontinental migration was the Chinese diaspora of the nineteenth and early twentieth centuries, especially into Southeast Asia.

Since World War II, the pattern of intercontinental migration has changed. Instead of the dominant pattern of heavy migratory flows from Europe to the New World, the tide of migrants today is overwhelmingly from developing to developed countries (Figure 8.20). Migration into industrial Europe and to North America has been spurred partly by widening global economic inequality and by rapid rates of population increase in the developing world. Immigrants thus form significant populations in many countries around the world, particularly in the United States, western Europe, Australia, and South Asia. Some of the new arrivals are refugees, whereas others are unskilled workers seeking jobs outside of their native lands.

The era of heavy intercontinental migration is over. Mass external migrations still occur, but mostly at the intracontinental scale (within a continent). In Europe, forced and impelled movements of people in the aftermath of World War II have been succeeded by a system of migrant labor. Just as thousands of Latin Americans travel to the United States each year

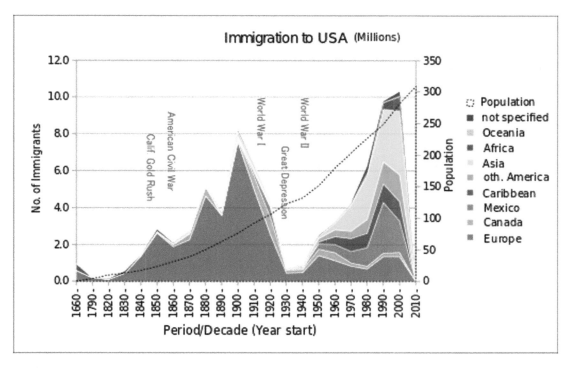

Figure 8.19: Immigration levels to the United States reflect two distinct periods. The first, stretching from the late nineteenth century to the 1920s, saw a heavy influx from southern and eastern Europe. The second, beginning in the 1960s and 1970s, continues today with large numbers of immigrants from Latin America and Asia.

to work, workers from the economic periphery of Europe migrate to the most prosperous west European countries. France and Germany are the main receiving countries of labor migrants to Europe. France attracts numerous workers from North Africa. Germany draws workers from Italy, Greece, Turkey, and eastern Europe. Britain has a sizable population of Polish immigrants. Migrant workers from eastern Europe usually have low skills and perform jobs unacceptable to indigenous workers.

International migrant labor also exists in the developing world. In Africa, laborers move great distances to work in mines and on plantations. In West Africa, the direction of labor migration is from the interior to coastal cities and export agricultural areas. In East Africa, agricultural estates attract international labor, typically refugees. In southern Africa, migrants focus on the mining-urban-industrial zone that extends

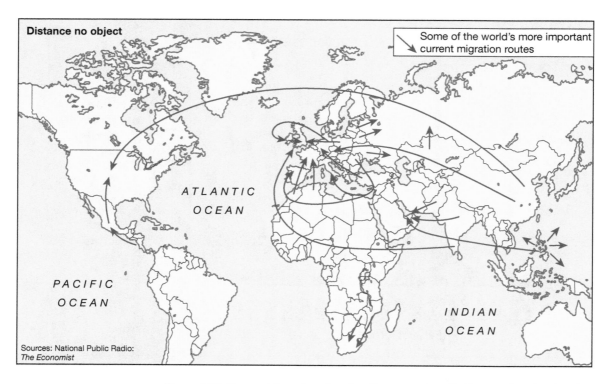

Figure 8.20: The world's major migration flows are primarily from the less developed to more developed countries as millions seek work and better lives. These flows include the movements of Arabs and Africans to Europe, Latin Americans and East Asians to the United States, and South Asians and Filipinos to the Middle East.

from southern Zaire in the north, through Zambia's Copper Belt and Zimbabwe's Great Dyke, to South Africa's Witwatersrand in the south.

Today, refugee-generating and receiving countries are concentrated in Africa (six million people), Southeast Asia (four million), and Latin America (two million). The causes of refugee movement typically include wars (e.g., Vietnam, World War II, Iraq, Afghanistan); racial and ethnic persecution (e.g., South Africa, Bosnia); economic insufficiency increased by political turmoil (e.g., Sudan); and natural and human-caused disasters (e.g., Central American hurricanes).

Colonizing migration and population drift are two types of interregional migration. Examples of colonizing migration include the nineteenth-century spontaneous trek westward in the United States and the planned eastward movement in Russia beginning in 1925. General drifts of population occur in almost every country, and they accentuate the unevenness of population distribution. Between the two world wars, there was a drift of African Americans from the rural South to the cities of the nation's industrial heartland in the Northeast and Midwest. Since the 1950s, there has been net out-migration from the center of the United States to both coasts and a shift of population from the industrial northeastern and midwestern states (Rustbelt) to southern and western states (Sunbelt), primarily because of economic stagnation in the former and rapid job growth in the latter. Today, the majority of Americans live in the South and West, as opposed to the North and Midwest, although the vast expanses of land in the Sunbelt states generate lower population densities than in the Northeast.

The most common type of internal migration is rural-urban migration, which is usually for economic motives. The relocation of farm workers to industrial urban centers was prevalent in developed countries during the nineteenth century. Since World War II, migration to large urban centers has been a striking phenomenon in nearly all developing countries. Burgeoning capital cities, in particular, have functioned as magnets attracting migrants in search of "the good life" and employment.

In highly urbanized countries, migration between cities is increasingly important. Although many migrants to cities come from rural areas and small towns, they form a decreasing proportion. Employment opportunities contribute to intercity migration. So, too, does ease of transportation, especially air transportation. For intermetropolitan migrants from New York, the two most popular destinations are Miami and Los Angeles.

■ KEY TERMS

Baby boom: the generation born during the post–World War II boom, i.e., roughly 1946 to 1964

Birth rate: the number of birth per 1,000 people per year; beyond the crude birth rate, age-specific death rates refer to births per 1,000 people in a given age category

Death rate: the number of deaths per 1,000 people per year; beyond the crude death rate, age-specific death rates refer to deaths per 1,000 people in a given age category

Demographic transition theory: a model of how birth, death, and natural growth rates change over time as a population goes through successive stages of economic development

Diminishing marginal returns: the tendency for additional inputs to yield ever-smaller volumes of output

Doubling time: the amount of time a population needs to double at a given growth rate

Emigration: movement out of one country to another

Epidemiological transition: the shift in causes of death that accompanies economic development, from infectious diseases to environmental and behavioral causes

Fertility rate: the average number of children to which a woman gives birth over her lifetime

Green Revolution: the worldwide effort that began in the 1960s to develop genetically engineered crops to enhance agricultural productivity in the developing world

Immigration: movement into one country from another

Infant mortality rate: the proportion of babies who die before their first birthday

Migration: the movement and permanent relocation of people over long distances

Mortality: death, or demographically, the incidence of death in a population

Natural growth rate: the birth rate minus the death rate

Net migration rate: the difference between in-migration and out-migration in a region

Nuclear families: families in which several generations and extended relatives do not live together

Opportunity cost: what one gives up to obtain something; the opportunity cost of having a child is largely felt through the income the mother sacrifices to stay home with children

Pandemic: an epidemic of disease that reaches worldwide proportions

Population pyramids: the distribution of a population by age and gender as represented in a graphic with age on the vertical axis

Sex ratio: the ratio of males to females in a society

■ STUDY QUESTIONS

1. What are the major world regions containing large shares of the planet's population?
2. Define crude birth rates and mortality rates. How does fertility relate to the birth rate?
3. What are the four major components of population growth? What mathematical function describes the relationships among them?
4. Why did Malthus have such a gloomy view of the future?
5. How do neo-Malthusians resemble and differ from Malthusians?
6. Is it fair to blame poverty on overpopulation, or is this view simplistic?
7. What is the Green Revolution?
8. What are the four stages of the demographic transition?
9. What explains the world map of fertility rates?
10. What were the most important direct contributors to the decline in mortality rates in developed nations?
11. How do the most common causes of death differ between developed and developing nations?
12. Why do the poorest countries have the most rapid natural growth rates?
13. Why do birth rates decline as countries become wealthier?
14. What is a population pyramid, and how does it vary between developed and developing countries?
15. Summarize the major causes of international migration.
16. What are some consequences of international migration? How do these resemble or differ from consequences of internal migration?
17. What were the two largest periods of immigration in the United States? How have the major source areas changed?

BIBLIOGRAPHY

Anderson, B. (2014). World Population Dynamics: An Introduction to Demography. *New York: Pearson.*

Caldwell, J., Caldwell, B., Caldwell, P., McDonald, P., & Schindlmayr, T. (2006). Demographic Transition Theory. *New York: Springer.*

Galor, O., & Weil, D. (2000). Population, technology, and growth: From Malthusian stagnation to the demographic transition and beyond. American Economic Review, 90*(4), 806–828.*

Hardin, G. (1968). The Tragedy of the Commons. Science, 162, *1243–1248.*

Larkin, R., & Johnson-Webb, K. (2013). Population Geography: Problems, Concepts, and Prospects. *Sunnyvale, CA: Kendall Hunt.*

Newbold, K. (2007). Six Billion Plus: World Population in the Twenty-first Century *(2nd ed.). Washington, DC: Rowman and Littlefield.*

Newbold, K. (2013). Population Geography: Tools and Issues. *Washington, DC: Rowman and Littlefield.*

Peters, G., & Larkin, R. (2010). Population Geography. *Sunnyvale, CA: Kendall Hunt.*

Robles, E. (2009). The Malthusian Catastrophe. *New York: Loyal Dog Publishing.*

Trewavas, A. (2002). Malthus Foiled Again and Again. Nature, 418, *668–670.*

Weeks, J. (2011). Population: An Introduction to Concepts and Issues. *Boston: Cengage Learning.*

IMAGE CREDITS

- Figure 8.1: El T, http://commons.wikimedia.org/wiki/File:Population_curve.svg. Copyright in the Public Domain.
- Figure 8.2: http://commons.wikimedia.org/wiki/File:Population_density.png. Copyright in the Public Domain.
- Figure 8.3: Copyright © Benjamin D. Hennig (CC BY-SA 4.0) at https://commons.wikimedia.org/wiki/File:World_Population_Cartogram_Map_2002.tif.
- Figure 8.4: Copyright © TastyCakes (CC by 3.0) at http://commons.wikimedia.org/wiki/File:China_Pop_Density.svg.
- Figure 8.5: Copyright © Jcostell4 (CC BY-SA 3.0) at http://commons.wikimedia.org/wiki/File:United_States_Population_Density.svg.
- Figure 8.6: Thomas Malthus, http://commons.wikimedia.org/wiki/File:Malthus_PL_en.svg. Copyright in the Public Domain.
- Figure 8.7: Mr Larry, "Club of Rome Projections," https://www.flickr.com/photos/4dtraveler/5967580802.

- Figure 8.9a: Copyright © Mestiso (CC by 3.0) at https://commons.wikimedia.org/wiki/File:Farming_-_panoramio_-_Mestiso.jpg.
- Figure 8.9b: Copyright © Sur Chakrabarty (CC BY-SA 4.0) at https://commons.wikimedia.org/wiki/File:Children_studying_and_farmers_working.jpg.
- Figure 8.9c: William Creighton, "Drying Peppers," https://commons.wikimedia.org/wiki/File:Drying_peppers_(5761949979).jpg. Copyright in the Public Domain.
- Figure 8.9d: Copyright © Depositphotos/dyvan82.
- Figure 8.10: Copyright © Roke (CC BY-SA 3.0) at http://commons.wikimedia.org/wiki/File:Birth_rate_Figureures_for_countries.PNG.
- Figure 8.11: Copyright © Roke (CC BY-SA 3.0) at http://commons.wikimedia.org/wiki/File:Death_rate_world_map.PNG.
- Figure 8.12: Copyright © Nummies (CC BY-SA 3.0) at http://commons.wikimedia.org/wiki/File:Life_Expectancy_Map_CIA_2013.svg.
- Figure 8.15: Kisipila, http://commons.wikimedia.org/wiki/File:Population_growth_rate_world_2011.svg. Copyright in the Public Domain.
- Figure 8.16: Securiger, http://commons.wikimedia.org/wiki/File:World_population_growth_rate_1950%E2%80%932050.svg. Copyright in the Public Domain.
- Figure 8.17: Tga.D, http://commons.wikimedia.org/wiki/File:World-Population-1800-2100.svg. Copyright in the Public Domain.
- Figure 8.18a: Delphi234, http://commons.wikimedia.org/wiki/File:Somaliapop.svg. Copyright in the Public Domain.
- Figure 8.18b: The World Factbook., "Population Pyramid Sweden," https://commons.wikimedia.org/wiki/File:Population_pyramid_of_Sweden_2015.png. Copyright in the Public Domain.
- Figure 8.19: Copyright © Masaqui (CC BY-SA 3.0) at http://commons.wikimedia.org/wiki/File:US_immigration_1660_to_2009_EN.svg.
- Figure 8.20: Copyright © Imma Moles (CC BY-SA 3.0) at http://commons.wikimedia.org/wiki/File:Report_on_Immigration.png.

Chapter Nine

POLITICAL GEOGRAPHY

MAIN POINTS

1. The nation-state, which is central to the world's politics, emerged from the sixteenth to the twentieth centuries on the heels of global capitalism.

2. Nationalism is the ideology that celebrates the nation-state as given and eternal, and is often coupled with suspicions of those from other countries.

3. Geopolitics explores how international politics and geography are intertwined. Whereas its classical version was largely a matter of imperial strategy, more recent critical geopolitics focuses on the ways in which nation-states are constructed symbolically.

4. World systems theory holds that although the economic geography of the planet is defined by the world economy, politically its structure is manifested through numerous competing nation-states.

5. Geographies of war and conflict take many forms, including fights between states, civil wars within them, and terrorism, as well as struggles over natural resources and nuclear weapons.

6. Electoral geography analyses the spatial dimensions of elections, including electoral strategies and outcomes as well as the processes by which electoral districts are created, such as redistricting and gerrymandering.

Although you might not be interested in politics, politics is interested in you. Politics is all about power—who has it, who doesn't, who controls what, who gets to shape the rules. Politics is half of the notion of political economy introduced in Chapter 1. Politics means much more than elections; it deeply shapes everyday life, including the politics of the family or the workplace. People who ignore politics are people who lack power, and suffer for it. Politics is also deeply geographical, because power is unevenly distributed over space. Seen in this way, politics takes a variety of forms that links people's everyday lives with the dynamics of the world economy.

If human geography is about the role played by space in the conduct of human affairs, political geography is about the power to exercise control over people and the spaces they occupy. The creation of specific territories is the basis for political organization and political action. The political partitioning of space creates a fundamental geographic division, the sovereign state, which is the dominant political actor of our time. The spatial manifestation of power, however, is not confined to the nation-state, for it also plays out at the international level (e.g., transnational corporations) and at scales below that of the state, such as in counties and cities, and even within local neighborhoods.

This chapter summarizes some of the broad outlines of political geography. First, it explores the origin and meanings of the nation-state. Next, it turns to the ideological counterpart of the nation-state, nationalism, which produces moral geographies of sameness and otherness. Third, it focuses on classical and critical geopolitics, or different attempts to theorize relations among states. Fourth it addresses world-systems theory, which views the world's countries as part of an integrated totality. Fifth, it addresses issues of war and conflict, including their frequency and severity, especially civil wars and struggles over resources. Finally, it points to electoral geography, the spatial distribution of voting patterns, which closely relates to how people are represented in the machinery of state power.

9.1 THE NATION-STATE

Capitalism is not simply an economic system for organizing the production and consumption of resources through markets. It is also a political one that heavily involves the state, that is, governments and governance. Thus, the emergence of capitalism witnessed a series of political changes that accompanied the rise of market-based societies.

Today, the nation-state is the most important political actor in the international arena. Many people confuse nation-states with nations. In fact, the terms have quite distinct meanings. It is important, therefore, to understand what a nation is and how it relates to the nation-state.

Popular usage of the word "nation" is broad and often vague. It frequently refers simply to a country, such as in sentences such as "It is raining across the nation." The term can also refer to entire cultures or tribes, such as the Spanish, Arabs, or Pawnee. The formal term **nation**, however, has a specific meaning: a group of people who share a common culture, ethnicity, heritage, language, religion, history, territory, and identity. Nations are simultaneously social and psychological constructs. For many people, nations are the primary form of community, the definition of a "we" to whom they belong.

In places such as China, the nation is a very old entity. It is often held to date as far back as 221 BC, when the Ch'in (Qin) dynasty unified Chinese-speaking peoples for the first time. Similarly, nations have long existed in Japan, Korea, and other relatively culturally homogeneous non-Western places such as Iran. As we saw in Chapter 5, the political geography of feudal Europe consisted of large, multiethnic empires that contained many nations within them. In the medieval era, when the term nation first arose, it referred to a group united by blood or common ancestry, essentially referring to people bound by a common ethnicity or culture. In this sense, nations existed long before the nation-state, or capitalism, which is often held to be the context in which the modern nation-state was born. Feudal empires had many nations within their borders; the Holy Roman Empire and the Austro-Hungarian Empire, for example, contained dozens of different ethnicities, each with its own language, culture, and, often, loyalties. Individual and collective identity was largely defined in religious terms, ancestral lineage, or as subjects of a particular king. Thus, in the medieval era, ethnic identity was not the primary definer of political identity. In Europe before about 1600, the sovereignty of most individual rulers depended on the allegiance of people rather than on control over a clearly defined territory. However, the classical Greek and the medieval Italian city-states were exceptions.

By the eighteenth century, however, as these empires began to disintegrate, the nation had increasingly come to be the primary source of identity of many peoples. During the expansion of capitalism, the term *nation* expanded to include a more explicit political connotation, often defined as part of a collective effort to defend national territory. Thus, the modern idea of nation is closely coupled with that of the state, or government, an idea expressed in the ideology of nationalism, which celebrates the nation-state.

In contrast to the nation, which is relatively homogenous culturally, the nation-state refers to an independent country, typically a nation with independence or sovereignty. A **nation-state** is a clearly defined large group of people who self-identify as a group (a nation) and who occupy a spatially defined territory with the necessary infrastructure and social and political institutions (a state). Closely connected to the nation-state is the notion of **sovereignty**, in which authority over the territory and population of a state, vested in its government, is held to be the most basic right of a state as a political community. The nation-state is, therefore, fundamentally a *political* construction. Each nation-state is, in principle, a political territory including all members of one national group and excluding members of other groups. In practice, few nation-states fit that strict definition because the vast majority of them are not composed of just one national group. Not all nations are nation-states; for example, today several groups qualify as nations but do not have independence, including the inhabitants of Scotland, Québec, and Tibet, as well as the Kurds, Basques, Baluchis, and Palestinians (Figure 9.1).

The rise of the nation-state was a pivotal moment in world history. It formed an integral part of the development of capitalism, first in Europe and later in much of the rest of the world. Although some nation-states had roots that can be traced to the centralized monarchies of western Europe in the feudal era (for example, Britain and France), many social scientists and historians hold that the nation-state became the dominant actor in global politics with the signing of the Treaty of Westphalia in 1648. In ending the Thirty Years' War (1618–1648), which was part of the fratricidal conflicts of the Reformation and Counter-Reformation, and guaranteeing the independence of the Netherlands and Portugal, the treaty signaled the breakup of the vast religious empires that long dominated medieval Europe. It also allowed the individual states making up the Holy Roman Empire to choose their own religion, a power that had been usurped by the Holy Roman Emperor. Most importantly, it elevated the autonomous territorial state into the world's preeminent form of political organization and decisively shifted the scale at which territorial politics was constituted from city-states and feudal empires to the nation-state. Moreover, it marked the beginning of an era in which states were expected to respect one another's sovereignty, abide by noninterference in one another's internal affairs, and respect national borders as clearly defined, sacrosanct entities representing the limits of control and governance.

Thus, the **Westphalian system** of nation-states, which continues today, is characterized by borders that are closely monitored and policed, as well as restrictions on geographical movement such as passports (although supranational federations of states such as the European Union have relaxed internal controls in this regard). It is worth noting that the Westphalian ideal was always exactly that—an ideal. Complete noninterference, of course, has rarely been practiced, borders have always been porous, and states have always shaped, and been shaped, by events inside their neighbors.

In the nineteenth century, nationality often became conflated with the idea of race, as if the two were synonymous and had biological origins. More recently, it has become clear that nations are not so easily defined simply by language or religion. Indeed, nations are always more ethnically and culturally diverse than nationalists claim. On the eve of the French Revolution in 1789, for example, only one-half of the people of France spoke French. Indeed,

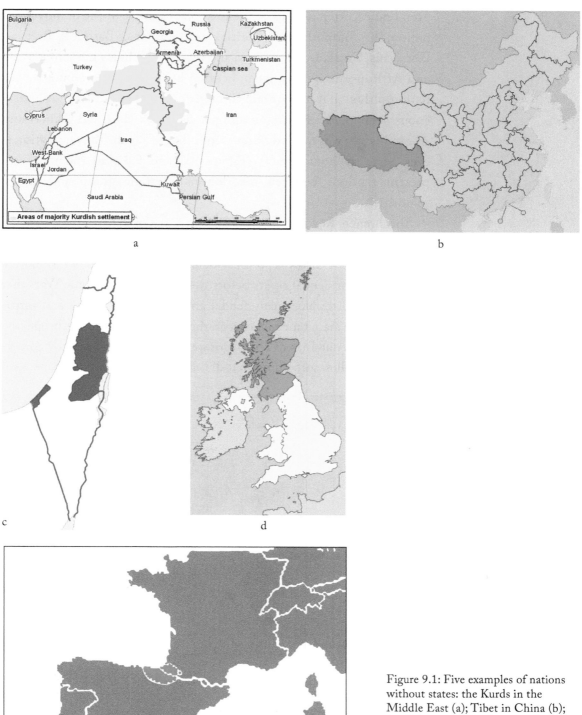

Figure 9.1: Five examples of nations without states: the Kurds in the Middle East (a); Tibet in China (b); the Palestinians (c); Scotland (d, in red); and the Basques in France and Spain (e).

even France, often held to be the textbook example of the homogenous nation, has important minorities of Germans, Basques, Bretons, Corsicans, Catalans, and others, and recently includes numerous Arab and African immigrants. Japan is often held up to be a homogeneous nation, although it too contains minorities of Koreans, Ainu (an indigenous people), *burakumin* (an indigenous underclass), and Okinawans, whose history was quite different from that of Japan. Thus no nation is ever perfectly homogenous (Figure 9.2 and 9.3); all include ever-present tensions between the dominant ethnic group and ethnic minorities.

The French Revolution of 1789 was a major turning point in the breakdown of the old feudal social order and the rise of the new modern one, creating new forms of political identity such as citizens who enjoyed, at least in theory, equal rights before the law. The Napoleonic Wars that ended in 1815 toppled many feudal aristocrats from power and introduced modern state bureaucracies in their place. In western Europe, the centralized monarchies of feudalism were gradually replaced by constitutional republics, something that did not occur in eastern Europe until

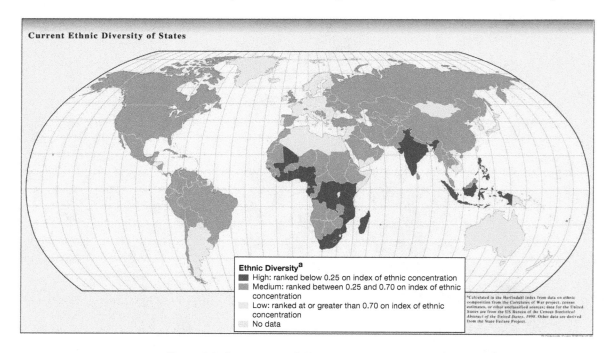

Figure 9.2: States vary widely in their degree of ethnic homogeneity; because many states in the developing world are products of colonialism, they consist of numerous tribes and ethnic groups.

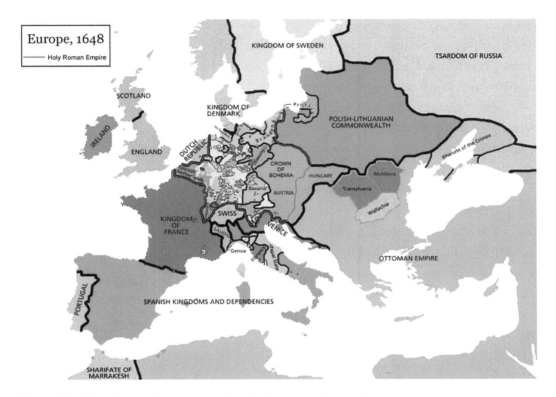

Figure 9.3: Today, all states have ethnic minorities in them, although the degree of cultural heterogeneity varies widely. Typically, former European colonies, with externally imposed boundaries, have the greatest range of ethnic groups, which often leads to civil wars.

after World War I. By the end of the twentieth century, the nation-state had become the dominant form of political organization throughout the world, including many former European colonies.

Ever since Westphalia, sovereign national states have become deeply ingrained not only in the world's multiple legal and economic systems, and the basis of international relations, but also in our ways of thinking. National social systems are so deeply equated with "society" in the minds of most people that national borders appear to many on world maps as natural as mountain ranges or rivers. Indeed, typically we define "a society" as a nation-state. However, it is worth repeating that, historically, the nation-state is a relatively recent product, in comparison with tribes or empires. In addition, some nation-states only emerged only recently. Italy was united in 1861, Germany in 1871, and with the collapse of European empires after World War II, many nation-states in the developing world are only two or three generations old. In fact, nation-states continue to be born today in former European colonies, including Namibia (1990), Eritrea (1992), and South Sudan (2011), although some of these are more collections of different tribes than ethnically similar nation-states.

The emergence of market societies facilitated the growth of nation-states in several ways. Key factors included rising wealth, the diffusion of mass literacy, growing cadres of intellectuals, and political parties demanding a role in newly democratic societies. Other institutions related to capitalism also supported the rise of the nation-state, such as national banks and currency, the military draft (which provided troops for the growing national armies and socialized young men from disparate backgrounds into a common national culture), the media (i.e., newspapers), and national rail systems, which tied together the diverse parts of the emerging nation-states and reinforced their sense of unity. In turn, the new nation-states facilitated markets in several ways, including, for example, the public construction of an infrastructure (including roads and railways), the provision of public services such as schools and transportation, the maintenance of national monetary supplies, and the protection of domestic producers by taxation of imported goods. These relations led some social scientists to discard the conventional view that markets were born free of the shackles of state in favor of a perspective that maintains markets *required* the state to survive.

It is important to remember that capitalism long preceded the nation-state and that there is no necessary correspondence between the two, an observation with important implications in the current age of globalization. Capitalism began in the city-states of northern Italy in the fifteenth and sixteenth centuries, and exhibited a political geography quite distinct from that of today. The nation-state arose out of the industrialization of the eighteenth and nineteenth centuries, long after capitalism had become firmly entrenched. The ascendancy of the nation-state marked a new scale at which capitalist social relations could be contained and managed, including trade and production networks. Simultaneously, a world economy was coming into being marked by numerous competing states. Capitalism thus produced the scale of the national at the same time as it produced the international one.

9.2 NATIONALISM

The ideological counterpart of the nation-state is **nationalism**, which celebrates the nation-state as "natural" and optimal—that is, as given, not made. Nationalism describes either the attachment of a person to a particular nation or a political action by such a group to achieve statehood (or national self-determination). As the state became the primary source of authority in the eighteenth and nineteenth centuries, ruling elites had construct a narrative that provided the often culturally diverse populations that inhabited the state territory with a single identity. Nation and state came to overlap in determination of many nations to acquire self-determination and sovereignty, or a country's right to determine its own fate.

Nationalism was originally a means for the dominant classes to exert control over the diverse and increasingly mobile populations within their respective territories. As the nation-state became the primary form of politics from the seventeenth through the nineteenth centuries, it became increasingly important for ruling elites to construct a narrative that provided their often culturally heterogeneous populations with a single identity, a sense of community typically defined by its opposition to "otherness," or "foreigners." Nationalism generally describes either an emotional and ideological attachment to a particular nation or a political action by a nation to achieve statehood (Figure 9.4). The political manifestation of national identity as nationalism is a relatively modern concept. Nationalism became an increasingly important political ideology in the twentieth century as the existence of a distinct national identity became a necessary condition for the creation of a sovereign state.

There are several types of nationalism. Perhaps the most common distinction is that between civic and ethnic nationalism. **Civic nationalism** developed first, and is represented by the model of national identity as a rational choice for those desiring self-determination and individual liberty. In this view, the nation is composed of individuals who occupy a common territory and willingly form bonds based on ideology and shared political institutions. This relatively inclusive model permits membership to anyone who voluntarily resides in the territory and recognizes the institutions and ideology of the group.

The goal of civic nationalism is generally statehood, as it is only at that level of sovereignty that ideology can be translated into fully independent institutions. Civic nationalism embodies the concept of sovereignty of the people. The nation cannot exist without the collective will of the people who support it. This model began in late medieval England but is best represented today in the United States, where the desire for self-determination was largely based in the ideological drive toward greater individual liberty and the need to create new institutions to enshrine popular sovereignty. After two centuries of absorbing immigrants who identify with countless different language, religious, ethnic and racial groups, the very strong nationalism of the United States remains almost entirely based on attachment to political symbols and institutions.

In the eighteenth and nineteenth centuries, however, a new form of national identity emerged, that of cultural or **ethnic nationalism**, primarily in Europe as well as Japan. This movement emphasized the bonds of language, perceived ethnicity, race, and ancestry. Rejecting the ideological and voluntary basis of French and English national models, this form of

Figure 9.4abc: Nationalism is arguably the most important form of political identity in the world today, celebrating the nation-state as given, not made, although it must compete with subnational tribal loyalties and transnational religious ideologies.

national identity emphasized genetics and birth. One can neither voluntarily join this national group nor leave it; national identity is imprinted through race and ethnicity and is expressed through language and religion. This form of nationalism was especially powerful in Germany, which viewed the large German populations of French Alsace-Lorraine and eastern Europe as members of the newly formed German Empire, an idea that culminated with the Nazis before and during World War II.

Throughout Europe, as the old feudal empires disintegrated and new nations took their place, culturally distinctive groups used literature, music, mythology, art, and language to define themselves, so that they could then press for self-determination. The rise of compulsory public education provided the stage for states to promote their vision of the nation. Simultaneously, however, it became a battleground for ethnic minorities attempting to preserve and promote their language and culture in the face of the drive toward homogeneity that characterized the emergence of state-led nationalism. For example, language became a highly politicized expression of nationalism for the Scots, Basques, Bretons, and others (Chapter 3). At the same time, the development of mass production and the mass media, at first in the form of newspapers and then in magazines, posters, and radio all permitted the widespread dissemination of these new ideas about national identity.

Ethnic nationalism likely reached its peak during and immediately following World War I (1914–1917), when the German, Austro-Hungarian, and Ottoman Empires collapsed, leaving a multitude of small European nation-states in their wake. Dominant ethnic groups often used the state to enforce cultural homogeneity. In the nineteenth and early twentieth century, the U.S. government tried to eliminate Native American languages by educating indigenous children in state boarding schools where tribal languages were illegal. Likewise, in the Soviet Union—a multination, multiethnic, multilingual, multireligious collection of territories founded shortly after the Bolshevik Revolution of 1917—the imposition of the Russian language and culture was widespread.

As decolonization of Asia, African, and the Middle East proceeded in the interwar and post-World War II periods, nationalism again flourished across the globe. New states, their borders drawn by colonial powers and rarely conforming to the ideal of having a nation synonymous with a state, readily embraced nationalism to unify their culturally diverse populations. Today, nationalism is as widespread and powerful in the developing world as in the developed one, perhaps more so. Many states, however, have been torn by nationalist tensions, often in the form of tribal conflicts and civil wars. This process occurred most forcefully in Africa, the political geography of which was constructed by the old colonial overlords in the Berlin Conference of 1884. African nationalism overlaps and contends with tribal loyalties that often undermine states there, such as the horrific violence in Rwanda in 1994, in which almost one million people were murdered in strife between Hutus and Tutsis.

9.3 GEOPOLITICS

The ways in which national power is constructed and projected across space is a major dimension of today's world system, which is comprised of roughly 200 independent states. There have been several attempts to theorize how nation-states interact with one another. Political geographers have often approached this topic through **geopolitics**, which consists of the older, classical form and new, critical geopolitics.

CLASSICAL GEOPOLITICS

First developed in the late 1890s by the Swedish political geographer Rudolf Kjellen, geopolitics in the first half of the twentieth century was primarily concerned with the geographical foundations of the state and national power with reference to natural resources, population, and geographical location. It is no accident that **classical geopolitics** flourished during the height of nationalism. Classical geopolitics viewed interstate relations as a science similar to biology and tended to be highly nationalistic in orientation. It was also largely concerned with the art of statecraft in the international arena during a period of major global upheaval, shifting alliances, world wars, and ethnic movements. Geopolitics was studied with great interest in Germany, Italy, Japan, and Britain to put forth "laws" about international politics based on a series of geographical facts such as the relationship between land and sea-based powers. Informed by social Darwinism, which viewed social relations entirely as a wilderness in which only the fittest survived, classical geopolitics emphasized the struggle among national states to secure the "fittest" states and peoples. In this line of thought, the state should be thought of as a superorganism, which exists in a world characterized by struggle and uncertainty. To survive, let alone prosper, in these testing circumstances, states need to acquire territory and resources.

An early, famous proponent of classical geopolitics was the German geographer Friedrich Ratzel (1844–1904), who wrote extensively on the links between the physical environment, society, and the state. Heavily influenced by Social Darwinism, Ratzel considered the state as an organism that must expand into space to grow, thus requiring *Lebensraum* ("living space"), additional territory. As different states attempted to expand, there would inevitably ensue a clash of cultures (*Kulturkampf*), in which the strong would triumph over the weak. This idea became politicized and adopted by the Nazis as a justification for Germany's expansionist policies (Hitler read Ratzel's works while in prison). Indeed, geopolitics became closely associated with Germany's view of the world as a state encircled by the Russians, French, and British, and one needing to break out of these confines.

A leading British geographer, Sir Halford Mackinder, writing in 1904, was another to formulate a geopolitical theory of international relations. Mackinder's **Heartland theory** (Figure 9.5), which attempted to explain how geography and history had interacted over the past 1,000 years, has strong environmental determinist overtones. It reflects British concerns about perceived Russian threats to British colonies in Asia, especially India. Mackinder created political geography's most famous global geopolitical model, the "Heartland" model, in which he divided the world into three main areas: (1) the *Heartland* or pivot area, analogous originally to much of the Asian part of Russia but later extended to include most of the Soviet Union, (2) the *inner crescent* or

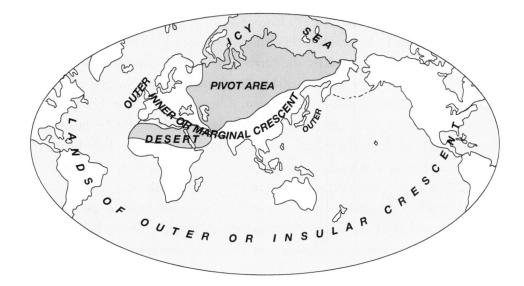

Figure 9.5: Halford Mackinder's Heartland theory was an enormously influential early manifestation of classical geopolitics that emphasized the strategic role of eastern Europe and Central Asia.

rimland, the European and Asian areas surrounding the Heartland, and (3) the *outer crescent* or everything else (Figure 9.4). Mackinder argued that the Heartland was the key to world power, that whatever country controlled the Heartland could control the "world island" (i.e., Eurasia) and ultimately the world itself. He based this argument on the historical record that repeated attempts throughout history to conquer Europe had originated in the Heartland. Moreover, he held that the Heartland was impenetrable from attack on three sides because of natural barriers, including the east (Gobi Desert), the south (Himalayas), and the north (Arctic Ocean), thereby making it invulnerable defensively. Thus,

> Who rules East Europe commands the Heartland;
> Who rules the Heartland commands the World Island;
> Who rules the World Island commands the World.

Mackinder emphasized the role of railroads in the ability of states to control terrestrial space, including the movement of troops. Mackinder's major interest was making sure that the Germans did not challenge the British for global supremacy. His main concern was that Germany would gain control over eastern Europe. Thus, this supposedly scientific view of the world was in reality a warning to Britain to look out for Germany and a possible alliance between Germany and Russia. He argued that a naval power—that is, Britain—would not be able to take over the Heartland. He suggested that the only way into the Heartland was through eastern Europe. Thus, controlling eastern Europe was the key to world domination.

In Germany after World War I, General Karl Haushofer became a leading Nazi geopolitical theoretician. He was bitterly disappointed by the territorial losses Germany experienced as a result of World War I. For Germany to prosper, Haushofer argued that interwar Germany needed to become a "space hopping" country rather than a "space bound" one. He contended that if the world were organized into a series of pan-regions then it might be possible for Germany, the United States, and Japan to exercise global leadership. His ideas have been credited with informing Hitler's plans of spatial expansionism in eastern Europe and Russia. By the end of World War II his influence was declining, and in 1946 he committed suicide.

Although the links between classical geopolitics and the Nazis were never as clear as some critics alleged, after World War II the school of thought lost much of its earlier influence. Geopolitics continued to influence political science conceptions of how countries interact, such as realism. It enjoyed a modest renaissance under U.S. Secretary of State Henry Kissinger in the 1970s, who applied it to justify the U.S. war in Vietnam. By the 1980s, however, classical geopolitics had given way to a new, more liberal variant.

CRITICAL GEOPOLITICS

A newer version of geopolitics, one that overcame the racist and imperialist legacy of its early twentieth century counterpart, is **critical geopolitics**. The rise of critical geopolitics allowed geographers to reject the baggage associated with German geopolitics and to offer a more liberal, emancipatory version. Central to critical geopolitics is the relation between knowledge and power, the view that all ways of looking at the world are tied to someone's interests and the ability to get others to think in the same way. Critical geopolitics relies heavily on the notion of discourse, a structured set of meanings, ways of knowing, or rules of knowledge that shapes how people view the world. For example, nationalism as a discourse plays a large role in shaping people's identities, political views, and how they interpret international events; nationalism produced nationalists. In this light, people are not simply the producers of ideas but also their products. Thus, to be critical is to expose the linkages between knowledge and power.

Rejecting the notion that geopolitics was an objective "science," this school of thought approaches geopolitics as a set of discourses and political practices—that is, as a multitude of representations about the world that reflect, and in turn shape, how people think. Discourses, in short, do not just mirror social reality, they help to construct it. Thus, rather than approaching power relations among states as objective and external to people, critical geopolitics emphasizes the subjective and psychological nature of power, its ability to make social relations seem natural rather than human constructions, and to motivate people to think and feel in some ways and not others. Rather than simply focusing on the relations *among* states, critical geopolitics turns to how politics is constructed *within* states by different groups for different purposes. Whereas classical geopolitics was motivated by concerns to advance the interests of one state or another, critical geopolitics is motivated by the desire to expose the power relations within countries that give rise to some worldviews and not to others.

Critical geopolitics draws attention to how geopolitics is performed and how it creates representations of the geographies of global politics in the media, speeches, film, newspapers, and cartoons. Critical geopolitics is interested, for example, in how elites, politicians, and policymakers construct visions of other countries and whose interests are served by those representations, how ideology and symbols can be used to assert and justify national power, and how language and thought reflect, and are in turn shaped by, foreign policy. Although much of critical geopolitics has been concerned with how elites have framed the world, it has also explored "banal nationalism"—that is, how ideas about national unity, status, and power are reflected in everyday life, including forms such as postage stamps and bumper stickers.

9.4 WORLD-SYSTEMS THEORY

An especially influential body of theory, **world-systems theory**, started by the sociologist Immanuel Wallerstein, has had enormous influence in human geography and is one of the primary means that geographers have of making sense of how the world is organized and functions. Unlike perspectives that see the world's countries as isolated units in chaotic competition with one another, world systems theory posits that the world's states operate in an integrated system of relations that binds them together into a whole. Its focus is on the entire world rather than individual nation-states. Fundamentally, this view maintains that one can't study the internal dynamics of countries without also examining their external ones. Thus the boundary between the foreign and the domestic effectively disappears.

World-systems theory distinguishes between large-scale, precapitalist world *empires*, such as the Romans, Mongols, or Ottomans, which appropriated surplus from their peripheries through the state (e.g., tribute and taxation), and the primarily Western capitalist world *system*, which arose in the "long sixteenth century" (1450–1650). Under global capitalism, no single political entity rules the world. Instead, there is one global market but multiple political centers, with no effective control over the market. The political geography of capitalism is thus not the nation-state but the interstate system, which has taken various forms over time, including multiple city-states. Occasional attempts to reassert a world empire included the Hapsburgs, Napoleon, and Germany in World Wars I and II, all of which failed.

World-systems theorists maintain that capitalism takes many forms and uses labor in different ways in different regions. In the core, labor tends to be for wages (i.e., organized through labor markets), whereas in developing countries there is considerable use of unfree labor, ranging from slavery to indentured workers to landless peasants working on plantations. The world economy structures places in such a way that high-valued goods are mostly designed and produced in the core and low-valued ones in the periphery. It is the search for profits through low-cost labor that drives the world system forward to expand into uncharted territories.

Unlike the split between developed and less developed countries that both modernization and dependency theories advocate, in world-systems theory there is a three-part division among core, periphery, and semiperiphery (Figure 9.6). The core and periphery are the developed and developing countries, respectively; one is wealthy, urbanized, industrialized, and democratic, the other rural, impoverished, agriculturally based, and dominated by authoritarian governments. The semiperiphery has characteristics of both core and periphery and includes states at the upper tier of the less developed countries, such as the newly industrializing countries. For example, the rapidly industrializing countries of East Asia, including China, South Korea, Thailand, and Malaysia, are important parts of the semiperiphery. Processes in core countries

Figure 9.6: World-systems theory divides the world into a core, periphery, and semiperiphery, based on countries' relative levels of wealth and power.

generate high wages, high levels of urbanization, industrialism and postindustrialism, advanced technology, and a diversified product mix. Processes in the periphery, in contrast, generate low wages, low levels of urbanization, preindustrial and industrial technology, and a simple production mix. Consumption is low. In between are states that are part of the semiperiphery where both sets of processes coexist to a greater or lesser degree. The theory suggests that the core countries exploit the semiperiphery countries with regard to raw materials and product flow at the same time the semiperiphery countries exploit periphery countries.

World systems advocates say that **hegemony** exists when one core power enjoys supremacy in production, commerce, and finance and occupies a position of political leadership. The hegemonic power owns and controls the largest share of the world's production apparatus. It is the leading trading and investment country, its currency is the universal medium of exchange, and its city of primacy is the financial center of the world. Because of political and military superiority, the dominant core country maintains order in the world system and imposes solutions to international conflicts that serve its self-interests. Britain played this role in the nineteenth century and the United

States has done so since World War II. Consequently, hegemonic situations are characterized by periods of relative international peace (for example, during the nineteenth century). During a power's decline from hegemony, rival core states, which can focus on capital accumulation without the burden of maintaining the political and military apparatus of supremacy, catch up and challenge the hegemonic power. Thus, in the early twentieth century, Germany challenged Britain for global leadership, with catastrophically violent results.

World-systems theory pays particular attention to the control a dominant nation (hegemon) has over the global political and economic system at different historical moments. During the period of Spanish dominance in the sixteenth and seventeenth centuries, for example, mercantilism was the dominant ideology. Under the British Empire of the nineteenth century, free trade was the norm. And, since the rise to dominance of the United States, especially since World War II, neocolonialism, in which core countries exert economic but not political control over the developing world, has been the typical pattern (although during the Cold War there were two superpowers and a split world system). Dominant powers may overextend themselves militarily, leading to an erosion of their economic base. When powers in the core have conflicts among themselves, they open opportunities both for new world leaders and for countries on the periphery; the Napoleonic Wars of the early nineteenth century and World War II in the twentieth century were thus openings for nationalist anticolonial movements worldwide, first in Latin America and later in Asia and Africa.

9.5 THE GEOGRAPHY OF WARS AND TERRORISM

Starting in the eighteenth century, when capitalism and nation-states were becoming dominant in Europe, war became a state activity with the creation of professional armies and the elimination of private armies. Periods of interstate conflict became a temporary state of affairs, with long periods of peace between conflicts. Between the French Revolution (1789) and World War I (1914–1918), wars were waged between nation-states but were almost always territorial in nature. After 1918 conflicts became more ideological—variously between communism, fascism, and liberal democracy. Since 1945, most conflicts have been centered in the less developed world. Before 1945, most casualties were caused by global wars; after 1945, most casualties were caused by relatively local and civil wars.

The contemporary world contains many traditional hostilities. Most of these relate to ethnic rivalries (e.g., in Kenya, Iraq, India, and Pakistan), contending religions (e.g., in Sri Lanka and Nigeria), and/or competitions for territory and resources (e.g., between Sudan and South Sudan). Conflicts between countries have become increasingly formalized and structured over time as the number of independent states has increased. Since 1945, the United Nations has offered member states the opportunity to work together to avoid conflict, but not always with success.

Political struggles can found in numerous countries around the world today (Figure 9.7). Typically, these can be grouped in five categories: (1) traditional conflicts between states; (2) independence movements against foreign domination or occupation; (3) secession conflicts; (4) civil wars that aim to change regimes; and (5) military actions taken against states that support terrorism.

Conflicts between states since 1945 include three India–Pakistan wars (1949, 1965, 1971), four wars involving Arab states and Israel (1948, 1956, 1967, and 1973), the Vietnam War (1954–1975), and China's invasion of Vietnam (1979). Struggles for independence were an integral part of decolonization (Chapter 5). Examples include the conflicts in Indonesia (1946–1949), the Belgian Congo (1958–1960), Angola and Mozambique (1964–1974), and Algeria (1961). Conflicts over secession include the attempt to carve a state of Biafra out of Nigeria (1967–1970), Eritrea's independence from Ethiopia (1961–1991), and in 2011, the creation of the world's most recent state, South Sudan. Civil wars occurred in China (1945–1949), Bolivia (1967),

Figure 9.7: The geography of (ongoing) armed conflicts points to the instability of states around the developing world. Almost all wars today are within, not between, countries.

Colombia (1970s–2010), Lebanon (1975–1990), Angola (1975–2002), Sri Lanka (1983–2009), Yugoslavia (1989–1994), Rwanda (1994), Sierra Leone, Liberia, and Congo (1994–2011), Somalia (1990–2012), Libya (2011–2015), and Syria (2011–2015). Armed invasions against states suspected of harboring terrorism include the U.S.-led invasion of Afghanistan (2001–2013) and more recent drone attacks against suspected terrorist sites in Pakistan and Yemen.

CIVIL WARS

Although conflicts between states have the greatest potential to disrupt the human world, most wars today are civil wars—there were dozens of ongoing conflicts as of 2018—and most have been going on for years. Although the disruption caused by civil wars may be somewhat less than that caused by violent conflicts between states, there is good reason today to be especially concerned about them. Governments in the economically developed world often assume that such wars are rooted in ancestral ethnic and religious hatreds and that little can be done to prevent them (a view that led to international refusal to intervene in the 1994 genocide in Rwanda). Yet, this assumption is not necessarily valid. A principal cause of civil war is lack of development, in which different groups struggle over scarce resources, such as land or water. As countries develop economically, they tend to become progressively less likely to suffer violent conflict, and this in turn makes further development easier to achieve. By contrast, when efforts at development fail, a country usually is at high risk of civil war, which will further damage the economy, increasing the risk of further war. The fact that many civil wars are linked to natural resources, as discussed below, suggests that resources can be a curse rather than a blessing. Much national wealth that might be used to foster human development instead funds wars as groups within a country fight over access to oil, metals, minerals, and timber. Prime examples include Angola, where groups fought over oil and diamonds from 1975 until 2002, and Sudan, where fighting from 1983 to 2005 was related to oil, a struggle that continues today in its conflict with South Sudan.

Civil wars cause significant population displacement, mortality, and poverty among local populations. They have spillover effects on neighboring countries. For example, the civil war in Somalia spread into Ethiopia and Kenya; Rwanda's tribal conflict spilled over into Congo; Iraq's tensions between Sunnis and Shiites, and between Arabs and Kurds, has fostered unrest in Turkey. But they also have global impacts. For example, in a country engaged in civil war, some areas are impossible to control for illegal activities such as drug production or trafficking—most of the world's hard drugs are produced in countries experiencing civil wars. For example, Afghanistan is the world's leading producer of opium and heroin, and a long-running civil conflict in Colombia was centered in its coca-producing regions. Tragically, civil wars may be prolonged because a few people benefit financially from such activities. This is one of the reasons why a state might be appropriately described as failed.

FAILED STATES

It has become commonplace, at least since the end of the Cold War, for analysts to suggest that some countries have failed because they are either critically weak or no longer functioning effectively—they are either ungoverned or misgoverned (Figure 9.8). There are three types of state in the contemporary world; premodern states are those that have failed; modern states are governed effectively; postmodern states are those where national sovereignty is often being voluntarily dissolved, as in the member countries of the European Union, although this process is often resisted.

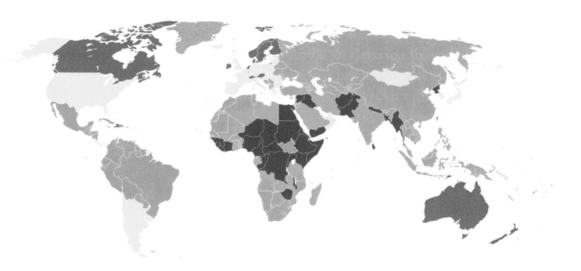

Figure 9.8: Failed states are those with dysfunctional governments incapable of providing their citizens with even basic public services. They are invariably very poor, typically highly corrupt, and frequently wracked by tribal conflicts or rebellion against the government. (Red: Alert, Orange: Warning, Yellow: Moderate, Blue: Sustainable, Grey: No information / Depending Territory)

Failed states are those incapable of providing even minimal levels of security, stability, and services for their populations. They pose both national and global problems. The global security implications of state failure are apparent. Such states may serve as safe havens for terrorists and illicit drug production (Afghanistan), generate numerous refugees (Congo), and/or allow pirates to operate freely in busy shipping lanes (Somalia).

DRUG PRODUCTION, MOVEMENT, AND CONFLICT

The international illegal drug trade is also a source of violence. Several Latin American countries are struggling to cope with drug production and movement. Large quantities of cocaine are produced in and regularly shipped from the Andean region, forcing peasant farmers from the land, prompting gang wars, and compromising state institutions (Figure 9.9). In Mexico roughly 60,000 people were killed in drug-related conflict between 2005 and 2016, some of it fueled by arms imported from the United States. Drug-related violence has also wracked Colombia (cocaine), Myanmar (Burma), Thailand, Pakistan, and Afghanistan, where opium production is widespread.

A recent development in drug trafficking is movement from Colombia in South America to West Africa and then to Europe. This route is favored because trafficking through the Caribbean has become more difficult as a result of more intensive policing. In West Africa, the small country of Guinea-Bissau might be labeled the world's first narco-state after the arrival of Colombian drug cartels in 2005. Already a failed state following a series of conflicts, Guinea-Bissau, with no prisons and few police, was a logical location for the storage of cocaine prior to shipment to Europe. Of course, the demand in Europe and North America fuels most of the production, movement, and related conflict.

TERRORISM

Terrorism is the use or threat of violence by a group against a state or other group, with the general goal of intimidation designed to achieve some specific political outcome. Difficulty in defining "terrorism" reflects a fundamental contradiction, namely, that many people labeled as terrorists by those whom they attack self-identify as freedom fighters and are considered as such by those whose interests they represent. The incidence of terrorism varies widely around the world (Figure 9.10). In recent decades, most terrorist

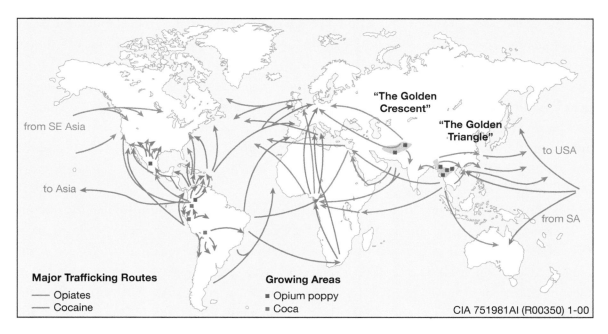

Figure 9.9: Like many industries, the drug trade has become globalized. The largest flows include shipments of cocaine from Latin America to the United States and Europe and heroin from Afghanistan to the West. "Narcodollars" flooding the producing countries have important economic and social effects, including increased crime and drug addiction. The geopolitics of illegal drugs demonstrates how globalization and the nation-state are intertwined.

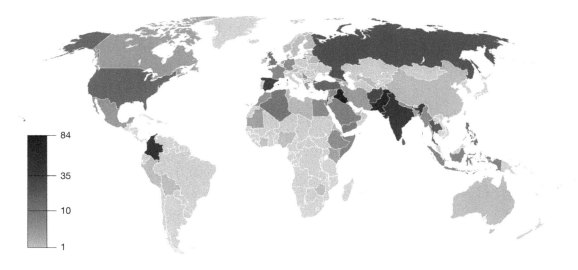

Figure 9.10: The map of terrorist incidents fluctuates over time, but reflects a variety of global processes (e.g., Muslim fundamentalism) and national policies. In the map, darker areas note increased terrorist incidents.

activities have been a response to national policies. For example, the Irish Republican Army (IRA) conducted a campaign against British rule in Northern Ireland, various Palestinian groups continue to fight against Israeli occupation of Palestine and the West Bank, and ETA (*Euskadi ta Askatasuna*, meaning "Basque Fatherland and Liberty") has fought for an independent Basque homeland in northern Spain and southwest France.

Many commentators consider that the nature and goals of terrorism changed with the 2001 attacks on the United States by the militant Islamic group Al-Qaeda. These attacks and the organization behind them are clearly of a different order than previous terrorist activities. In 1988 Al-Qaeda was founded by Osama bin Laden (who was killed by the U.S. forces in 2011) in Afghanistan. The organization has links with many other groups worldwide and is responsible for numerous attacks around the world. Of these, the most devastating, both in terms of the numbers killed and in terms of the response, were the September 11, 2001, attacks in the United States, which killed roughly 3,000 people. More recently, several groups have arisen that are affiliated with Al-Qaeda, such as Boko Haram in Nigeria, Al-Shabaab in Somalia, Lashkar e-Taiba in Pakistan, and Al-Qaeda in Mesopotamia (i.e., Iraq). In Iraq and Syria, the Islamic State (ISIS) has attempted to turn a broad area into a Muslim caliphate governed by sharia law (Chapter 4).

Much of the immediate reaction to these attacks, employing the "clash of civilizations" terminology, was couched in the context of the long history of conflict between Islam and Christianity. Some critics, however, find that view superficial and argue that "Islamic" terrorism is the work of individuals and small extremist groups—not representative of most Muslims. Certainly, organizations like Al-Qaeda and the Islamic State do not reflect Muslims as a whole, who form the vast bulk of their victims, and it is a serious error to associate terrorism with Islam in general.

Coping with terrorist organizations is especially difficult because—unlike states—they are not tied spatially to the political landscape. This does not mean, however, that they function in a geographic vacuum. Most terrorist groups are supported, at least informally, by states—hence the U.S.-initiated invasion of Afghanistan following the 2001 attacks. The United States also invaded Iraq in 2003 under the banner of combating terrorism or preventing it from developing nuclear or chemical weapons, although no linkages between Iraqi leader Saddam Hussein's government and Al-Qaeda were ever proven, nor were any weapons of mass destruction discovered. Certainly, the recent upsurge of extremist Islamic terrorism has dramatically changed global geopolitics, and it is not surprising that some commentators see this as a principal feature of the world today and one likely to continue for the foreseeable future, especially as many analysts in the United States and throughout the West point out that the American "War on Terror" in fact has created terrorists and the environment conducive to terrorism.

THE GEOGRAPHY OF NUCLEAR WEAPONS

As of 2017, there are nine official nuclear powers: the United States, Russia, China, France, Britain, India, Pakistan, and North Korea. Israel is widely known to possess nuclear bombs but does not officially admit to having them. South Africa developed nuclear bombs but gave them up with the collapse of the apartheid regime in 1994. There is much concern today about the nuclear intentions of North Korea and Iran (Figure 9.11). Algeria, Saudi Arabia, and Syria are suspected of having nuclear intentions, but no weapons programs have been identified. Two countries, Iraq and Libya, recently ended their nuclear programs, while Belarus, Kazakhstan, and Ukraine all scrapped the weapons inherited from their inclusion in the former USSR.

Nuclear power is an increasingly important source of energy. Unfortunately, however, it is also the first step along the road to the production of nuclear weapons. Thus, as more countries develop the technology for nuclear power, it seems likely that the world will confront more crises about the nuclear weapon intentions of those countries. This issue has been at the heart of the international community's attempts to limit or stop Iran's development of enriched uranium

Figure 9.11: Seven countries are officially known to possess nuclear weapons: the United States, Russia, China, Britain, France, India, and Pakistan. Israel is widely thought to have perhaps a hundred nuclear bombs, and North Korea has a few. Several others, including Iran, are suspected of attempting to construct them. (Red: Major nuclear weapons power, Orange: Average nuclear weapons power, Blue: Low nuclear weapons power, Black: Presumed nuclear weapons ability)

and the capacity to produce nuclear bombs. Of course, it is important to ask: Do countries that currently have nuclear weapons have any moral authority to say that other countries cannot have those weapons?

Nuclear war would mean national suicide for any country involved, and a large-scale nuclear war would affect all environments and peoples. Large areas of the northern hemisphere would likely experience subzero temperatures for several months, regardless of the season, in a scenario known as "nuclear winter." Low temperatures and reductions in sunlight would adversely affect agricultural productivity and lead to mass starvation. A nuclear winter would result in many deaths from hypothermia as well.

9.6 ELECTORAL GEOGRAPHY

In addition to analyses of relationships among states in the world, political geography is also concerned with how political processes play out within them, particularly in democratic societies in which the populace selects its leaders. Because democracy is a relatively recent historical product, emerging only after the Enlightenment in the eighteenth century (Chapter 5), this branch of political geography is concerned with more contemporary events than, for example, world-systems theory.

Electoral geography is the analysis of the geographic patterns of elections, referenda, and legislative votes. Electoral geographers have studied the spatiality of elections at the national and local levels; redistricting and gerrymandering (discussed below); shifts in voter preferences, turnout rates, and correlations with various socioeconomic variables; and neighborhood effects on political behavior. Traditional electoral geography focused on such spatial patterns of votes. More recently, the field has also included other concerns such as how elections reflect global processes (e.g., the worldwide spread of democracy), changes in voting technologies, and how elections are intertwined with the spatial dynamics of political life such as class, ethnicity, and gender.

Producing a map of election results is the most basic technique in electoral geography. These maps reveal the relative strength of support for candidates, parties, or referenda in different regions, and they are now often a prominent feature of election coverage in the popular media. The map of the 2012 presidential election in the United States, for example, shows Republican support stretching across a vast area of the country, whereas Democratic strength was concentrated in the Northeast, the West Coast, border regions, and scattered counties in the Midwest (Figure 9.12).

Spatial patterns of voting can trace the emergence and evolution of political parties. When old parties decline and new ones appear, electoral geographers trace the way in which voters

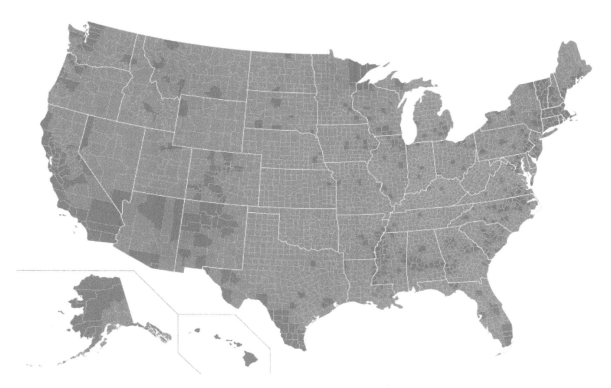

Figure 9.12: Electoral politics is an important part of political geography. This map of the 2016 presidential contest reveals that Democratic candidate Hillary Clinton carried densely populated counties with large cities, giving her the majority of the popular vote, while Republican candidate Donald Trump carried less populated counties in rural areas. Riding a wave of populist anger against elites and globalization (Chapter 6), Trump carried many blue-collar areas in the industrialized Midwest that had formerly leaned Democratic.

transfer membership and loyalty, or in some cases, discover which voters display little party loyalty and are willing to switch between elections. Such transitions are typically tied to changes in regional economies, demographic and migration trends, and regionally based public investment.

REDISTRICTING

The practice of **redistricting** is an important application of electoral geography. In political systems where candidates run for office in a specific geographic constituency, the location and pattern of voters are

Figure 9.13: Gerrymandering is the practice of drawing political boundaries to create districts that heavily favor one political party or candidate. This famous cartoon stems from a Massachusetts state senate race in 1812 in which critics argued that the districts favored the party of Governor Gerry.

often of concern for the organization responsible for drawing new districts. Political parties attempt to use redistricting to distribute their likely voters so as to create as many districts as possible in which the party has a majority or plurality. Redistricting is the process of changing the boundaries of electoral districts to maintain equal populations. It is a critical part of the maintenance of constituency-based systems, especially those with single-member districts.

In international contexts, redistricting is often called boundary delimitation, and in commonwealth countries, redistribution. (Redistricting should not be confused with reapportionment, the redistribution of congressional seats among states following the census every 10 years.) Manipulation of redistricting for electoral advantage is called **gerrymandering** (Figure 9.13). The party in power in each state attempts to redraw congressional districts to maximize its advantage, sometimes leading to districts with bizarre shapes (Figure 9.14). Occasionally gerrymandering can assume a racial dimension, as when it is used to dilute the voting power of African Americans. Today, using detailed spatial data and Geographic Information Systems, districts can be artfully and precisely gerrymandered to maximize the chances of reelection. This is one of several factors that reinforce the advantage of incumbents.

The way district boundaries are drawn can have a far-reaching impact on the distribution of political power. Placing large numbers of one group of voters in a single district and spreading them out over many districts are just two of the strategies used by politicians who are intent on tilting the balance in favor of one side or the other. Bias in drawing electoral district boundaries can undermine the basic fairness of an electoral system. Notably, careful redistricting has led to the creation of many safe congressional seats in which incumbency all but assures victory. The creation of safe districts—a usual result of gerrymandering—is

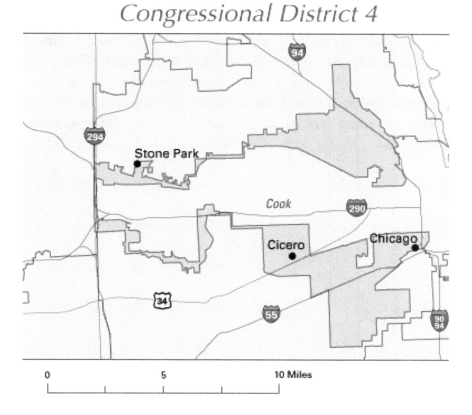

Figure 9.14: Gerrymandering today is very sophisticated, using Geographical Information Systems that draw boundaries to protect incumbents from serious challenges to reelection. For example, the Fourth Congressional District in Illinois illustrates the bizarre shape that gerrymandered districts can assume.

often blamed in the United States for contributing to the polarization of the two main parties.

After the U.S. constitutional requirement for equal population size among districts, redistricting in the United States is driven by the intersecting interests of partisanship and race. It is also heavily litigated in the courts. Although a few states involve nonpartisan commissions in their redistricting process, most allow the party in control of the state legislature and the governor's mansion to use the redistricting process to boost their political fortunes. The U.S. Supreme Court has thus far refused to intervene although it has not completely closed the door on the possibility that a partisan gerrymander might violate the Constitution. Judges have been much more active, however, in redistricting cases in which racial or ethnic minority groups are alleged to have been disadvantaged. The Voting Rights Act, originally passed in 1965 and renewed repeatedly by Congress, allows the federal government and private plaintiffs to challenge state and local redistricting plans that they allege dilute the ability of minority voters to elect candidates of their choice. Many states of the former Confederacy, for example, must have their congressional boundaries approved by the U.S. Department of Justice to prevent attempts at racialized gerrymandering designed to minimize the representation of minority voters.

■ KEY TERMS

Civic nationalism: nationalist ideology that is not focused on a particular ethnic identity, such as that found in the United States

Classical geopolitics: various theories about international relations popular in the early twentieth century that viewed states as similar to organisms competing in an ecosystem, thus inevitably leading to conflict

Critical geopolitics: a contemporary form of geopolitics concerned about the ideological and representational nature of states as constituted through discourse and ideology

Electoral geography: the branch of political geography that examines the spatial dimensions of elections and voting

Failed states: countries that are unable to provide even minimal government services for their citizens, including security

Geopolitics: a branch of political geography concerned with international relations among states

Gerrymandering: the process of drawing congressional districts designed to amplify the power of one political party, sometimes with the result of discrimination against minorities

Heartland theory: put forth by geopolitician Halford Mackinder, a view of international relations that focused on the region of eastern Europe and western Russia as key to controlling the "world island," or Eurasian landmass

Hegemony: in general, a synonym for dominance; in world systems theory, the unique advantages enjoyed by the dominant superpower at a given historical moment

Nation: a group of people with shared culture, ethnicity, history, territory, and sense of collective identity

Nationalism: the ideology that upholds the nation and nation-state as the proper and appropriate vehicle for the expression of national pride and power

Nation-state: a nation that enjoys independence and sovereignty, that is, it has its own state

Redistricting: the process of redrawing congressional boundaries following the U.S. census every 10 years

Sovereignty: independence and self-control over a state's affairs

Terrorism: the systematic use of violence against noncombatants to achieve political goals

Westphalian system: the model of international affairs based on independent nation-states that emerged after the Treaty of Westphalia in 1648

World systems theory: a model of international relations that focuses on the world as a whole so that no part can be viewed in isolation from others. It holds the world is constituted through a single world economy and numerous states.

■ STUDY QUESTIONS

1. Are nations exclusively European or Western creations?

2. What were the links between capitalism and the rise of the nation-state?

3. What was the Treaty of Westphalia and why was it so important?

4. What is the difference between a nation and a nation-state?

5. Does nationalism overlook or celebrate ethnic diversity? Why or why not?

6. Contrast civic and ethnic nationalism.

7. Can relations among states be reduced to a "survival of the fittest," as implied by classical geopolitics?

8. How does nationalism produce nationalists? Who benefits from this process?

9. What is the difference between world empires and the world system?

10. In world systems theory, what are the core, semiperiphery, and periphery?

11. What have been some dominant powers throughout the history of capitalism? What new power appears to be lurking on the horizon?

12. What are five reasons for wars among and within states?

13. Where are most civil conflicts located today? Why?

14. What are the linkages between the global drug trade and political conflicts?

15. Is terrorism qualitatively different from state-led warfare, or is it simply the extension of politics by another means?

16. What are some examples of resource-fueled conflicts among or within states?

17. How does redistricting affect the borders of congressional districts? Who stands to benefit?

BIBLIOGRAPHY

Agnew, J., & Corbridge, S. (1995). Mastering Space: Hegemony, Territory and International Political Economy. *New York: Routledge.*

Agnew, J., Mitchell, K., & Ó Tuathail, G. (Eds.). (2003). A Companion to Political Geography. *Oxford: Blackwell.*

Anderson, B. (1983). Imagined Communities: On the Origins and Spread of Nationalism. *London: Verso.*

Archibugi, D. (2008). The Global Commonwealth of Citizens: Toward Cosmopolitan Democracy. *Princeton, NJ: Princeton University Press.*

Fowler, R., Hertzke, A., Olson, L., & den Dulk, K. (2004). Religion and Politics in America: Faith, Culture, and Strategic Choices. *Boulder: Westview Press.*

Frank, T. (2004). What's the Matter with Kansas? How Conservatives Won the Heart of America. *New York: Holt.*

Giddens, A. (1987). The Nation-State and Violence. *Berkeley: University of California Press.*

Gilpin, R. (1987). The Political Economy of International Relations. *Princeton, NJ: Princeton University Press.*

Hobsbawm, E. (1990). Nations and Nationalism Since 1780. *Cambridge: Cambridge University Press.*

Modelski, G. (1978). The long cycle of global politics and the nation-state. Comparative Studies in Society and History, 20, *214–235.*

Ó Tuathail, G. (1996). Critical Geopolitics. *Minneapolis: University of Minnesota Press.*

Ruggie, J. (1993). Territoriality and beyond: Problematizing modernity in international relations. International Organization, 47, *139–174.*

Shannon, T. (1996). An Introduction to the World-System Perspective *(2nd ed.). Boulder: Westview Press*

Wallerstein, I. (1979). The Capitalist World-Economy. *Cambridge: Cambridge University Press.*

Warf, B., & Leib, J. (Eds.). (2011). Revitalizing Electoral Geography. *London: Ashgate.*

IMAGE CREDITS

- Figure 9.1a: Copyright © Ebrahimi-amir (CC BY-SA 3.0) at http://commons.wikimedia.org/wiki/File:Kurd_hafeznia.jpg.
- Figure 9.1b: Copyright © ASDFGH (CC by 3.0) at http://commons.wikimedia.org/wiki/File:Tibet_CN.png.

- Figure 9.1c: Copyright © Oncenawhile (CC BY-SA 3.0) at http://commons.wikimedia.org/wiki/File:Palestinian_Territories,_1948-67.svg.
- Figure 9.1d: Copyright © UKPhoenix79 (CC BY-SA 3.0) at http://commons.wikimedia.org/wiki/File:Uk_map_scotland.png.
- Figure 9.1e: Nafar, http://commons.wikimedia.org/wiki/File:Basque_language_location_map.png. Copyright in the Public Domain.
- Figure 9.2: National Intelligence Council, "Ethnic Diversity 2000," http://www.lib.utexas.edu/maps/world_maps/ethnic_diversity_2000.jpg. Copyright in the Public Domain.
- Figure 9.3: Copyright © Roke (CC BY-SA 3.0) at http://commons.wikimedia.org/wiki/File:Europe_map_1648.PNG.
- Figure 9.4a: Copyright © Egon Tintse (CC BY-SA 2.0) at https://commons.wikimedia.org/wiki/Eesti#/media/File:XVIII_tantsupidu_(2)_(1).jpg.
- Figure 9.4b: Copyright © Yuri Palamarchuk (CC BY-SA 3.0) at https://commons.wikimedia.org/wiki/File:Antijapanese_demonstration4.jpg.
- Figure 9.4c: Copyright © Lemsipmatt (CC BY-SA 2.0) at https://www.flickr.com/photos/lemsipmatt/3539292936.
- Figure 9.5: Copyright © Arnopeters (CC BY-SA 3.0) at http://commons.wikimedia.org/wiki/File:Map_Geopolitic_Mackinder.gif.
- Figure 9.6: Naboc1, http://commons.wikimedia.org/wiki/File:World_trade_map.PNG. Copyright in the Public Domain.
- Figure 9.7: Eszett, http://commons.wikimedia.org/wiki/File:Map_of_sites_of_ongoing_armed_conflicts_worldwide.PNG. Copyright in the Public Domain.
- Figure 9.8: Copyright © Tiiliskivi / Quintucket (CC BY-SA 3.0) at http://commons.wikimedia.org/wiki/File:Failed-states-index-2012.png.
- Figure 9.9: http://commons.wikimedia.org/wiki/File:Drugroutemap.gif. Copyright in the Public Domain.
- Figure 9.10: Emilfaro, http://commons.wikimedia.org/wiki/File:Number_of_Terrorist_Incidents.png. Copyright in the Public Domain.
- Figure 9.11: Copyright © Lokal_Profil (CC BY-SA 2.5) at http://commons.wikimedia.org/wiki/File:Localisation_nuclear_weapons.svg.
- Figure 9.12: Copyright © Ali Zifan (CC BY-SA 4.0) at https://commons.wikimedia.org/wiki/File:2016_Presidential_Election_by_County.svg.
- Figure 9.13: Elkanah Tisdale, "The Gerry-Mander Edit," https://commons.wikimedia.org/wiki/File:The_Gerry-Mander_Edit.png. Copyright in the Public Domain.
- Figure 9.14: National Atlas of the United States, "Illinois District 4 2004," https://commons.wikimedia.org/wiki/File:Illinois_District_4_2004.png. Copyright in the Public Domain.

Chapter Ten

INDUSTRIAL AGRICULTURE

MAIN POINTS

1. Industrial agriculture is usually synonymous with commercial farming, or agribusiness, which entails production for sale on the market, large quantities of machinery, little labor, and monoculture crop yields.

2. U.S. government agricultural policy is largely a reflection of the political influence of agribusiness. It includes subsidies to farmers to compensate for low market prices, often leading to large surpluses of grains and other foods.

3. Unlike commercial agriculture, sustainable agriculture attempts to achieve an ecological balance, with minimal inputs of energy, water, and chemicals and the lowest possible impacts on the environment.

4. Because food is not simply an economic phenomenon but also very much a social and cultural one, food consumption is closely linked to people's self-image, status, and identity.

Food is hugely geographical. Next time that you eat at a fast-food place, think about where your food comes from. The meat likely originated in Kansas, Australia, or Costa Rica. The burger buns were created out of processed wheat flour grown in North Dakota. The fries were made from genetically engineered potatoes raised in Idaho. The ketchup was an amalgam of tomatoes from California; vinegar, salt, and onions from various locations; and high-fructose corn syrup from corn grown in Iowa or Illinois. The soft drinks were sweetened with government-subsidized sugar grown in Louisiana and Florida. The whole fast-food experience, which started after World War II, ties agribusiness firms, food-processing wholesalers, franchised retailers such as McDonald's, and consumers together in a complex web of geographic relations. Knowing this, you may look at your meal in a new light: The simple act of eating a meal embeds you in networks that dissolve the boundaries between your personal life and the broader world.

Agriculture is intimately bound up with the cultural evolution of human beings since the Neolithic Revolution (Chapter 2). Several types of preindustrial agriculture, which rely on animate sources of energy, emerged in different parts of the world and continue to be practiced today, largely on a subsistence basis. In the nineteenth and twentieth centuries, as part of the Industrial Revolution (Chapter 5), an agricultural revolution took place. Today, industrial agriculture dominates agricultural activities in most developed countries, and many developing countries are now adopting it. In contrast to preindustrial agriculture, industrial agriculture involves extreme capital intensity (use of machinery and equipment instead of labor) and high levels of energy use. The high degree of mechanization is evident in the enormous farms found throughout the Midwest, in which thousands of acres can be used for crops but workers are almost absent, except for the occasional operator of a combine or tractor. Although industrial agriculture has dramatically increased the amount of food produced by a given amount of land and labor, it has also depleted water and soil resources, polluted the environment, and destroyed a way of life for millions of farm families.

The industrialization of agriculture drastically reduced the number of farmers in North America. In the United States, the number of farmers declined from 7 million in 1935 to around 1.7 million in 2012. In Canada, 600,000 farm operators existed in 1951 but less than one-third that number were still in operation in 2010. Europe witnessed similar trends. In Britain, for example, a decline in the number of farm workers has occurred for decades. Today, the percentage of a labor force engaged in agriculture is a useful measure of economic development around the world (Chapter 7). In parts of the developing world, several types of preindustrial agricultural systems can be found, all of which are labor-intensive (Figure 10.1). Thus, significant proportions of workers are engaged in farming. In most of Africa and East Asia, for example, more than 60% of the employed population works as peasant farmers. In the developed world, relatively few workers (e.g., 2%) engage in agriculture.

Figure 10.1: Types of agricultural systems in the world. Industrialized agriculture predominates in the temperate regions of the developed world but is growing in the developing world.

The emergence of agriculture and its subsequent spread throughout the world has meant that little, if any, land still can be considered "natural" or untouched. Almost everywhere, nature has been modified extensively by human beings, making it difficult to speak of nature as a phenomenon separate from human impacts (Chapter 1). Agriculture is perhaps the most widespread and visible way in which humans have transformed the planet's environments, a process that extends back to the Neolithic Revolution. In addition to industrialized agriculture, millions of people still rely on subsistence methods such as slash-and-burn (Chapter 2), which require much more space per person. Virtually all of the world's major vegetation zones show signs of extensive clearing and burning, as well as the grazing of domestic animals. The destruction of vegetative ground cover has led to the creation of productive agricultural and pastoral landscapes, but it has also led to reduced land capability.

In general, industrialized farming practices pose more danger to the environment than preindustrial farming. **Monocultures** (the cultivation or growth of a single crop) and the breaking down of soil structure by huge machines are a few factors that may destroy the topsoil. Repeated droughts and dust storms in the twentieth century on the

Great Plains of the United States gave testimony of how nature and industrial agriculture can combine to destroy the health of grasslands. Industrial agriculture reflects the emphasis under capitalism to maximize profits. Corporate producers want to make land use more efficient and productive; thus, farming is often viewed as just another industry. However, we must remember that land is more than a means to an end: it is finite, spatially fixed, and ecologically fragile.

10.1 COMMERCIALIZED AGRICULTURE

Commercial agriculture produces crops and livestock for sale off the farm rather than for immediate use by the farmer (as in **subsistence agriculture**). Regions or nations that use commercial agriculture share several major features:

1. Only a small proportion of the population is engaged in agriculture. Populations fed by commercial agriculture are urban populations engaged in other types of economic activity, such as manufacturing, services, and information processing.
2. Machinery, fertilizers, and high-yielding seeds are used extensively, with high-energy inputs such as petroleum. Output per unit of labor is very high, but output per unit of energy input is very low.
3. Farms are extremely large, and the trend is toward even larger ones.

Commercial agricultural areas include the United States and Canada, Argentina and portions of Brazil, Chile, Europe, Russia and Central Asia, South Africa, Australia, New Zealand, and portions of China.

Agriculture in the United States epitomizes the contemporary capitalist system of food production. The American agricultural system developed in the nineteenth century as European farmers settled the Midwest and West. Railroads and steamships dramatically lowered transport costs to the markets along the East Coast, and agricultural trading centers and ports, such as Chicago, St. Louis, and New Orleans grew rapidly. By the turn of the century, the United States had become a major supplier of wheat and other agricultural products to Europe. The only other major producer, Russia, effectively withdrew from world markets following the Bolshevik revolution in 1917. Consequently, the vast agricultural region stretching across the Ohio and Mississippi River basins into the Great Plains, and extending into central Canada, became the core of the North American food-producing system that fed other parts of the world.

Today, a handful of large firms dominate the tremendously productive U.S. agriculture industry. Giant food companies such as ConAgra, Bunge, Cargill, Dole, Nabisco, General Mills, General

Foods, Hunt-Wesson, Archer Daniels Midland, and United Brands control the whole food chain from "seedling to supermarket." These companies dictate what crops will be grown and sell the seeds to farmers. Whereas the popular imagination clings to the stereotype of the small family farmer, in reality most American agriculture is organized around the needs of a handful of large firms. Families still own and operate most farms, but the firms mentioned above control the food production and processing (e.g., slaughter, canning), distribution to retailers (e.g., supermarkets), price and cost information, and marketing. Most farmers have little power in the agricultural production process and are often left to take the risks (e.g., droughts) while corporations reap the profits.

Agribusiness is extremely capital-intensive and energy-intensive. Farmers rely on copious quantities of chemical fertilizers and pesticides, tractors, harvesters, and other equipment—most of it very sophisticated, computer-controlled, and expensive (Figure 10.2). This equipment requires a large amount of oil and other fossil fuels. Because

Figure 10.2: Industrialized agriculture utilizes large quantities of equipment, but very little labor. Although very efficient, this system also relies heavily on petroleum for fertilizers and farm machinery. It is thus very energy-intensive.

of these expenses, farms must be large to be profitable. In this way, farmers keep human labor low and productivity high. Agriculture employs only 2% of the U.S. labor force, and it not only feeds the other 98% of the populace but exports vast amounts of food as well. The very high per capita productivity has led people to migrate from rural to urban areas in pursuit of work. The use of tractors worldwide is a measure of the capital-intensity of agricultural production: Countries with largely commercial agricultural systems rely extensively on this technology, freeing people from the farm. In contrast, developing nations that employ large numbers of people in subsistence farming usually tend to lack tractors.

Industrial agriculture is also practiced in the form of market gardening, which is sometimes called **truck farming**. Crops are grown in relatively small, intensively cultivated space close to metropolitan areas. Industrial food-production truck farms specialize in intensively cultivated fruits and vegetables (those that use lots of inputs and generate high outputs per unit area), and they depend on migratory seasonal farm laborers to harvest their crops. California is the largest center of fruit and vegetable farms in the United States, although such farms are widespread in Florida as well.

Agribusiness has also extended livestock farming immensely, including the mechanized raising and slaughter of cattle. Other examples of modern food production include poultry ranches and egg factories. These animals tend to live and die under inhuman conditions, including being confined in extremely close quarters with insufficient room to move, which also makes them susceptible to disease. At one time, livestock farmers raised combinations of crops and animals on the same farm. In recent decades, livestock farming has become highly specialized and separated from crop farming. Factory-like feedlots raise thousands of cattle and hogs on purchased feed and generate huge quantities of animal waste. Feedlots exist mainly in the western and southern states, in part because of the mild winters. These feedlots raise more than 60% of the beef cattle in the United States.

As an industry, corporate agriculture resembles the production of other goods such as cars. As discussed, agriculture's major inputs include petroleum and machinery, with labor only a small part of the production costs. The sales of this sector include the food-processing sector (which transforms crops into foods that consumers can eat) and meat production; a large share of cereals, especially corn, becomes animal feed.

Modern American farming responds quickly to new developments, such as new production techniques. Consequently, farmers with sizable investments of money, materials, and energy can create drastic changes in land-use patterns. For example, farmers in the low-rainfall areas of the western United States have converted large areas of grazing land to forage and grass production with the use of center-pivot irrigation systems, in which a centralized source sprays water in a large circle. Other western farmers grow sugar beets and potatoes through federally subsidized irrigation projects.

American corporate farming is becoming a worldwide food-system model. Poultry-raising operations in Argentina, Pakistan, Thailand, and Taiwan increasingly resemble those in Alabama and Maryland. Enormous, politically connected enterprises such as United Brands, Del Monte, Archer Daniels Midland, and Unilever divert food production in developing countries toward consumers in developed countries. Often, demand exists in developed countries for products that are out of season or that cannot be grown locally.

The percentage of laborers in developed countries working in commercial agriculture is less than 5% overall. In contrast, some portions of the developing world that practice intensive subsistence agriculture (such as swidden or Asian rice paddy farming; see Chapter 2) employ 90% of the population in farming. On average, 60% of the populace in developing nations works as farmers. Today, U.S. farmers on average produce enough food for themselves and 70 other families.

In 2015, the United States had approximately 2.1 million farms, compared with 5.7 million in 1950 (Figure 10.3a). This reduction in the number of farm families comes from waves of corporate consolidation and mergers, leading to a growth in the average size of farms (Figure 10.3b). Factors that have driven families off farms include the high cost of equipment, low prices, or high interest rates. Farming is difficult, often dangerous work. Farmers often borrow money for seed at the beginning of the growing season and repay the bank after the harvest and sale of the crops. Therefore, low crop prices can be ruinous. Meanwhile, opportunities for college education and higher-paying occupations in the cities have long lured farm children off the land. The decline in farms also stems from the encroachment of the metropolitan area onto farmland. Suburban sprawl has replaced rich topsoil and farmland with housing subdivisions and shopping centers. Brought by interstate highways, sprawl and rural-to-urban land conversion surround many metropolitan areas in the United States and Europe.

Although it employs relatively few people to produce a large amount of products, commercial agriculture relies heavily on expensive tractors, combines, trucks, diesel pumps, and other heavy farm equipment. All of this equipment needs fuel in the form of petroleum and gasoline. In addition, farmers purchase miracle seeds that are hardier than their predecessors and that produce more impressive tonnages. Commercial agriculture is also fertilizer intense.

Improvements in transportation to the market have resulted in less spoilage. Products arrive at the canning and food-processing centers more rapidly than they did previously. By 1850, many American farms were well connected to cities by rail transportation, which allowed the output of the midwestern Great Plains to be shipped to the markets in Chicago and the East Coast. In the mid-nineteenth century, long cattle drives connected cattle-fattening areas in

Figure 10.3a: Number of U.S. farms.

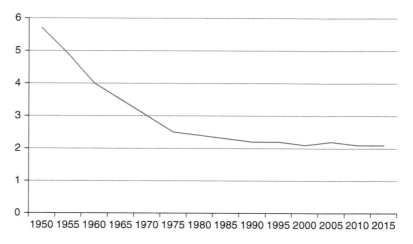

Figure 10.3b: Size of U.S. farms (acres).

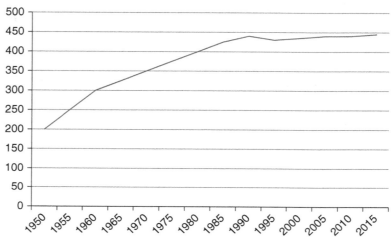

Figure 10.3ab: As farming in the United States has become increasingly consolidated, the average size of farms has increased and the number has dropped. This has led to the decline of the traditional family farm.

Texas, Oklahoma, and Colorado with the Union-Pacific rail line stretching from St. Louis to Kansas City to Denver. More recently, trucks have supplanted rail transportation, and the refrigerator car and truck preserve freshness. Cattle also arrive at packing houses by truck as fat as when they left the farm.

Every state now has agricultural research stations, usually affiliated with public land-grant universities (those that started out with agricultural schools and originally emphasized the teaching of farming and livestock raising). These stations have greatly improved agricultural techniques with in improved fertilizers and hybrid plant seeds, hardier animal breeds, and better insecticides and herbicides. In addition, local and state government farm advisors can provide information about the latest techniques, innovations, and prices so that the farmer can make wise decisions concerning what, when, and how much to produce.

TYPES OF COMMERCIAL AGRICULTURE

Commercial agriculture falls into six main categories: mixed crop and livestock farming, dairy farming, grain farming, cattle ranching, Mediterranean cropping, and horticulture and fruit farming.

Mixed crop and livestock farming is the principal type of commercial agriculture. It appears in Europe, Russia, Ukraine, North America, South Africa, Argentina, Australia, and New Zealand. In mixed crop and livestock farming, the main source of revenue is livestock, especially beef cattle and hogs. Other income may come from milk, eggs, veal, and poultry. Although the majority of farmlands are devoted to the production of crops such as corn, most of the crops are fed to the cattle. Cattle fattening intensifies the value of agricultural products and reduces bulk (the farmer raises fewer, fatter cows). Because of the developed world's preference for meat as a major food source, mixed crop and livestock farmers have fared well during the past 100 years.

In developed nations, the livestock farmer maintains soil fertility by using a system of crop rotation in which different crops are planted in successive years. Each type of crop adds different nutrients to the soil. The fields become more efficient and naturally replenish themselves with these nutrients. Farmers today use the four-field rotation system, wherein one field grows a cereal, the second field grows a root crop, the third field grows clover as forage for animals, and the fourth field lies fallow, resting the soil for that year.

Most cropping systems in the United States rely on corn because it is the most efficient food source for fattening cattle. American corn is highly subsidized by the federal government, largely owing to the political clout of agribusiness. Thus, farmers produce in much larger volumes of corn for lower prices than would be the case if it were a "free market." Most corn goes to feed cattle or hogs, even though cows evolved to eat grass, not corn, which may cause problems with their digestive and immune systems. The population consumes corn in the form of corn on the cob, corn oil, or margarine. Corn appears in corn sweeteners in a large variety of processed foods, including soft drinks and ketchup. The second most important crop in mixed crop and livestock farming-regions of central North America and the eastern Great Plains is the soybean. The soybean has more than 100 uses (e.g., soy oil, which is widely used in processed foods), but most soybeans are used for animal feed. In China and Japan, tofu from soybean milk represents a major food source high in protein and low in fat.

Dairy farming accounts for the most farm acreage in the northeastern United States and northwestern Europe and accounts for 20% of the total output by value of commercial agriculture. Ninety percent of the world's milk supply is produced in these areas. Most milk is consumed locally because of its weight and perishability. Some dairy farms produce butter and cheese as well as milk. In general, the farther the farm is from an urban area, the more expensive the transportation of fluid milk, and the greater proportion of production in high-value-added (i.e., profitable) commodities such as cheese and butter. For example, the Swiss discovered ways of transforming their milk products into chocolates, cheeses, and spreads that are distributed worldwide. These processed products are not only lighter but also less perishable. On the other hand, in the United States, most farms produce liquid milk because of their proximity to major cities including Boston, New York, Philadelphia, Baltimore, and Washington, DC, on the East Coast; and Chicago and Los Angeles in the Midwest and West. Farms throughout the rest of the United States primarily produce butter and cheese. Worldwide, remote locations such as New Zealand devote three-fourths of their dairy farms to cheese and butter production. In contrast, three-fourths of the farms in Britain produce fluid milk because of the much higher population density and thus markets at close proximities.

Dairy farms are relatively labor-intensive because cows must be milked twice a day. Most of this milking is done with automatic milking machines. However, the cows still must be herded into the barn and washed, the milking machines must be attached and disassembled, and the cows must be herded back out and fed. The difficulty for the dairy farmer is to keep the cows milked and fed during winter, when forage is not readily available and must be stored.

Commercial grain farms usually appear in drier territories not conducive to dairy farming or mixed crop and livestock farming. Most grain, unlike the products of mixed crop and livestock farms, is produced for sale directly to consumers. Only a few places in the world can support large grain-farming operations. These areas include China, the United States, Canada, Russia, Ukraine, Argentina, and Australia. Important grains include wheat, barley, oats, rye, and sorghum. These grains are not particularly perishable and can be shipped long distances. Wheat is the most important crop for world food production, as the primary grain used to make flour and bread (Figure 10.4). Global wheat yields increased markedly between 1970 and 2010, whereas cropland area has increased only slightly. However, production per capita is much more disappointing in developing countries, particularly Africa. Nonetheless, the world's food-producing system, however constrained and imperfect, has allowed the global food supply to keep pace with world population increases. In this way it has denied, or at least forestalled the Malthusian prophecy (Chapter 8).

Wheat constitutes the leading international agricultural commodity transported among nations. The United States and Canada lead as export nations for grains and together account for

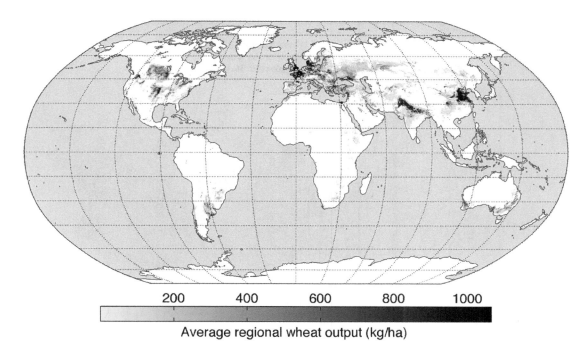

Figure 10.4: Wheat is the major crop produced by industrialized agricultural systems (but second to rice in terms of total world output). Wheat production is concentrated in the midlatitude grasslands common in North America, Australia, and Europe, where it is grown and harvested mechanically.

50% of wheat exports worldwide. The North American wheat-producing areas have been called the world's breadbasket. They provide the major source of food to many areas, including several hungry countries in Africa. The United States stands as the world's major agricultural superpower and plays a unique role in the global food production system. American agricultural exports generate more than $110 billion in export revenues annually, and the United States exports 20% of all food traded internationally. As with other economic sectors, agriculture has become thoroughly globalized. Farmers in Nebraska and Kansas know that next year's revenues will depend on weather and political events in European and Asian markets.

In the United States, the Winter Wheat Belt is west of the mixed crop and livestock farming area of the Midwest and is centered in Wyoming, Kansas, and Oklahoma (Figure 10.5). Major cities in the Great Plains, such as Minneapolis and St. Paul, were often established

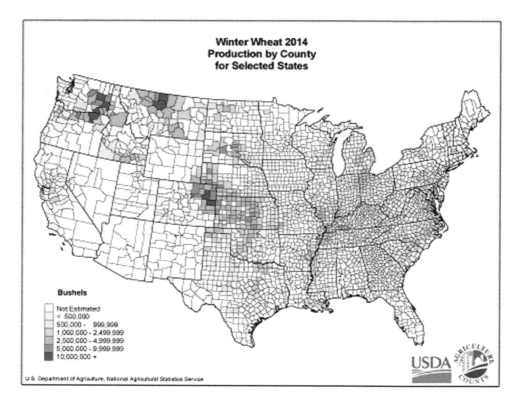

Figure 10.5: The map of U.S. wheat production reveals two major concentrations, including spring wheat in the north and winter wheat in the central Great Plains.

as centers of flour milling and distribution. Because winters are harsh, the seeds would freeze in the ground. For this reason, spring wheat is planted in the spring and harvested in the fall; the fields are fallow in the winter. Winter wheat, however, is planted in the fall, and moisture accumulation from snow helps fertilize the seed. It sprouts in the spring and is harvested in early summer. Like corn-producing regions, wheat-producing areas are heavily mechanized and require high inputs of energy resources. Today the most important machine in wheat-producing regions is the combine, which not only reaps but also threshes and cleans the wheat. Large grain elevators to store wheat and other cereals appear prominently in the landscape of the Great Plains of the United States and Canada. To some extent these reflect the enormous surpluses that farmers have accrued, in part due to government subsidies (like those for corn).

Figure 10.6: Enormous feedlots to fatten cattle before slaughter are major contributors to the gasses that cause global climate change as well as local pollution of groundwater and streams.

Regions of the developed world that are too dry to support crop farming often practice cattle ranching. Raising cattle is an extensive (using lots of land per unit of output) agricultural pursuit because it requires many acres. In some instances, cattle live near cities. Forage is trucked to cattle-fattening pens called **feedlots** (Figure 10.6). Because cattle are fed crops that require immense amounts of energy to produce (notably corn), meat production is a major contributor to the release of carbon dioxide and other gases associated with human-induced climate change.

10.2 U.S. AGRICULTURAL POLICY

Agricultural landscapes reflect not only the invisible hand of the market but the very visible hand of the government. Far from being a "free market," agribusiness has long seen active government intervention of different types. In the historical development of America, farms were family-owned, small, and served local markets. Farm prices were relatively stable and predictable, although persistent overproduction in the late nineteenth century gradually depressed prices and bankrupted many farmers. The early twentieth century was a prosperous time for U.S. farmers, especially during World War I, when they provided a large amount of food for Allied troops. However, many farmers lost their fortunes during the 1920s and 1930s with the twin economic and environmental catastrophes of the Great Depression and the Dust Bowl, which caused prices to drop and eroded the soils of the Midwest. Many farms ceased to operate as a result. World War II created another upswing for agricultural pursuits as farmers once again provided food for a much larger Allied army, and a hungry nation.

After World War II, farmers' fortunes dwindled in the 1950s and 1960s. The U.S. government agreed to major grain trade agreements with the Soviet Union in the 1970s, which increased American exports. Since then, world markets for U.S. grain have diminished as many foreign countries have improved their own food production. For example, as a result of Green Revolution technology, India, formerly a net food importer, is now a net food exporter. Relatively high interest rates made the cost of borrowing fertilizer, seeds, and equipment frequently prohibitive, and periodic increases in the value of the U.S. dollar relative to other currencies made American exports uncompetitive. At the same time, the prices of farm operation—including machinery, fertilizer, land, and transportation—have increased drastically. These less profitable times for farmers, in which costs have outrun income, have continued. Technological and mechanical improvements and hybrid seeds have increased yields so much that U.S. farm productivity is the highest in the world. The quantity of farm products has increased much more rapidly than has demand. The result has been relatively low returns to farm families and has spelled disaster for many farmers. With lower prices and increased quantities, more and more farmers cannot afford the rapidly rising costs for machinery, fertilizer, transportation, and labor. World farm prices have likewise fallen over the long run. The result to U.S. farmers has been a continuing reduction in return for their investment. As a result of persistent overproduction and low prices, there has been a large movement of farm families and labor away from the farm. In 1910, 35% of the U.S. population lived and worked on farms. By 2010, this figure dropped to less than 2% of the total population. However, considering present production, prices, and consumption, America today still has too many farmers. In a normal market, resources would have shifted away from agriculture into other economic activities. However, because of U.S. government price supports for farms, this has not been the case.

In 1933, with farming in deep crisis, Congress passed the Agricultural Adjustment Act to aid American farmers. This act was designed to help a large proportion of the population (up to 33%) who lived in rural areas. The act artificially raised farm prices so that farmers could enjoy a "fair price" for their products, defined as "equality between the prices farmers could sell their products for, and the price they would spend on goods and services to run the farm." The period selected to determine parity prices was from 1910 to 1914, when farm prices were relatively high in comparison with other products. Since 1933, however, the ratio between farm prices and all consumer goods declined until 2010, when it was approximately 30% of the original 1914 parity established in 1933. In other words, without parity, farmers could sell products and purchase only 30% of what they could in the earlier period. Admitting that markets had created widespread instability in agriculture, the federal government stepped in through the Agricultural Adjustment Act to establish a subsidy program, or a price floor, for key agricultural commodities, a guaranteed price above the market price, to insure that farmers had an adequate income. These supports were minimum prices that the government could ensure farmers received. For example, the government bought all corn and wheat from farmers and sold it at what the market would bear. It stored many of these commodities in its own storage facilities. Thus public funds were used not only to encourage farmers to grow more than consumers could consume, but also to store the surplus.

Over the past 80 years, these price supports have provided U.S. farmers with artificially high compensation for their products. Lawmakers had hoped, of course, that market prices would rise to parity, and they did during World War II and during the Soviet grain trade agreements of the 1970s. However, most of the time, the price of farm products has been much less than the parity price. Today, these subsidies amount to more than $10 billion annually. The government also attempted to reduce production with the Soil Bank program, which paid farmers to keep acreage out of production. Initially, this approach worked, but the per-acre yields increased amazingly and completely overshadowed the lost acreage in terms of total yield.

The small American farmer currently exists as an endangered species. Although the government **price-support programs** kept less efficient small farmers in business for years, government subsidies based on a farm's output actually favor large corporate farms over small, family-owned ones. The U.S. corn industry receives the largest share of government subsidies, including $20 billion in 2010. In essence, the large corporate agribusiness farms have become richer and, with the lion's share of U.S. government subsidy, forced food prices even lower. This has continued to force small farmers off the land. The political clout of agribusiness has kept the federal government from trimming agricultural subsidies. However, doing so would improve the country's agricultural land, eliminate overproduction, and reduce the enmity toward the United States that subsidized agricultural exports generate.

The United States is not alone in subsidizing its farmers. Almost one-half of the budget of the European Union goes toward the Common Agricultural Policy, which subsidizes farmers in France, Germany, and other countries. Japan heavily subsidizes its farmers in a country with relatively little arable land. Japanese consumers pay prices well above the world average as a result. Farming often appears central to a nation's interests and identity, making politicians everywhere reluctant to reduce agricultural subsidies. In 2010, the world's developed countries spent more than $600 billion in agricultural subisides, flooding the world with cheap food and bankrupting farmers throughout the developing world. In this way, agriculture exemplifies the powerful role of the state in almost all market-dominated societies. There is certainly no "free market" in farming.

10.3 SUSTAINABLE AGRICULTURE

Industrialized agriculture has profoundly affected the natural world over the twentieth and twenty-first centuries. An agribusiness grows only one crop and relies heavily on mechanical, biological, and chemical inputs (e.g., tractors, hybrid seeds, herbicides, and pesticides). Industrial agriculture also requires fossil fuels for tractors and irrigation, petrochemical-derived fertilizers to increase yields, and the energy required to transport food globally. The use of pesticides around the world has grown since World War II, costing billions of dollars annually and causing thousands of human deaths and adverse health risks each year in the form of cancer, acute poisonings, and neurological damage. Pesticides also harm a wide variety of wildlife, including microorganisms, fish, birds, and mammals. Soil health has also been damaged by industrial agriculture due to increasing farm specialization, ever-larger farms, and monoculture-cropping practices. With inorganic fertilizer use, the natural organic matter of the soil is not replenished. Despite chemical substitutes, many areas have seen a decline in soil productivity. Moreover, with the increased use of pesticides and herbicides, some insects, weeds, and fungi eventually evolve resistance. Similarly, overuse of antibiotics in industrial livestock production leads to antibiotic-resistant bacteria that pass to humans via the food chain. This development causes health risks and reduces the efficacy of important antimicrobial drugs. Many observers question the long-term ecological sustainability of this system, a given its contributions to environmental destruction, global climate change, and the worldwide depletion of fossil fuels. Another option, **sustainable agriculture**, promotes social, ecological, and economic health for food- and fiber-producing land and communities (Figure 10.7). Numerous locally based, low-energy input and knowledge-intensive farm systems follow the principles of sustainable agriculture. Sustainable agricultural systems are often referred to as "alternative," yet these approaches are actually quite traditional, and are practiced by millions of farmers in the developing world.

Approaches that follow the principles of sustainable agriculture include organic farming, agroecology, holistic management, urban gardening, community-supported agriculture, and natural systems agriculture. These forms of farming arose in different environments, including cities, but all attempt to grow a diverse array of crops with relatively little machinery, fertilizer, or pesticides. Sustainable agricultural systems all rely on local knowledge and have minimal ecological impacts. The knowledge that local farmers possess about local climate, topography, soils, water runoff, and plants represents a rich and reliable source of information regarding **agroecosystems** (ecosystems that include crops), and helps to replace some of the costly fertilizers, machinery, and hybrid seeds found in industrialized agriculture. In sustainable agricultural systems, farmers' local knowledge can improve nutrient cycling; crop productivity; profitability; and the conservation of soil, water, energy, and biological resources. Some of these

Figure 10.7: Sustainable agriculture follows rules that use the fewest inputs such as energy or water and attempts to have the lowest possible environmental impacts. This Oregon winery uses solar panels as power.

savings occur with the reduction or elimination of external, nonlocal and nonrenewable inputs (e.g., oil, pesticides), which often have financial and environmental costs. Sustainable agricultural systems are often relatively labor-intensive, and covering the costs of labor can make their crops more expensive than those produced through commercial means. However, almost any farmer worldwide can incorporate sustainable agricultural approaches into existing operations. Indeed, many people believe that if future supplies of petroleum dwindle, sustainable agriculture will be necessary for future food production.

■ KEY TERMS

Agroecosystems: ecosystems that include crops grown for consumption

Feedlots: large facilities for the storage and fattening of cattle before slaughter

Monoculture: agricultural systems that rely on one crop over extensive areas

Price-support programs: government policies to subsidize farmers by paying higher than market prices

Subsistence agriculture: farming for consumption rather than sale of output on a market for a profit

Sustainable agriculture: farming systems that have a minimum of negative environmental impacts

Truck farming: commercial farming near urban areas that grow specialized fruits and vegetables for local consumption

■ STUDY QUESTIONS

1. How do preindustrial and agricultural systems differ from each other?

2. What proportion of the U.S. population works in agriculture? Why is it so low?

3. What are the defining characteristics of commercial agriculture?

4. How have the number and size of U.S. farms changed over time?

5. Describe the geography of U.S. agriculture: which region lies at the heart of American agribusiness?

6. What are the causes and impacts of government agricultural subsidies?

7. How does sustainable agriculture differ from industrialized forms of farming?

BIBLIOGRAPHY

Bowler, I. (2014). The Geography of Agriculture in Developed Market Economies. *London: Routledge.*

Kimbrell, A. (2002). Fatal Harvest: The Tragedy of Industrial Agriculture. *Sausalito, CA: Foundation for Deep Ecology.*

Kub, E. (2012). Mastering the Grain Markets: How Profits Are Really Made. *Charleston, SC: CreateSpace Independent Publishing.*

Mandelblatt, B. (2012). Geography of food. In J. Pilcher (Ed.), The Oxford Handbook of Food History. *Oxford: Oxford University Press.*

Marsden, T., Munton, R., Ward, N., & Whatmore, S. (1996). Agricultural geography and the political economy approach: A review. Economic Geography, 72(4), 361–375.

Morgan, D. (2000). The Merchants of Grain. The Power and Profits of the Five Giant Companies at the Center of the World's Food Supply. *Bloomington, IN: iUniverse.*

Niles, D., & Roff, R. (2008). Shifting agrifood systems: The contemporary geography of food and agriculture; an introduction. Geojournal, 73(1), 1–10.

Patterson, M., & Pullen, N. (2014). The Geography of Beer: Regions, Environment, and Societies. *New York: Springer.*

Winter, M. (2003). Geographies of food: Agro-food geographies—making reconnections. Progress in Human Geography, 27(4), 505–513.

IMAGE CREDITS

- Figure 10.1: Copyright © Miyuki Meinaka (CC BY-SA 3.0) at http://commons.wikimedia.org/wiki/File:Agricultural_Map_by_Whittlesey,_D.S.png.
- Figure 10.2: Copyright © Hinrich (CC BY-SA 3.0) at https://commons.wikimedia.org/wiki/File:Claas-lexion-570-1.jpg.
- Figure 10.4: Copyright © AndrewMT (CC BY-SA 3.0) at http://commons.wikimedia.org/wiki/File:WheatYield.png. \
- Figure 10.5: http://www.nass.usda.gov/Charts_and_Maps/Crops_County/ww-pr.asp. Copyright in the Public Domain.
- Figure 10.6: Gene Daniels, "Aerial Of Cattle on Huge Feedlot," https://commons.wikimedia.org/wiki/File:AERIAL_OF_CATTLE_ON_HUGE_FEEDLOT._(FROM_THE_DOCUMERICA-1_EXHIBITION._FOR_OTHER_IMAGES_IN_THIS_ASSIGNMENT,_SEE_FICHE..._-_NARA_-_553047.tif. Copyright in the Public Domain.
- Figure 10.7: Copyright © Eyeliam (CC by 2.0) at http://commons.wikimedia.org/wiki/File:Solar_panels_in_Oregon_vineyard.jpg.

Chapter Eleven

MANUFACTURING

MAIN POINTS

1. The geography of the world's manufacturing is dominated by a few major industrial core regions in North America, Europe, and Russia. Increasingly, East Asia has become an important center of industry, particularly Japan and China.

2. Manufacturing geographies are constantly in flux. In the West, this means that many traditional forms such as steel production have declined because of deindustrialization.

3. Major manufacturing sectors such as textiles/garments, steel, automobiles, and electronics each have their own, unique geographies that have evolved over time in response to changes in markets, technologies, and government policies.

Think about your cell phone: Where was it made? Who made it? How did it get from the workers to you? You can ask similar questions about your clothing, iPod, laptop, or car. Our lives revolve heavily around the consumption of goods, yet few people take the time to inquire where their things came from. But everything we buy—from furniture to underwear, shoes to televisions—was made by somebody. This fact locks commodities into complex webs that involve workers, companies, transportation, and intermediaries such as wholesalers, many of which span the globe today. Studying the geography of manufacturing helps us uncover these linkages from producers to consumers. It also sheds light on the spatial structure of capitalism as it constantly floods the planet with every manner of product.

To manufacture is to make things—to transform raw materials into goods that satisfy needs and wants. Manufacturing plays a crucial economic role because it produces goods that sustain human life, provides employment, and generates economic growth. It has done so since the Industrial Revolution began in Britain in the late eighteenth century. Over the following century, manufacturing generated the working classes of Europe, North America, and Japan (Chapter 5), and is doing so today in many developing countries.

Geographers who study manufacturing emphasize how firms decide where to locate factories and the nature of the places they create. Manufacturing does not simply respond to differences among locations: It helps to create them as firms generate local environments, including labor markets and investments in the built environment. First, this chapter examines the major regions in the world that produce goods—North America, Europe, and East Asia. Manufacturing is highly unevenly located around the world. How did these clusters come to exist? Second, it explores four crucial industries in more depth: textiles, steel, automobiles, and electronics. Third, it introduces the notions of flexibility, post-Fordist production, and just-in-time systems, which have revolutionized the manufacturing production and delivery process throughout the world.

Manufacturing involves deciding what is to be produced, gathering together the raw materials and semifinished inputs at a plant (assembly), reworking and combining the inputs to produce a finished product (production), and marketing the finished product (distribution). The assembly and distribution phases require transportation of raw materials and finished products, respectively. The production phase—changing the form of a raw material—involves combining land, labor, capital, and management, factors that vary widely in cost from place to place. Each of the steps of the manufacturing process has a spatial or a locational dimension.

Changing a raw material into a usable good increases its use or value. Flour milled from wheat is more valuable than raw grain. Bread, in turn, is worth more than flour. Economic thinkers refer to this process as **value added** by manufacturing. The value added by manufacturing is quite low in an industry engaged in the initial processing of a raw material. For example, turning sugar beets into sugar yields an added value of about 30%. In contrast, changing a few ounces of steel and glass into a watch yields a high added value—more than 60%. The cost, productivity, and skill level of labor play important roles in high-value-added manufacturing.

11.1 MAJOR CONCENTRATIONS OF WORLD MANUFACTURING

Manufacturing capacity and employment cluster unevenly around the world, and this fact explains much of the global economy's uneven spatial development. Three major areas account for approximately 80% of the world's manufacturing jobs and capacity: North America, Europe (including Russia and Ukraine), and East Asia (including Japan, China, and South Korea). Starting with the Industrial Revolution (Chapter 5), these regions have evolved over time, creating regional historical geographies that have intertwined and shaped one another.

NORTH AMERICA

North American manufacturing largely centers on the northeastern and midwestern United States and southeastern Canada. This region, the North American **Manufacturing Belt**, accounts for one-third of the North American population, two-thirds of North American manufacturing employment, and one-half of its industrial output. The belt extends from the northeastern seaboard along the Great Lakes to Milwaukee, where it turns south to St. Louis, then extends eastward along the Ohio River valley to Washington, DC (Figure 11.1).

This area grew rapidly in the late nineteenth century, with access to raw materials from the continental interior as well as European markets. As capital flooded in, much of it foreign, the Manufacturing Belt developed a large labor pool, often made up of immigrants from southern and eastern Europe. The transportation system included the St. Lawrence River and the Great Lakes, which were connected to the East Coast and the Atlantic Ocean by the Mohawk and Hudson Rivers. This transportation system allowed the easy movement of bulky and heavy materials. Later, canals and railroads supplemented the river and lake system. The first major factories in the Belt—the textile mills of the 1830s and 1840s—clustered along the rivers of southern New England. Between 1850 and 1870, mills replaced water with coal as a power source (leading to the rise of the Appalachian Coal Belt), and railroads replaced rivers as a

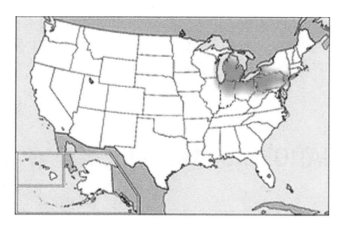

Figure 11.1: Encompassing most of the northeastern United States and much of the Midwest, the Manufacturing Belt began to form in the late nineteenth century as waves of industrialization swept the county. The site of the nation's steel, rubber, plastics, machinery, and automobile industries, it propelled the U.S. economy to world leadership. Since the 1970s, however, it has lost enormous numbers of jobs. Many communities within it have seen economic and social devastation.

means of transport for raw materials and manufactured commodities. Freed from the riverbanks, industrial production expanded rapidly in many urban areas. Manufacturing employment in New York City, Philadelphia, and Chicago soared more than 200% between 1870 and 1900. Factories concentrated in large cities for a combination of the following reasons: (1) They could be near large labor pools, including unskilled and semiskilled immigrants; (2) they could secure easy railroad and waterway access to major resource deposits, such as the Appalachian coalfields and the Lake Superior–area iron mines; (3) they could be near suppliers of machines and other intermediate products, which lowered transport costs; and (4) they could be near major markets for finished goods. This highly concentrated pattern of industrial production served the nation (and much of the world) well for about 100 years—roughly the century between the 1870s and 1970s. In the late twentieth century, this pattern was disrupted by a massive wave of economic, technological, and geographic changes that reflected the combined impacts of globalization and the microelectronics revolution.

Canada's most important industrial region stretches along southern Ontario near the St. Lawrence River valley, on the north shore of the eastern Great Lakes. This area has access to the St. Lawrence River–Great Lakes transportation system as well as proximity to the largest Canadian markets, abundant skilled labor, and inexpensive electricity from Niagara Falls. Its products include iron and steel, machinery, chemicals, processed foods, pulp and paper, and primary metals, especially aluminum. For example, Toronto is a leading automobile-assembly location in Canada, whereas Hamilton is Canada's leading iron and steel producer.

The 1960s marked the start of the steady migration of manufacturing jobs from the above regions to the U.S. South and parts of the Southwest. Southern states often had lower labor costs because workers were less skilled and less unionized. (Southern states are "right to work" states, meaning that unions cannot force employees at an establishment to join.) The southeastern manufacturing region of the United States stretches from central Virginia through North Carolina, western South Carolina, northern Georgia, northeastern Alabama, and northeastern Tennessee. It wraps around the southern flank of the Appalachian Mountains because of poor transportation across the mountains.

The region is the primary producer of American textiles. This industry moved from the Northeast to take advantage of less expensive, nonunion labor. Transportation equipment, furniture, processed foods, and lumber constitute other important products. Aluminum manufacturers moved to this region because of the inexpensive electricity produced by the more than 20 dams built by the Tennessee Valley Authority during the Great Depression. Birmingham, Alabama, was long the iron and steel center of the southeastern United States because of the plentiful iron ore and coal supplies nearby, although it never rivaled the large centers of the north such as Pittsburgh. Like its northern counterparts, it has lost much of its iron and steel industry. The Gulf Coast manufacturing region stretches from Southeastern Texas (i.e., Houston) through the petroleum complex in southern Louisiana, to Mobile, Alabama. Nearby oil and gas fields support petroleum refining and chemical production. The region also produces primary metals, including aluminum, as well as electrical machinery and electronic products.

EUROPE AND RUSSIA

Europe has a number of the world's most important industrial regions (Figure 11.2). They are located in a north–south belt that starts in Scotland and extends through southern England. The belt continues from the mouth of the Rhine River in the Netherlands, passes through the Ruhr region in Germany as well as through northern France, and concludes in northern Italy. Good supplies of iron ore and coal, as well as a highly skilled labor force, fuel the success of these industrial regions.

The Industrial Revolution started in the Britain in the mideighteenth century (Chapter 5). It had its basis in textile and woolen manufacturing. Because many other countries have since learned to produce their own iron, steel, and textiles, the world currently has an oversupply of these items. Therefore, the market for British goods decreased substantially over time. Britain's outmoded factories, high labor costs, slow productivity growth, and deteriorating infrastructures reduced its global competitiveness. In contrast, Germany rebuilt after World War II, modernizing its plants and industrial processes with new equipment. As a result, Germany

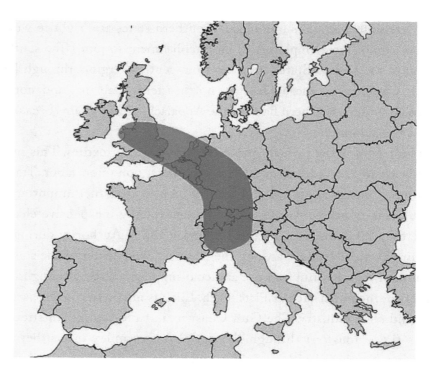

Figure 11.2: In Europe, manufacturing is concentrated in a belt stretch from the British Midlands across Belgium, northern France, western Germany, and northern Italy. This region developed during the Industrial Revolution but has been supplemented or displaced by newer centers such as northern Italy.

became a major industrial success in the late twentieth century. In contrast, Britain has suffered persistent industrial decline to a greater extent than any other modern industrialized country.

The largest European manufacturing region today lies in the northern European lowland countries of Belgium and the Netherlands, northwestern Germany, and northern France. In this region, the Rhine and Ruhr Rivers meet, offering cheap access to the oceans and the world market. This region's backbone has been the iron and steel industry because of its proximity to coal and iron ore fields. Production of transportation equipment, machinery, and chemicals helped lead western Europe into the industrial age long before the rest of the world. The Rhine River, which is the largest waterway of European commerce, empties into the North Sea in the Dutch city of Rotterdam. Because of its excellent location, Rotterdam became one of the world's largest ports, although recently East Asian ports such as Hong Kong and Singapore have surpassed it.

The Upper Rhine–Alsace-Lorraine Region lies in southwestern Germany and eastern France. Its central location makes it well situated for distribution to population centers throughout western Europe. The main cities on the German side include Frankfurt, Stuttgart, and Mannheim. Frankfurt became the financial and commercial center of Germany's railway, air, and road networks. Stuttgart produces precision-engineered goods (e.g., medical equipment) and high-value manufactured goods, including the Mercedes Benz, Porsche, and Audi automobiles. Mannheim, located along the Rhine River, is noted for its chemicals, pharmaceuticals, and inland port facilities. The western side of this district, in the Alsace-Lorraine section of France, produces a large portion of the district's iron and steel.

The Po River valley in northern Italy contains Turin, Milan, and Genoa. It includes only one-fourth of Italy's land, but more than 70% of its industries and 50% of its population. This region specializes in iron and steel, automobiles, textile manufacture, and food processing. The Emilia-Romagna region is Italy's high-technology center, generating a number of high-value-added textiles, industrial ceramics, processed foods, and electronic equipment. The Alps, a barrier to the German and British industrial regions, give the Italian district a large share of the southern European markets as well as cheap hydroelectricity.

The Ukraine industrial region relies on the rich coalfield deposits of the Donets Basin. The iron and steel industry base centers on the city of Krivoy Rog, with nearby Odessa serving as the principal Black Sea all-weather port. The area is collectively known as Donbass. Like the German Ruhr area, Ukraine's industrial district lies close to iron ore and coal mines, a dense population, a large agricultural region, and good transportation facilities.

In Russia, the Moscow industrial region takes advantage of a large, skilled labor pool as well as a large market. Its best known products include textiles: linen, cotton, wool, and silk fabrics. East of Moscow, the linear Volga region extends northward from Volgograd (formerly Stalingrad) astride the Volga River. The Volga, a chief waterway of Russia, is linked via canal to the Don River and thereby to the Black Sea. This industrial region developed during and after World War II because it was just out of reach of the invading Nazi army that occupied Ukraine. Substantial oil and gas production and refining occur here. Recently, a larger oil and gas field was discovered in West Siberia. Nonetheless, the Volga district supplies most of Russia's oil and gas, chemicals, and related products.

Just east of the Volga region, the Ural Mountains separate the European and Siberian parts of Russia (Figure 11.3). The Urals have large deposits of industrial minerals, including iron, copper, potassium, magnesium, salt, tungsten, and bauxite. Although coal must be imported from the nearby Volga district, the Urals district provides Russia with iron and steel, chemicals, machinery, and fabricated metal products. The Kuznetsk Basin centered on the trans-Siberian railroad is the chief industrial region of Russia east of the Urals. Again, as in the case of the Ukraine and the

Figure 11.3: The Soviet Union industrialized extremely rapidly during the 1920s. This development made Russia and the surrounding states significant centers of heavy industry, including steel and petrochemicals. With vast natural resources, Russia's industrial base remains important in the region. However, it has been hobbled during the transition to a capitalist economy following the disintegration of the USSR in 1991 by widespread corruption and mismanagement.

Urals districts, the Kuznetsk industrial district relies on an abundant supply of iron ore and coal—it has the largest supply of coal in the country. The Kuznetsk Basin resulted from the grand design of former Soviet city planners. These planners poured heavy investments into this region, hoping that it would become the industrial supply region for Soviet Central Asia and Siberia. Following the disintegration of the Soviet Union in 1991, lack of innovation and investment has led to stagnation and decline in this region.

EAST ASIA

East Asia has emerged as the world's largest manufacturing region, with numerous subregions (Figure 11.4). This part of the world includes long-established economies such as Japan; more recently

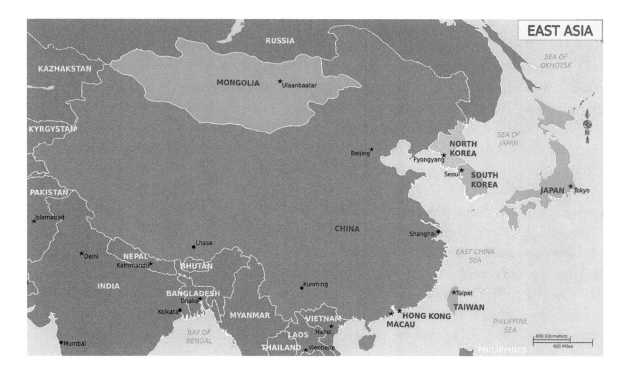

Figure 11.4: East Asian economies have grown rapidly since World War II. This region has become the world's largest producer of many manufactured goods, including textiles and electronics. Japan pioneered the way, with South Korea, Taiwan, and Hong Kong following in a "flying geese formation." More recently, China's explosive growth has made it the second-largest economy in the world. China's manufacturing is clustered in several districts, mostly along the Pacific Coast. They include the Guangdong region in the south as well as several regions along the Yangtze River, such as the greater Shanghai metropolitan area.

industrialized ones, such as South Korea, Taiwan, and Singapore; and latecomers such as Malaysia, Indonesia, and Thailand. And of course, it includes China, the world's second-largest economy and largest exporter of manufactured goods.

Starting in the late nineteenth century, Japan, set the model for industrialization in East Asia. With the collapse of the feudal Japanese society in the 1870s, Japan embarked on a program of rapid industrialization that involved imitating the West in many ways. Within two decades Japan was producing steel, railroad cars, and military equipment, becoming the first non-Western country to industrialize. However, Japan's manufacturing base has stagnated recently as its economy has moved steadily into services. The traditional notion of the "flying geese formation" compares countries with a flock of birds flying in a V-formation, with Japan at the head, the newly industrialized countries (NICs) such as South Korea, Taiwan, Hong Kong, and Singapore closely behind, and other countries (Thailand, Indonesia, Malaysia) forming a third generation. China, of course, is the giant looming over all of them today. As the NICs industrialized, they acquired a capacity first in unskilled, light forms of manufacturing such as textiles and **garments**, then in successively more skilled, higher value-added industries such as ships, steel, automobiles, and electronics.

Following World War II, Japan set about rebuilding itself to become a potent economic power. In the late twentieth century, Japan's output in several manufacturing sectors increased dramatically. Today it has the third-largest GNP in the world and is the world's leading producer of electronics, steel, commercial ships, and automobiles. Compared with the United States and Britain—countries with the physical resources to sustain an industrial revolution—Japan is much less well-off. Except for coal deposits in Kyushu and Hokkaido, Japan has almost no significant raw materials. It depends on imported raw materials for its industrial growth. Human resources, however, are not scarce. There are 126 million Japanese, one-quarter of whom live crammed into an urban-industrial core near Tokyo. Japan's combination of a strong work ethic, a high level of collective commitment, a first-rate educational system, and government support has produced a world economic power.

Japanese manufacturing is dominated by large groups of firms called *keiretsu*, which typically consist of a large multinational corporation (e.g., Toyota, Mitsubishi) and closely related banks and parts suppliers. Although permanent workers in large firms receive good salaries, Japan has a lower relative cost of labor than many other developed countries. The state directs savings and profits toward the goals set forth by a unique collaboration between Japan's Ministry of International Trade and Industry (MITI) and private enterprise. Under MITI's guidance, Japan has subsidized research and relocated industries such as steel making and shipbuilding "offshore" in the newly industrializing countries NICs, where labor costs are even lower. Thus, Asian manufacturing typically involves much higher levels of government intervention than found in the West. However, Japan has suffered from persistent economic stagnation for the past 20 years in what is commonly called the end of the "bubble economy." Japanese labor costs are high for the region, its financial markets remain stagnant, and its political system has been unable to deal with the shifting necessities of remaining competitive in a rapidly changing global economy. Although Japan remains the second-largest economy in Asia (after China), its growth rate has been stagnant compared with its more dynamic neighbors to the west and south.

Now, countries such as South Korea, Taiwan, and Singapore have developed their own higher value-added industries (e.g., consumer electronics). South Korea, with a manufacturing capacity centered on Seoul, has become the eighth-largest economy in the world, a major producer of electronics goods, steel, autos (e.g., Hyundai), and ships. Similar to the Japanese model, most Korean manufacturing occurs through interlocking combinations of giant firms and their suppliers known as **chaebols** (e.g., Samsung, Hyundai). These "tigers" industrialized in the 1970s and 1980s just as American manufacturing began to decline. Many started in light industry, such as garments, and moved swiftly into more skilled, more profitable sectors such as electronics. As with Japan and Korea, their growth was largely driven by exports rather than domestic demand.

Of course, China, which began to industrialize in the 1980s, has become a major world powerhouse in its own right. Following the collapse of communism in China in the late 1970s, the government moved aggressively to attract foreign investment capital and started a series of industrial districts. With light manufacturing centered on Hong Kong and Guangdong province in southern China and stretching along the Pacific Coast (e.g., Shanghai), China is the world's largest producer of steel, concrete, ships, apparel, toys, and many other manufactured goods. With a population of 1.3 billion people, it has a vast reservoir of cheap labor, although growth in the southern regions has slowly lifted wages there. Its transition to economic superpower status continues unabated. The economy has grown at an annual clip of about 9% to 12%. China is becoming the world's factory, with enormous trade surpluses to prove it.

The U.S. trade deficit with China has increased steadily and stood at about $300 billion in 2012, more than one-third of the American total. In the near future, China is expected to dislodge Japan as the country giving America its biggest trade deficit. Imports to the United States from China have grown 30% per year and exports by 15%. Part of this imbalance occurs because China maintains an intricate system of import controls that restrict the influx of foreign goods. Most products are subject to quotas, licensing requirements, or other restrictive measures. In contrast, three out of four toys sold in America's more liberalized market come from abroad, with 60% of those imports coming from China. Sixty percent of all shoes sold in America come from China. Other major goods imported into the United States from China include clothing, telecommunications equipment, household appliances, televisions, computer chips, computers, and office equipment.

If current rates of economic growth continue, China may be the world's largest economy by 2040 (Figure 11.5). Already, China's rapidly growing economy has led it to import vast quantities of raw materials (Table 11.1), including Middle Eastern oil, iron ore and wheat from Australia, soybeans from Brazil, and numerous minerals and oil from Africa. Chinese

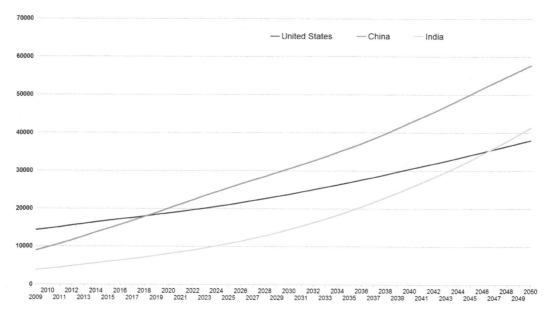

Figure 11.5: With annual growth rates of 9% to 12%, China has the world's most rapidly expanding economy. If current growth rates continue, it will surpass the United States as the world's largest economy around the year 2040.

competition has helped to drive up prices of all of these goods worldwide. To fuel its growth, China uses enormous amounts of coal and is now the world's leading producer of gases linked to global climate change.

TABLE 11.1: CHINA AND U.S. SHARES OF GLOBAL CONSUMPTION, 2015

	CHINA	U.S.
Cement	40	6
Cotton	33	7
Steel	27	12
Lead	23	21
Copper	20	16
Aluminum	19	24
Soybeans	18	23
Wheat	18	6
Oil	12	25

11.2 DEINDUSTRIALIZATION

As capitalism underwent one of its periodic restructurings in the late twentieth century, a new international division of labor arose quite different from the world order that had existed until then. Similar to many earlier moments of transition, this one was born out of an economic crisis. In the 1970s, the average rate of world economic growth declined, and the long boom period following World War II drew to a close. A deep recession in 1974–1975 followed the first oil shock in 1973, and Asian countries began to compete with North America and Europe in various manufacturing sectors. One of the most visible effects of this new order was widespread deindustrialization in economically developed countries, a decline in manufacturing capacity, and a loss of manufacturing jobs. As corporations restructured or closed in a climate of intense international competition, millions of workers were laid off. Deindustrialization affected all of the world's developed economies to some extent (Figure 11.6). Britain, France, Germany, Canada, the United States, and to a lesser extent, Japan, all lost millions of well-paying jobs. In this context, globalization led to widespread job displacement for unskilled workers.

In the United States, almost all states in the Manufacturing Belt experienced manufacturing job loss, and virtually all states in the South and the West registered manufacturing job gains. As manufacturing plants closed, once-vibrant cities in the Northeast and Midwest became economically stagnant with devastating impacts on their communities (Figure 11.7). Today, many of the Manufacturing Belt's inner-city areas are littered with closed factories, bankrupt businesses, high unemployment, depressed real estate markets, and struggling blue-collar neighborhoods. Victims of plant closings sometimes lost not only their current incomes but often their accumulated assets as well. When savings run out, people lose their ability to respond to life crises and often suffer depression and marital problems. Although job losses occurred in many occupations, unskilled workers were especially vulnerable. Unskilled and semiskilled workers, particularly African Americans, were especially hard hit. By driving up unemployment, the deindustrialization of the inner city helped create the impoverished ghetto communities there.

Some old industrial cities such as Pittsburgh successfully built new bases for employment in services (e.g., health care). Southern New England actually underwent a new round of industrial expansion based on electronics, drawing on its pool of highly skilled workers. Some of the jobs from the Manufacturing Belt moved to new industrial areas of the Sunbelt (South and West), reflecting another round of uneven spatial development as capital abandoned some areas for others.

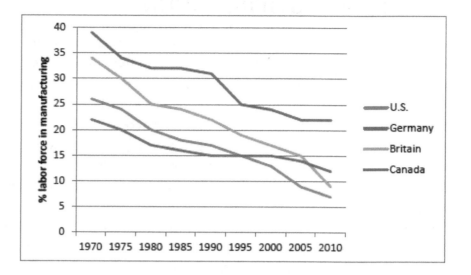

Figure 11.6: Throughout the developed world, manufacturing employment has been on the decline since 1970. This stems from automation as well as the shift of many industrial plants to the developing world. As manufacturing has declined, services have grown as the largest part of the economy (Chapter 12).

Several factors underlie the decline in manufacturing in the industrialized world. First, wage costs in Europe and North America are much higher than in the developing world, creating an incentive for firms to save by relocating abroad. In addition to wages, industrial firms faced high pension costs and, without universal health insurance in the United States, significant health coverage costs. (The average automobile company, for example, spends more on health insurance per car than it does on steel). Second, technological changes in manufacturing reduced the number of workers needed per unit of output. Thus, although the United States lost numerous manufacturing jobs, its manufacturing output stayed almost constant. Third, although industrial firms reinvested in new production techniques, they often failed to reinvest sufficiently in research and development to remain competitive on the world stage. Rather, many firms spent their revenues on high stock dividends, that is, returns to shareholders rather than investments to raise productivity. Finally, inadequate public investment in education and infrastructure has hurt countries such as the United

Figure 11.7: The decline of manufacturing jobs devastated many once-vibrant cities, turning them into deindustrialized wastelands. As job losses reverberated throughout the local economy, they depressed local retail sales and property values.

States. Too few workers have the skills needed for a postindustrial economy, and transportation costs have risen.

Conversely, manufacturing employment and output have increased sharply in lower wage industrializing countries; the flip side of deindustrialization is the rapid economic growth of parts of the developing world, particularly Asia. Between 1974 and 2010, the advanced industrial countries lost 35 million jobs, whereas the newly industrializing countries gained 29 million. The most rapid growth of manufacturing output occurred in East and Southeast Asia, the world region with the fastest rate of economic growth since World War II. This region attracted many industries from North America. The "sunrise" industries of the developing world thus corresponded with the "sunset" industries of the West (e.g., textiles and garments). Deindustrialization in some places has translated into industrialization in others. Massive social changes have occurred in newly industrializing nations, reshaping the lives of hundreds of millions of people. Many have been lifted out of poverty and catapulted from rural villages into industrial cities. Although beneficial in many ways, the social dislocation often leads to great personal cost and suffering.

11.3 MAJOR MANUFACTURING SECTORS

Globalization occurs differently in different industries. Various forms of manufacturing have their own specific technical, labor, and locational requirements. Additionally, nation-states regulate different industries in a variety of ways. Therefore, each industry has evolved in a unique manner on the global stage. Four industries—textiles and garments, steel, automobiles, and electronics—dramatize the similarities and differences among manufacturing sectors as they create their own geographies, reflecting and producing globalization in sector-specific ways.

TEXTILES AND GARMENTS

The textile and garment industries dramatically reveal the globalization of manufacturing. Textile manufacture creates cloth and fabric, whereas clothing manufacture uses textiles to produce wearing apparel. About one-half of textile output is used in apparel; the rest goes to sectors such as carpets and industrial fabrics (for example, upholstery in cars). Textiles and garments make up a classic low-tech industry: unskilled, labor-intensive, with relatively little technological sophistication and small firms. The industry is relatively easy to enter and exit, making it highly competitive. Producers therefore seek savings by paying very low wages. Thus, textiles and garments have long been among the most brutally exploitative industries.

Textiles and garments were the leading sector of the Industrial Revolution in Britain, fueling the rise of a manufacturing base in cities such as Manchester and providing grist for the novels of Charles Dickens. Then, as today, the industry relied heavily on the labor of young women children. By the early nineteenth century, it expanded in continental Europe, including northern Italy and Lyon in France. In the United States, textile production generated the manufacturing cities of southern New England. The industry later moved to the Piedmont states in the South, particularly North and South Carolina, which remains the core of American textile production. Concentrations of garment production exist in New York, a major center in the early twentieth century, as well as southern California, where sweatshops using immigrant (and often illegal) labor are common.

Textile and garment production in the developed world declined steadily from the 1970s onward as the industry faced a wave of cheap imports. Lured by significantly lower wages in the developing world, especially in East Asia, the industry moved abroad. Today, major cotton fabric manufacturers include India and China, the world's largest producer. In both nations, working conditions resemble those of Britain 200 years ago (Figure 11.8): large groups of young women working for long hours for very low wages under horrific conditions, including poor lighting, respiratory problems from dust and small threads circulating in the air, and limited restroom breaks. Bangladesh has emerged as a significant garment producer, accompanied by disastrous collapses of factories that killed dozens of workers. A small industry has also sprouted up in Central America, where sweatshops are found in Honduras, Guatemala, and El Salvador. Everywhere, the industry uses young, predominantly female workers, who are unlikely to organize or resist their working conditions. By continually looking for new sources of cheap, easily exploitable labor, this industry has exhibited a very fluid geography over time, a pattern that continues in the twenty-first century.

Figure 11.8: Garment and textiles production have long been among the most brutally exploitative industries on the planet. Even today, working conditions in many sweatshops do not differ markedly from those described by Charles Dickens in nineteenth-century England.

STEEL

The steel industry has played an enormously important role in the development of industrialized societies. Iron and steel production generate a wide variety of outputs essential to many other sectors. Examples include parts for automobiles, ships, and aircraft; steel girders for buildings and dams; industrial and agricultural machinery; pipes, tubing, wire, and tools; furniture; and many other uses. Steelmaking requires relatively simple inputs including iron ore, which is purified into pig iron; large amounts of energy, generally in the form of coal; and limestone, which is used in the ore purification process. Steel production is a very capital-intensive process, that is, it uses immense amounts of equipment and relatively little labor per unit of output. It demands a huge financial investment. This requirement raises high barriers to entry, so a small group of large firms have generally dominated the sector. Steelmaking uses a highly skilled workforce that is almost entirely male (Figure 11.9). Transport costs have traditionally been high in this sector, necessitating investments in railroads and ships and limiting the market for steelmaking firms.

The historical geography of steel production includes Britain's Midlands cities such as Sheffield and Birmingham (Chapter 5). These areas played a critical role in the early Industrial Revolution. A similar complex of steel production arose in the Ruhr region in western

Figure 11.9: Steel production uses immense amounts of equipment and relatively little labor per unit of output, that is, it is capital-intensive. This feature shapes the behavior of steel firms over time and their location in space.

Germany on the banks of the Rhine River. In the northeastern United States, the earliest iron and steel producers were very small and localized. They used wood and charcoal as fuel and served local markets.

However, many changes in the late nineteenth century dramatically reshaped the industry. The Bessemer open hearth furnace made the production of steel relatively cheap, using huge amounts of energy in plants open 24 hours per day. Geographically, the industry came to center on the steel towns of the Manufacturing Belt. Pittsburgh became the largest steel-producing city in the world. Other centers within this complex included Hamilton, Canada; Buffalo, New York; Youngstown and Cleveland, Ohio; Gary, Indiana; and Chicago. These locations allowed easy access to coal from Appalachia, iron ore from Minnesota and upper Michigan, and cheap transportation via the Great Lakes and the railroads. By clustering there, they had ready access to specialized pools of labor and suppliers of parts and services. The industry became increasingly controlled by few giant firms. The rise of the U.S. Steel Company under Andrew Carnegie saw 30% of the nation's steel output in the hands of one firm in 1900.

The United States led the world's steel industry in the early and midtwentieth century, producing as much as 63% of the world's total output after World War II. However, beginning in the 1960s, other nations claimed more of a share in the world's steel market. New competitors arose first in the rebuilt factories of Japan and Europe (particularly Germany, France, and Spain). Later, developing countries (e.g., Brazil, South Korea) joined in. The United States produced 8.3% of the world total in 2005 (Figure 11.10). Conversely, the industry has flourished in East Asia. China is now by far the world's largest producer of steel, with Japan in second place and the United States in third. These shifts reflect the continual reproduction of uneven development under capitalism.

The decline of the U.S. steel industry generated enormous problems for former steel towns. Waves of plant closures in the 1970s and 1980s generated high unemployment, rising poverty, depressed property values, and out-migration. Today, the Manufacturing Belt produces very little steel, and Pittsburgh produces none at all. Between 1970 and 2007, the North American and western European proportions of global steel production declined from 67% to 41%, whereas developing countries' production levels increased from 10% to more than 33%. The overall amount of steel produced has risen. Many steel-manufacturing firms have gone out of business as the global steel-production capacity exceeds global demand. The industry's response to crisis, other than plant closures, was both to call for protectionism from imports and to restructure. Because of government subsidies (particularly in Europe), steel mills in some countries have remained open in the face of dwindling quotas. The introduction of computerized technology led to widespread changes in steel production, including the emergence of highly automated **minimills** that use scrap metal as inputs, are not generally unionized, and

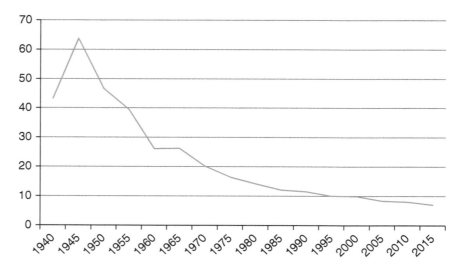

Figure 11.10: The steep and steady decline in the U.S. share of world steel output reflects the ongoing globalization of the industry as it has shifted to the developing world, notably China.

produced specialized outputs for niche markets. Since the 1970s, U.S. production has decreased 33%, whereas employment in the steel industry has declined by 66%.

AUTOMOBILES

From its inception at the dawn of the twentieth century, the automobile industry has unleashed major effects on cities and everyday life throughout the world. Originally the automobile was made on a craft basis, that is, by small groups of skilled workers, and was affordable only to the wealthy. Henry Ford standardized the European invention of auto production around the turn of the twentieth century. He introduced the moving assembly line and a highly detailed division of labor to cut costs, making automobiles affordable to the middle class. Ford also paid high wages to reduce turnover rates for his workers. So successful was his model of production that it became known as "Fordism" and was widely imitated by other sectors. Ford's success in this sector centered on the Detroit region, adding another layer of industrial investment to the Manufacturing Belt. Other than being

Ford's hometown, Detroit was the previous center of the wagon and buggy industry and had a supply of skilled labor accustomed to working with machines. Its other locational advantages included numerous parts-producing firms, the rubber-producing center in nearby Akron, and numerous rail lines and shipping routes that brought raw materials across the Great Lakes.

Throughout the twentieth century, the automobile market expanded dramatically as incomes rose and as western cities increasingly developed around low-density suburban forms. Many auto companies purchased mass transit lines and destroyed them, as in Los Angeles, to force people to buy cars. In the 1950s and 1960s, the construction of the Federal Interstate Highway System greatly enhanced the demand for cars. As it matured, the auto industry became steadily concentrated in a handful of highly capital-intensive transnational corporations. For example, the world's 10 leading automobile manufacturers produce over 70% of the world's automobiles. In the United States, General Motors, Chrysler, and Ford account for 95% of national automobile output. However, the U.S. auto industry has been steadily besieged by competition from abroad and has suffered declining market share, partly because of its reluctance to produce smaller, fuel-efficient vehicles. The German firm Daimler-Benz bought Chrysler, only to sell it entirely later as performance declined. In the wake of the financial crisis that began in 2007, General Motors and Chrysler went bankrupt. Only massive federal government bailouts saved them from obliteration. The industry laid off tens of thousands of workers, mostly in the greater Detroit area as well as parts of Ohio and Kentucky.

Worldwide automobile production extends to many countries across the globe. Its capacity has expanded as the growing middle class in many developing countries seeks out cars. Three major nodes of automobile production exist—Japan, the United States, and western Europe. In 2010, Europe accounted for 25% of the world's automobiles; Japan, 20%; and the United States, 19%. The three developed regions of the world, East Asia, North America, and Europe, accounted for 72% of the automobiles produced. Japanese firms also invested heavily in the United States in the late twentieth century. They built factories in much of the Midwest that allowed them access to the American market, low transport costs, and freedom from fluctuating exchange rates and threats of U.S. protectionism. Smaller production centers exist in Brazil, Russia, South Korea, and India, each of which has started its own national automobile industry, sometimes with foreign investment.

ELECTRONICS

Although its roots extend into the nineteenth century, the electronics industry underwent enormous changes with the microelectronics revolution of the late twentieth century.

Microelectronic technology is the most significant technology of the present, transforming all branches of the economy and many aspects of society.

The radio was invented and produced as early as 1901, but the modern electronics industry was not born until America's Bell Telephone Laboratories built the transistor in 1948. The transistor supplanted the vacuum tube, which had been used in most radios, televisions, and other electronic instruments. A solid-state device made from silicon, the microelectronic transistor acted as a semiconductor of electric current. By 1960, the production of the integrated circuit allowed transistors to be connected on a single small silicon chip. By the early 1970s, a computer came into production that was tiny enough to fit on a silicon chip the size of a fingernail. Each of these microprocessors could do the work of a roomful of vacuum tubes. By the 1980s, the use of semiconductors replaced transistors, making computers even faster, cheaper, and more powerful. With these changes, it became possible to process information in digital format (i.e., binary code zeros and ones) rather than analog format (as electrical waves), which made it much easier for computers to use. The microprocessor made possible the contemporary, small, powerful computer such as laptops and desktops, which in turn revolutionized the collection and analysis of information, particularly office work. As the industry generated wave upon wave of improvements, computing power and memory rose exponentially (Figure 11.11).

Although chips became increasingly smaller and more powerful, new applications for the electronics industry emerged. These included calculators, electronic typewriters, computers, industrial robots, aircraft-guidance systems, and combat systems. New discoveries were applied to automobile construction for guidance, safety, speed, and fuel regulation. An entire new range of consumer electronics became available for home and business use. The electronics industry, like textiles, steel, and automobiles before it, has come to be regarded as the modern touchstone of industrial success. Hence all governments in the developed and rapidly emerging economies operate substantial support programs for the electronics industry, including tax breaks, worker training, and export promotion.

Japan and the United States produce most of the world's electronic components, including semiconductors, integrated circuits, and microprocessors. Other significant production exists in western Europe and Southeast Asia. From the 1960s through the 1970s, the United States dominated the field of semiconductor manufacture. However, by the 1990s, Japan took over this role. In 2007, Japan accounted for 40% of the world production of semiconductors; the United States, 21%; and Europe, 11%. Southeast Asia, South Korea, Malaysia, Taiwan, Thailand, and China were significant manufacturers.

The world manufacture of consumer electronics such as televisions, computer peripherals, and microwave ovens has shifted steadily to developing countries, especially in East and

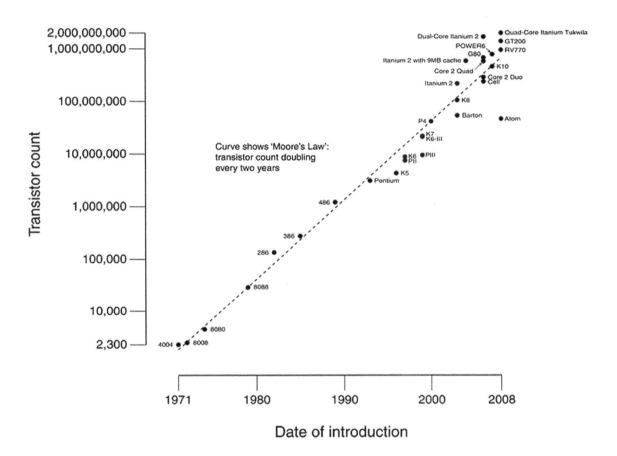

Figure 11.11: The microelectronics revolution unleashed enormous rounds of improvements in computing power and memory over time. Each new generation of electronic technology was markedly faster and cheaper than its predecessors. As the cost of computers declined, they became ubiquitous, with enormous impacts on work, education, entertainment, travel, information access, and daily life.

Southeast Asia. Much of the television production that formerly occurred in the United States, Germany, and the United Kingdom now takes place in China and Malaysia. In Latin America, Brazil has a small but growing electronics industry that generates aerospace equipment and televisions.

■ KEY TERMS

Chaebol: a Korean term for large complexes of industrial firms surrounded by dedicated networks of suppliers

Garments: clothing; the garment industry is related, but not identical, to the textile industry

Manufacturing Belt: the region of the United States stretching from southern New England along the shores of the Great Lakes, which was long the dominant concentration of U.S. manufacturing

Minimills: small, technologically sophisticated steel mills that arose in the late twentieth century that use scrap metal rather than iron ore as their input

Textiles: cloth, as distinct from clothing; the textile industry tends to be more capital-intensive than garment production and have slightly different locational requirements

■ STUDY QUESTIONS

1. What are four major world regions of manufacturing?

2. Summarize the historical development of the U.S. Manufacturing Belt.

3. When and why did the Manufacturing Belt begin to lose industry? Where, specifically, did it go?

4. What is the "flying geese formation?" Who is the lead goose?

5. What explains China's enormous economic growth?

6. What are the causes and consequences of deindustrial-ization?

BIBLIOGRAPHY

Amin, A., & Thrift, N. (1992). Neo-Marshallian nodes in global networks. International Journal of Urban and Regional Research, 16, 571–587.

Bluestone, B., & Harrison, B. (1984). The Deindustrialization of America: Plant Closings, Community Abandonment, and the Dismantling of Basic Industry. *New York: Basic Books.*

Essletzbichler, J. (2004). The geography of job creation and destruction in the U.S. manufacturing sector, 1967–1997. Annals of the Association of American Geographers, 94(3), 602–619.

Gertler, M. (1995). Being there: Proximity, organisation, and culture in the development and adoption of advanced manufacturing technologies. Economic Geography, 71, 1–26.

Gertler, M. (2003). Tacit knowledge and the economic geography of context, or the undefinable tacitness of being (there). Journal of Economic Geography, 3, 75–99.

Hughes, A., & Reimer, S. (2004). Geographies of Commodity Chains. *London: Routledge.*

Krugman, P. (1992). Geography and Trade. *Cambridge, MA: MIT Press.*

Linge, G. (1991). Just-in-time: More or less flexible? Economic Geography, 67, 316–332.

Pacione, M. (2014). Progress in Industrial Geography. *London: Routledge.*

Saxenian, A. (2006). The New Argonauts: Regional Advantage in a Global Economy. *Cambridge, MA: Harvard University Press.*

Tickell, A., & Peck, J. (1995). Social regulation after Fordism: Regulation theory, neo-liberalism, and the global-local nexus. Economy and Society, 24, 357–386.

Walker, R., & Storper, M. (1991). The Capitalist Imperative: Territory, Technology and Industrial Growth. *Hoboken, NJ: Wiley-Blackwell.*

IMAGE CREDIT

- Figure 11.1 Copyright © Psemper (CC BY-SA 3.0) at http://commons.wikimedia.org/wiki/File:Rust-belt-map.jpg.
- Figure 11.2: Copyright © ArnoldPlaton (CC BY-SA 3.0) at http://commons.wikimedia.org/wiki/File:Blue_Banana.svg.
- Figure 11.3: U.S. Central Intelligence Agency, "Rs-map," https://commons.wikimedia.org/wiki/File:Rs-map.png. Copyright in the Public Domain.
- Figure 11.4: Copyright © Cacahuate (CC BY-SA 4.0) at http://commons.wikimedia.org/wiki/File:Map_of_East_Asia.png.
- Figure 11.5: Copyright © Srikar Kashyap (CC BY-SA 3.0) at http://commons.wikimedia.org/wiki/File:US,China_and_India_projected_GDP_growth_2009-2050_Pwc.PNG.

- Figure 11.7: Copyright © andrew (CC by 3.0) at https://commons.wikimedia.org/wiki/File:Abandoned_factory_in_Kirovsk.JPG.
- Figure 11.8: Copyright © Marissaorton (CC BY-SA 2.0) at http://commons.wikimedia.org/wiki/File:Lindintracuay.jpg.
- Figure 11.9: Copyright © Joost J. Bakker (CC by 2.0) at https://commons.wikimedia.org/wiki/File:Corus_Tata_steel_Velsen_IJmuiden.jpg.
- Figure 11.11: Copyright © Abovedrew23 (CC BY-SA 3.0) at http://commons.wikimedia.org/wiki/File:Transistor_Count_and_Moore%27s_Law_-_2008_1024.png.

Chapter Twelve

SERVICES

MAIN POINTS

1. Services—the production of intangible things such as banking, air travel, and haircuts—include many different industries and occupations. They comprise the bulk of employment in all industrialized countries.

2. Services have grown for several reasons: (1) changes in household spending, (2) the rise of health-related services, (3) an increasingly complex corporate environment, (4) a long-term trend of increasing government employment, (4) and interregional and international trade in services.

3. The labor markets of services differ from those in manufacturing: services tend to be more labor-intensive, include large numbers of female workers, and often (but not always) demand higher levels of education.

4. The geography of services includes (1) the locations of consumer services, which follow the distribution of people with the ability to buy, and (2) the locations of producer services, which serve other firms and tend to concentrate in large urban areas.

5. Financial services include many types of institutions that deal with money, such as banking, mortgages, and insurance. Following deregulation, this sector has mushroomed in size and power over the past half century, becoming thoroughly internationalized.

6. With the increasing importance of information in business, telecommunications form a critical part of services. This includes the Internet, now used by one-half of the planet's

people. Social media networks have become increasingly widespread. Telecommunications have reshaped finance through offshore banking centers and electronic commerce.

What type of occupation do you wish to pursue after graduation? Lawyer? Teacher? Nurse? Engineer? Physician? Dentist? Airline pilot? Banker? Social worker? Computer programmer? Entertainer? Real estate agent? Manager? All of these fall into the broad category of services, as do hundreds of others. We live today in a service-dominated world, which differs from the earlier environment of manufacturing in terms of skills, opportunities, salaries, constraints, and geography. There is a very high probability that your career, or careers, will be spent in one type of service or another. So, it's a good thing to know a little bit about services, what they are, their diversity, how they came to be, and their effects on the world around us.

"Services" encompass an enormous diversity of occupations and industries, ranging from professors to plumbers to prostitutes. A service is the production and consumption of an intangible product: a haircut, credit, a museum visit, or a vacation. In contrast, manufacturing creates a tangible product (e.g., computers, washing machines, clothing). Major service industries include retailing, banking, real estate, finance, law, education, and health care. Government also provides services including education, defense, and some types of health care. Economists speak of agriculture, manufacturing, and services as the primary, secondary, and tertiary economic sectors, respectively. In postindustrial economies, services predominate. However, the term "*the* service sector" oversimplifies because it masks a huge array of job types among and within different service industries.

Many observers anticipated the shift in developed nations from a manufacturing-based economy to a service-based economy. They tended to see services as information-processing activities, including clerical work, executive and management decision making, legal services, telecommunications, and the media. Such a view heralded information processing as a qualitatively new form of economic activity, different from manufacturing-based economies in terms of their labor markets, skills, workplaces, inputs, and outputs. Thus services supposedly represented a historically new form of capitalism. Unfortunately, this view of services was simplistic and mistaken. Although many service jobs do involve the collection, processing, and transmission of data, clearly others do not: The trash collector, restaurant chef, security guard, and janitor all work in services, but their activities have little to do with information processing. Rather, services represent another form of capitalist production, involving the same constraints to location, production, and consumption as other industries. The chef sells his labor for a wage; the restaurant has startup costs and overhead; it requires raw materials (ingredients) and must sell to consumers (diners) at a price they will pay.

Throughout most of the economically developed world, various forms of services have replaced manufacturing employment. This trend reflects the shift of manufacturing from more economically developed countries to low-cost less developed ones. The percent of national output (gross domestic product) generated by services varies considerably around the world (Figure 12.1); generally, the more economically advanced countries have the highest proportions. In the United States,

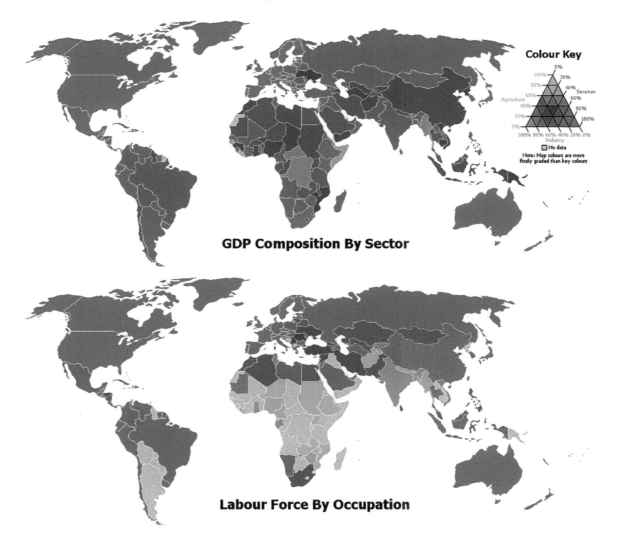

Figure 12.1: The proportion of national output and employment generated by services is closely associated with level of economic development. In wealthy countries, the bulk of activities and output is generated by services, whereas in poorer countries most people are employed in agriculture.

more than 80% of people work in services activities, whereas in poorer countries most people work in agriculture and the share of service workers is relatively low.

This chapter begins by pointing out problems in the definition of services and explaining some of the ways in which various service industries differ from one another. Next, it turns to the forces that underlie the growth of the service economy. Third, it discusses labor markets in services, contrasting them with those of manufacturing. Fourth, it summarizes the factors that affect the location of different service industries. Fifth, it addresses two important industries: **financial services** (including the recent financial crisis) and legal services. Sixth, it explores how telecommunications and the Internet have changed the operations of two key service industries: finance and clerical services.

12.1 DEFINING SERVICES

Services may be understood as the production and consumption of intangible inputs and outputs and thus stand in contrast to manufacturing, whose tangible product can be "dropped on one's foot." The definition of services as the production and consumption of intangibles makes measurements of output difficult. What, for example, is the hourly or annual output of a lawyer? A teacher? A doctor? It is impossible to measure these outputs accurately and quantitatively, yet they are real nonetheless. To complicate matters, many services generate both tangible and intangible outputs. Consider a fast-food franchise: The output is tangible, yet restaurants are generally considered a service; the same is true for a computer software firm, in which the output may be stored on CDs or as files on a server's hard drive.

The difficulty of measuring output in services complicates their analysis. For example, some critics of the service sector argue that the slowdown in productivity growth in many countries since the 1970s reflects the growth of services. Yet if output in services cannot be adequately measured, how can one argue that output per employee in services is high or low, rising or falling?

Figure 12.2 indicates how services have grown, while employment in manufacturing and agriculture has declined over time. Services now make up more than 90% of all jobs in the United States.

There are several major components of services, including the following:

1. **Producer services** are those primarily sold to and consumed by corporations rather than households. Many producer services also serve individual consumers; for example, some

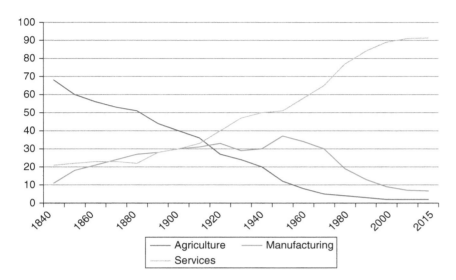

Figure 12.2: Services of different types make up the overwhelming majority of employment in the United States and the bulk of new jobs. Causes of the growth of services include changing consumer behavior, the rising demand for health care, government services, an increasingly complicated legal, financial, and technological environment, and domestic and international trade in services.

attorneys cater to both commercial clients and individuals. Producer services are commonly divided into two major groups:

a. The finance, insurance, and real estate (**FIRE**) **sector** includes commercial and investment banking, insurance of all types (e.g., property, medical, life, casualty), and the commercial and residential real estate industry.

b. **Business services** consist of legal services, advertising, computer services, engineering and architecture, public relations, accounting, research and development, and consulting that are sold mostly to other firms.

2. Transportation and communications industries move information, goods, and people. They include electronic media (film, radio, television, Internet), trucking, shipping, railroads, airlines, and local transportation (taxis, buses, and so forth).

3. Wholesale and retail trade firms are the intermediaries between producers and consumers, including, for example, warehouses, grocery stores, department stores, and large outlets such as Wal-Mart and Target.

4. **Consumer services** include as eating and drinking establishments, personal services (e.g., barbers, nail salons, interior designers), and repair and maintenance services, which, in contrast to business services, are sold primarily to individuals and households. All of these need to be located where their target customers can access them, either near their customers' homes or on transit networks. Various parts of services such as transportation, restaurants, and hotels, and motels that are oriented toward entertainment comprise elements of tourism, the world's largest industry in terms of employment (Chapter 6).
5. Government at the national, state, and local levels includes civil servants, the armed forces, and all those involved in the provision of public services (e.g., public education and health care, police and fire departments, social workers, and so forth).
6. The **nonprofit sector** includes charities, churches, museums, zoos, membership organizations, and private, nonprofit health care agencies, many of which play influential roles in local economies.

Every definition of services is slippery, however. For example, does the term *services* refer to a set of industries or occupations? (The U.S. government uses industries.) Yet when measured on the basis of their daily activities, many workers in manufacturing are in fact service workers. They include, for example, managers and executives; researchers; and administrative, clerical, and maintenance workers. Is the secretary at an automobile company part of manufacturing, but the secretary at a bank part of services? The use of industrial versus occupational definitions is particularly critical given the growth of many of the above "**nondirect production**" positions within manufacturing firms; many workers in manufacturing companies do not engage in production per se, but in service occupations such as accounting. Clearly, different definitions of services affect assessments of the size and composition of the service economy.

Services make up the vast bulk of output and employment in all economically developed countries. Indeed, the service sector employs more than 80% of the labor force of the United States. Similar proportions hold in Europe, Canada, and Japan. Even as early as 1910, services exceeded manufacturing in the United States, indicating that it was a "postindustrial" economy before becoming an industrial one! Further, services make up the vast majority—often 90%—of all *new* jobs generated in these economies, indicating that they are not only predominantly service oriented but are becoming increasingly more so. Even in much of the developing world, services make up a large share of the labor force, including much of the "informal" (untaxed, unregulated) economy (Chapter 7). This fact undermines the simplistic idea that all economies move through a rigid series of stages (from agricultural to industrial to postindustrial).

12.2 THE GROWTH OF SERVICES

Why have services grown so rapidly? In economically developed countries, employment in services has increased steadily despite low population growth, sluggish increases in incomes and productivity rates, and significant manufacturing job loss. Although both services and manufacturing produce commodities, services clearly exhibit growth and dynamics that differ from those of manufacturing. Frequently cited reasons for the growth of services include rising incomes, growing demand for health care and education, the increasingly complex economic and legal environment facing many firms, the growth of the public sector, and service exports.

RISING INCOMES

One possible explanation for the growth of services claims that income per capita has gradually been rising, particularly in the industrialized world. The demand for many services tends to increase as household purchasing power rises. Services prone to particularly high growth as countries become wealthier include entertainment, health care, and transportation. For example, as most people's incomes rise, they spend more money dining out and taking vacations. Thus, today U.S. households spend more on services than they do on durable goods (e.g., cars) and nondurable goods (e.g., clothes and food) (Figure 12.3).

This growth stems in part from the increasing value of time to individuals with rising incomes (especially with two income earners per family). As the value of time (measured by income) climbs relative to other commodities, busy consumers attempt to minimize the time necessary for many ordinary tasks such as cooking, cleaning, or repair, preferring to use that time for leisure and recreation. Although this phenomenon also explains the demand for dishwashers, automobiles, and fast food, it especially promotes the growth of services. More people order fast food, for example, to save time, or have their car's oil changed rather than doing so themselves, or have their clothes dry-cleaned rather than cleaning them at home. Thus, the increasing value of time has led to a progressive increase in the services that households purchase, so that work people used to do for themselves becomes a commodity purchased through the market.

Although popular, this explanation does not entirely account for the recent increase of the service sector in developed nations and faces some factual difficulties. First, real median household incomes in most industrialized countries have essentially remained stagnant or even declined over the past 40 years from deindustrialization, the offshoring of jobs overseas, weak unions, and the creation of many low-paying jobs. Second, rising consumer incomes would only explain the growth of services that cater to individuals and households (consumer

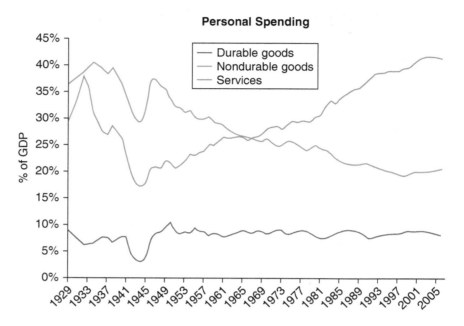

Figure 12.3: American households spend more today on services than durable or nondurable goods. As countries become wealthier, people allocate a larger share of their disposable incomes to the purchase of services, especially entertainment, travel, and health care.

services), not those that serve other firms (producer services); producer services, however, have grown the most rapidly. Clearly, then, something other than rising incomes is at work in driving the growth of the service economy.

DEMAND FOR HEALTH CARE AND EDUCATION

The rising demand for health and educational services has significantly fueled the growth of the service economy. Health services employment and expenditures have increased steadily throughout Europe, North America, and Japan. This increase has produced political conflicts about how to contain expenses, such as American debates about the rising costs of Medicaid and Medicare. In the United States, health care expenditures surpassed $2.7 trillion in 2010, roughly 17% of gross domestic product (GDP) and much higher than any other industrialized country

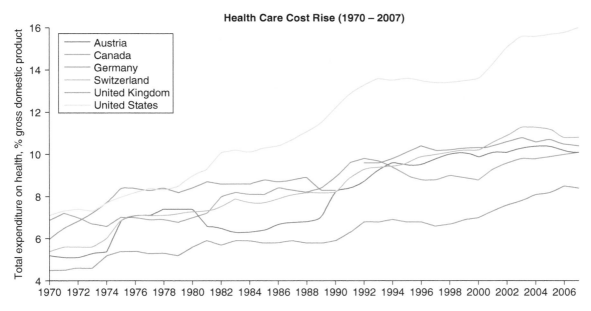

Figure 12.4: The only industrial country without national health insurance, the United States also has the most inefficient system, as measured by percent of GDP. Countries with government-sponsored health insurance tend to allocate a lower amount of national output to health care.

(Figure 12.4). The only developed country to lack a national health insurance ("single payer") system, the United States also has the most inefficient system. America spends far more per capita on health care than do countries with single-payer systems, and it devotes a larger share of its national output to health care than do Europe or Japan. Expenditures on medical emergencies constitute the leading cause of bankruptcy. Health care forms the largest and most important part of the local economy in cities including former steel capital Pittsburgh, Pennsylvania; Cleveland, Ohio; and Birmingham, Alabama.

The demand for health care has increased steadily in large part because of the changing demographics of industrialized countries. Throughout the more developed world, the most rapidly growing age groups today are the middle aged and the elderly (Chapter 8), precisely those that require high per capita levels of medical care. As the baby boom enters its retirement years and life expectancies increase, the demand for health services will only rise. Increased spending on medical equipment, medical insurance, litigation, and pharmaceutical research have also driven costs higher.

Similarly, globalization, increasing technological sophistication, and increasing demand for math and computer skills in the workplace have increased demand for educational services at all levels. University enrollments have soared as a college degree has become a requirement for most middle class jobs. Thus, whereas relatively few high school graduates in the United States attended college in the 1960s, today almost 40% do so.

INCREASINGLY COMPLEX ECONOMIC AND LEGAL ENVIRONMENT

Corporations today face a complex market and legal environment that includes specialized clients in niche markets, complicated tax codes, environmental and labor restrictions, technological changes, international competition, sophisticated financial markets, and real estate purchases and sales. Deregulation—the lifting of state controls in many industries—increased the uncertainty faced by many firms and had significant impacts on the profitability, industrial organization, and geographic location of numerous industries.

To succeed in this environment, firms require many service workers of various types. These include clerical workers to assist with mountains of paperwork; researchers to study market demand and create new products; advertisers, public relations professionals, and salespeople to market these products; lawyers and tax specialists to assist firms in litigation and following government regulations; accountants and financial experts to deal with banks and stock and bond markets; and legions of managers to make strategic decisions in a complicated environment. Similarly, sophisticated machinery requires maintenance and repair personnel, whereas offices or industrial plants need security and building maintenance staff—all service occupations. Although income-based or demographic arguments about the rise of the service sector focus on consumer services, this explanation accounts for the rise of business services. Such producer services have become the most rapidly growing part of the economy of most developed countries.

GROWTH OF THE PUBLIC SECTOR

A fourth reason underpinning the growth of services is the increasing size and role of the public sector. Governments contribute to the growth of services in two ways. First, government employment has increased steadily, especially since the 1930s, because the public demands the services that it provides. Governments provide services that are socially necessary but not profitable, such as public transportation and education, libraries, police and fire services, social workers, disaster relief, postal services, public health, and health care for the poor. Today, the federal government is the largest single employer in the United States, employing more than four million people. However, as a whole, local and municipal governments across the United States have far more workers (Figure 12.5).

A second way that government contributes to the growth of services is by enacting regulation that contributes to the demand for business services (as discussed above). Today's labyrinthine web of restrictions includes tax, environmental, zoning, labor, and antidiscrimination laws. To negotiate this legal and regulatory environment, businesses need attorneys, accountants, consultants, and other specialists.

SERVICE EXPORTS

A fifth reason for the growth of services is the rising levels of service exports within and among countries. "Export" in this context refers to the location of those who pays the bills: If the client is located

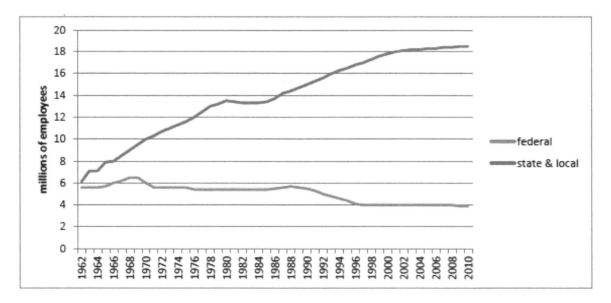

Figure 12.5: The growth of government, particularly at the local scale, has contributed to the expansion of services.

elsewhere, then sales of a service are exported. Many economists and development officials have long incorrectly believed that services always have local clients, and are thus of secondary importance to manufacturing. However, in reality the economies of many cities, regions, and countries derive a substantial portion of their incomes from the sale of services to clients located elsewhere. Many urban areas export services to other parts of the same country. Consider, for example, cities such as Las Vegas or New Orleans, whose economies revolve almost exclusively around exports of tourism and entertainment services to visitors. Whenever a television station in one city sells advertising time to a client in another, that city exports services. New York City depends heavily on exports of financial, accounting, advertising, and legal services domestically as well as globally, and Washington, DC, exports government services to the rest of the country.

Services also make up roughly 20% of international trade. Internationally, the United States is a net exporter of services, which form one-third of its foreign revenues, with an annual trade surplus of roughly $144 billion in 2016. However, the United States runs major trade deficits in manufactured goods of more than $840 billion. Exports of services have helped services employment expand domestically. Indeed, as the United States has lost much of its comparative advantage in manufacturing, it may have gained a new one in financial and business services. Sales of services among countries take many forms including tourism, fees and royalties from licensing of products and services (e.g., movies and music), business services, and profits from bank loans to overseas clients. Cities such as New York, Los Angeles, London, and Tokyo that are critical to global capital markets (e.g., markets in stocks and bonds) have benefited the most.

12.3 SERVICES LABOR MARKETS

In the economically developed world, the vast majority—often more than 80%—of all jobs involve services of one form or another. Further, new jobs—on the order of 90%—are overwhelmingly concentrated in services. Of course, these statistics include an enormously diverse array of industries and occupations. Nonetheless, despite these variations, the diverse labor markets in which most workers find employment demonstrate several common characteristics. Services jobs exhibit properties that simultaneously resemble and differ from those of manufacturing. Among these are their relatively labor-intensive nature, income distribution, gender composition, relative lack of unionization, and educational requirements.

LABOR INTENSITY

In contrast to manufacturing, most services tend to be relatively labor-intensive: They use relatively more labor per unit of output (e.g., workers per million dollars in revenue). Accordingly, the costs of wages and salaries for most services firms range from 70% to 90% of the total, compared with 5% to 40% in most manufacturing firms. Of course, some services, such as finance, can be quite capital-intensive (i.e., using large quantities of machinery, notably computers) and generate huge profits with minimal workers. However, services such as education and medical care require large numbers of employees. Manufacturing output has been relatively easy to mechanize—witness the remarkable changes since the Industrial Revolution (Chapter 5). In services the replacement of workers with machines is much more difficult and costly, particularly for jobs that involve variations in tasks, judgment, or dexterity. Some services, obviously, have exhibited enormous technological change. Examples include personal computers in the office and automatic scanners in retail stores. In poorly paying jobs, however, firms' incentives to replace workers with machines may be relatively low.

INCOME DISTRIBUTION

The distribution of incomes in services occupations has troubled many observers. The standard argument holds that industrial economies created societies with a large middle class and relatively few rich or poor (Figure 12.6). In contrast, services-based economies are frequently held by some observers to produce jobs into two categories: well-paying, white-collar managerial/professional jobs that require a college degree; and unskilled, low-paying jobs that require little to no higher education (for example, retail sales personnel, many health services, security guards, and so forth). Indeed, in contrast to early expectations that a service-based economy would eliminate poverty, today a large share of new service jobs pay poorly, offer few benefits, and are part time or temporary in duration. Many observers have described this concept as the "McDonaldization" or "K-Martization" of the economy. The most rapidly growing occupational groups in the United States, including over the next decade, include professionals but also low-wage service workers and retail trade employees.

The occupations with the greatest relative projected job growth include almost anything having to do with computers as well as skilled and unskilled health care professions. However, in absolute terms, the greatest number of new opportunities will be in low-wage, unskilled positions such as food preparation, customer service, retail sales, cashiers, clerks, and security

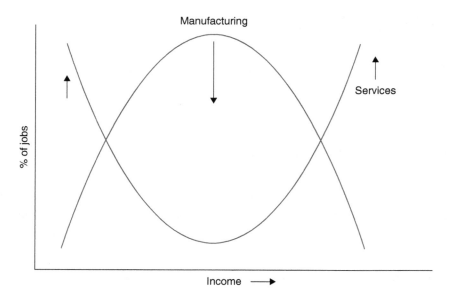

Figure 12.6: This graph illustrates a common view of the transition from a manufacturing-based economy to a service-based one. In this view, the decline of manufacturing erodes middle-class jobs. The growth of services increases jobs that are unskilled and poorly paid as well as those that are highly skilled and well paid.

guards. Further, the United States has witnessed a steady growth of unskilled part-time jobs typically filled by women and minorities (Table 12.1) who may be looking for full-time work. This process has heralded the birth of the **"working poor"** and the homeless caught between jobs that pay too little and housing that costs too much.

With deindustrialization reducing the available manufacturing jobs and service jobs becoming a larger share of the economy, concern has mounted that service professions produce income inequity. In general, most services have tended to pay poorly compared with manufacturing: The average clerical position in the United States pays only 60% of the annual salary of a blue-collar industrial worker, and the average retail trade job only 50% as much. These fears underlie current public discussion about the "declining middle class" and the polarization of postindustrial economies into distinct groups of "haves" and "have nots."

Statistically, however, little evidence suggests that the distribution of incomes among services occupations differs significantly from that in manufacturing (Figure 12.7). Indeed, services generate the vast majority of employment that allows millions of well-paid, middle-class

TABLE 12.1: TWENTY LEADING OCCUPATIONS OF EMPLOYED WOMEN IN THE UNITED STATES, 2015

OCCUPATION	% WOMEN
Secretaries	96.7
Elementary school teachers	96.7
Nurses	91.8
Home health aides	88.3
Cashiers	75.0
Administrative support	69.5
Retail sales staff	41.1
Customer service staff	70.1
Clerks	91.2
Accountants	60.8
Receptionists	93.9
Maids	89.7
Office clerks	83.8
Secondary school teachers	54.8
Waiters/waitresses	67.3
Financial managers	55.7
Teacher assistants	91.7
Preschool teachers	97.7
Social workers	76.1

suburbanites to live relatively comfortable lives. Although statistical evidence indicates mounting inequality in income distributions, particularly in the United States, this trend may reflect the changing of the U.S. tax code, which has greatly favored the wealthy, as well as the growth of unearned income (e.g., rents from property, royalties from patents, stock dividends), from which the rich derive vast bulk of their earnings. Thus, although the growth of low-wage services may have helped to keep family incomes stagnant; there are other drivers of income inequality.

GENDER COMPOSITION

A third dimension of labor markets in services concerns their gender composition. The developed world's previous manufacturing-based

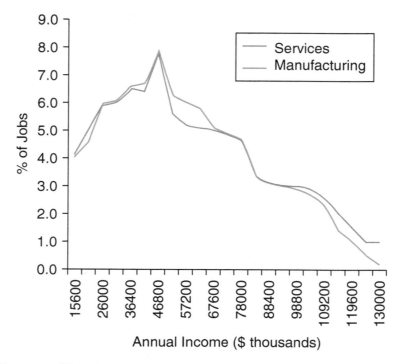

Figure 12.7: The U.S. wage distribution, 2015. Despite allegations that services cause income inequality, in fact the distribution of income in services resembles that of manufacturing. Income inequality is thus produced by other factors, including the growth of unearned income, or returns from assets such as stocks, which tend to be owned by the wealthy. The U.S. tax system also taxes such income at a lower rate than income from a job.

economies predominantly employed males. Although some women worked in factories (particularly textiles and garments), industrial economies traditionally saw relatively low rates of female labor force participation. Heavy manufacturing industries such as steel, automobiles, rubber, tool-and-die production, machine tools, and petrochemicals have long been heavily male-dominated domains. Thus, manufacturing-based economies also gave rise to a family structure in which men were the primary breadwinners. Most women who worked outside the home did so either for brief periods before getting married or in a few specialized occupations such as teaching or nursing.

In contrast, the growth of services in Europe, North America, and Japan since World War II has been accompanied by the steady increase in women working outside the home. The most rapid rise in women's labor force participation rates has been among married women with children. In the United States today, women make up 48% of all full-time employees. However, women's entry into services jobs has been in large part limited to so-called pink-collar jobs, including clerical and secretarial work, retail trade, health care (other than doctors), eating and drinking establishments, teaching, and child care. Most of these jobs pay relatively poorly. Indeed, until very recently, women's presence in well-paying occupations such as physicians, attorneys, or corporate management has been relatively low. Some women have complained about not being able to enter the highest levels of corporations, noting a "glass ceiling" that limited their upward mobility. This clustering in low-paying jobs contributes to the fact that U. S. women on average have incomes only 80% as high as those of men.

As the service sector expands, the rise of women's labor force participation carries social implications. The increase in women's work may compensate for declining male incomes in the face of deindustrialization. The rise of the two-income family has significantly changed gender roles at home and led to pressures to redistribute housework to men. In countries without national day care systems, such as the United States, the need for child care constrains women's job opportunities. Middle-class families capable of paying for private child care can generally overcome this constraint. However it represents a source of financial and emotional stress, complicating commuting patterns and changing the socialization patterns for the young. Most children today enjoy relatively little free time compared with earlier generations. For the poor, who typically cannot afford access to professional child care, adequate or otherwise, the rise in women's paid work has created a child care crisis.

LOW DEGREE OF UNIONIZATION

Since World War II, the percentage of workers belonging to unions has declined steadily (from 45% in the United States in 1950 to 12% today), largely due to deindustrialization. Despite their rapid growth, jobs in services are rarely unionized; the vast majority of workers in retail trade and in many skilled professions do not belong to unions. Very few restaurant workers, cashiers, security guards, or taxi drivers are unionized, for example. Neither are there many unionized attorneys, accountants, doctors, or professors. Among American workers, service employees are occasionally organized through unions in health care (e.g., Union 1199), teaching (e.g., American Federation of Teachers), or the public sector (e.g., American Federation of State, County, and Municipal Employees and Service Employees International). However, these are generally exceptions. Attempts to foment unionization in low-wage sectors such as large retail chain stores have been met with fierce resistance.

EDUCATIONAL REQUIREMENTS

Skilled managerial and professional service occupations require more education than do jobs in manufacturing. Employers in this part of the service sector usually require verbal, quantitative, and computer skills. Unemployment rates are almost always lower for well-educated workers compared with those without a high school degree. Moreover, the average income of workers with advanced degrees is considerably higher than those without (Figure 12.8). College degrees tend to pay for themselves many times over in terms of higher lifetime earnings. In the current financially depressed economy, however, many college graduates shoulder large quantities of debt and face dismal job prospects: In 2016, the average college graduate had $31,000 in student loan debt and total college student debt exceeded $1 trillion (larger than total credit card debt), and one-half of recent graduates were unemployed or underemployed (and as many as 85% moved back in with their parents).

The demand for higher educational services has risen throughout the industrialized world as a university degree has become a prerequisite for middle-class jobs. In the United States, for example, whereas

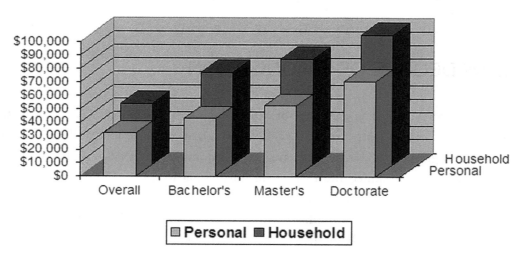

Figure 12.8: The growth of producer services has raised the importance of education. Those with more advanced degrees consistently earn higher salaries than those with less education. People who don't finish high school tend to be the poorest, most likely to be unemployed, and have few career prospects.

only 20% of high school graduates in the 1950s went on to attend universities, today roughly 40% do so (although many do not graduate). Accordingly, many industrial countries recognize that in the context of a knowledge-based economy, higher education bolsters national and local economic competitiveness. Within the United States, likewise, states that have invested in high-quality university systems have enjoyed a competitive advantage over those that have neglected educational funding. Such an advantage holds particularly in technology and business services. People who lack access to a university education—overwhelmingly the poor and minorities—face a lifetime of poverty. Whereas in the industrial economy of the early twentieth century, a strong back and arms ensured at least a lower middle-class lifestyle, today those who do not complete high school find few venues to compete in the labor market.

12.4 THE GEOGRAPHY OF SERVICES

Given their rapid growth, the location and functions of producer services are increasingly important. Generally, highly skilled, high-value-added functions such as financial and business services concentrate in metropolitan regions. Two-thirds of business and professional services jobs in the United States are located in large urban areas, which offer agglomeration economies in the forms of easy access to local suppliers and subcontractors (particularly in negotiating deals that require face-to-face contact), access to a pool of labor with a wider range of skills than is found in nonurban areas, superior transportation infrastructure and facilities, and a range of office accommodations and telecommunications facilities. Firms in the producer services sector, including finance and a numerous types of business services, benefit greatly by locating near other firms. Often the need for specialized information, which circulates through very narrow channels, constrains companies to locate in a certain part of a metropolis (e.g., New York City's Madison Ave. concentration of advertising firms or Hollywood film studios).

As discussed earlier in this chapter, a large part of the service sector concerns consumer services—those sold to individuals and households. We have seen that consumer services include a wide variety of industries and occupations, including retail trade and personal services, eating and drinking establishments, tourism, and sports and entertainment facilities. To some extent, the health care and education fields also provide consumer services.

The spatial distribution of consumer services largely reflects the degree of affluence of the local client base. As noted earlier, when disposable incomes rise, consumer services tend to thrive, although expenditures on goods and services do not rise equally for all types. The study of consumer services is thus intimately linked to the analysis of local demographics. Key factors include the associated tastes and preferences of the people in question; their inclination to save or spend; the advertising and other sources of information that shape their preferences; and their age, gender, and ethnic dimensions that affect spending and buying habits. Other forces that shape consumer services include changing household work patterns, the growth of chains and franchises, regulatory frameworks, and the relative prices of imports and other goods. (Import prices themselves reflect a galaxy of variables in the global economy.) The ever-changing geography of consumer services also depends on the location of purchasing power. For example, as suburbanization drew much of the middle class and its income to the metropolitan periphery, the urban geography of retail trade was reshaped, leading to an explosion of suburban malls (although the recent economic downturn has exposed a glut of malls and led some to close).

12.5 TELECOMMUNICATIONS AND GEOGRAPHY

Because large, complex economies depend on the circulation of vast quantities of information, the history of capitalism has produced wave after wave of innovations in communications.

Telecommunications began in 1844 with Samuel Morse's invention of the telegraph. The telegraph made possible the worldwide transmission of information concerning commodity needs, supplies, prices, and shipments—information that was essential to the efficient conduct of international commerce. Telegraphy grew rapidly in the United States, playing an important role in the American settlement of the West and allowing long-distance circulation of news, prices, stocks, and other information. In 1868, the first successful trans-Atlantic telegraph line was laid.

For decades after the invention of the telephone in 1876, telecommunications meant telephone service. Just as the telegraph facilitated the development of the American West, in the late nineteenth century the telephone became critical to the growth of American cities, allowing firms to centralize their headquarters functions (i.e., corporate decision making) while they spun off branch plants to smaller towns. Such multiestablishment corporations with offices in many cities used the telephone to coordinate production and shipments. In the 1920s, the telephone, like the automobile and the single-family home, became a staple of the growing middle class. It significantly affected friendship networks, dating, and other social ties. In 1956, the first international phone line was laid across the Atlantic Ocean. Today the telephone remains the most commonly used means of communication for businesses and households.

The most startling development in recent telecommunications has been the rise of the mobile or cellular phone. Since the late 1990s, when it was introduced, the cell phone has become the most rapidly diffusing technology in history. For developing countries, wireless technologies allow for "leapfrogging," i.e., moving directly into newer, lower cost forms of technology. Rapid decreases in size and cost of mobile phones, and the minimal infrastructure necessary to cultivate their networks, have made them increasingly affordable and convenient for vast numbers of people. In 2016, 98 percent of the world's inhabitants used mobile phones, giving almost everyone connectivity for the first time in human history (Figure 12.9). Mobile phones have become especially popular in the developing world, giving large numbers of the poor to have access to cheap and reliable telecommunications. Mobile telephony has had important repercussions for the contours and rhythms of daily life, including a mounting porosity between the private and public realms as once private conversations are held in public spaces. It has allowed for sharply improved coordination of mobility, improved use of time while driving or walking, security in case of emergencies, and its enhanced connectivity erodes the advantages of physical proximity. It has allowed mobile access to the Internet, and for many people is their only form of access. Texting has come to rival or surpass voice messages on many world telecommunications networks. Roughly 80% of the world's mobile phone users engaged in texting, sending a total of 6.1 trillion messages annually, or 200,000

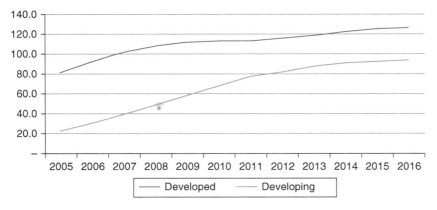

Figure 12.9: Cellular or mobile phone subscriptions per 100 people.
Source: International Telecommunications Union.

per second. Everywhere, teenagers took the lead in adopting texting as a means of telephony, although the precise applications range noticeably among different cultures.

The microelectronics revolution that saw the introduction of computers and digital information starting in the 1970s was particularly important in the telecommunications industry, which is arguably the world's most rapidly expanding and dynamic sector today. The ability to transmit vast quantities of information in real time over the planetary surface is crucial to contemporary capitalism. Large corporations need it to operate in multiple national markets simultaneously, coordinating the activities of thousands of employees within highly specialized corporate divisions of labor. Thus, telecommunications are important to understanding broader issues pertaining to globalization and the world economy, including the complex relations between firms and nation-states (see Chapter 6).

Today, two technologies—satellites and fiber-optic lines—underlie the global telecommunications industry. The transmission capacities of both grew rapidly in the late twentieth century. Multinational corporations in agriculture, manufacturing, and services (e.g., banks and media conglomerates) typically employ both technologies. These organizations either own private satellite and fiber-optic facilities or lease circuits from shared corporate networks. Roughly 1,000 fiber optic and two dozen public and private satellite firms provide international telecommunications services. The network of fiber lines linking the world constitutes the nervous system of the global financial and service economy, linking cities, markets, suppliers, and clients around the world (Figure 12.10).

There exists considerable popular confusion about the real and potential impacts of telecommunications on spatial relations, in part due to the long history of exaggerated claims made in the past. We often read, for example, that telecommunications will bring about "the end of geography, the "death of distance" or a "flat world": The erosion of the significance of place to the point where anything can occur anywhere. Often such views ignore the complex relations between telecommunications and local economic, social, and political circumstances. For example, repeated predictions that telecommunications would allow everyone to work at home via telecommuting, dispersing all functions and spelling the obsolescence of cities, have fallen flat. Rather, the past decade has seen persistent growth in densely inhabited urbanized places and global cities. In fact, telecommunications are usually a poor substitute for face-to-face meetings. Most sensitive corporate interactions still take place in person, particularly when the information involved is irregular, proprietary, unstandardized in nature, and requiring high degrees of trust. Most managers spend the bulk of their working time engaged in face-to-face contact such as meetings. No electronic technology can yet allow for the subtlety and nuances critical to such encounters.

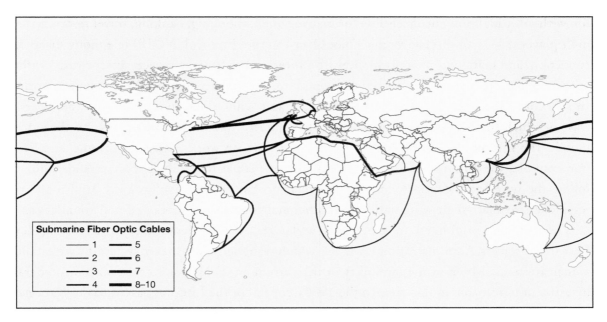

Figure 12.10: The world's fiber-optics networks form the backbone of the global telecommunications system. Fiber carriers much more information and is more secure than satellites, and transmits the vast bulk of information worldwide.

For this reason, a century of telecommunications has left most high-wage, white-collar, administrative functions clustered in downtown areas. In contrast, telecommunications are ideally suited for the transmission of routinized, standardized forms of data such as computerized files, medical records, spreadsheets, blueprints, and text documents. The processing of such information can easily be outsourced to low-wage regions. By allowing the decentralization of routinized functions, information technology actually contributes to the urban concentration of nonroutine, high-value-added functions that are performed face to face. This outsourcing frees land and labor that otherwise would be used for the processing of standardized data.

GEOGRAPHIES OF THE INTERNET

The most famous and significant telecommunications network today is clearly the Internet. The Internet is used for so many purposes that life without it is simply inconceivable for many: e-mail, shopping, banking, travel reservations, multiplayer video games, chat rooms, Voice Over Internet Protocol (VOIP) telephony, distance education, and searching for information. The Internet has become a necessity for vast swaths of the population in the economically developed world, and, increasingly, the developing world. In this context, simple divisions such as "offline" and "online" fail to do justice to the diverse ways in which the "real" and virtual worlds interconnect. Like the printing press and the automobile, the Internet has significantly changed the ways we work, interact, consume, and live. For example, the Internet allows large numbers of people to conduct their errands online; follow the news; and stay in touch with friends and family who live far away. It also allows researchers to share their work with one another and collaborate across national boundaries.

The Internet originated in the 1960s under the U.S. Defense Department's Agency Research Projects Administration (ARPA), which designed a computer system to withstand a nuclear attack. Much of the durability of the current system is due to the enormous federal investment in research in this area. In the 1980s, control of the Internet passed to the National Science Foundation, and in the 1990s control was privatized via a consortium of telecommunications corporations. The Internet emerged on a global scale through the integration of existing telephone, fiber-optic, and satellite systems. Technological innovations making this possible included packet switching and TCP/IP (Transmission Control Protocol/Internet Protocol), in which individual messages are decomposed, the constituent parts transmitted by various channels, and then reassembled at the destination. In the 1990s, graphical interfaces developed in Europe greatly simplified the use of the Internet and led to the creation of the World Wide Web.

The growth of the Internet has been phenomenal. Indeed, it is arguably the most rapidly diffusing technology in world history. In March 2017, an estimated 3.7 billion people, or roughly 50% of the world's population, in more than 200 countries, were connected to the Internet (Figure 12.11). However, the percentage of people with Internet access varies greatly among the world's major regions, ranging from as little as 0.2% in parts of Africa to as high as 100% in Scandinavia (Figure 12.13). In the United States, Internet penetration was 84% of the population. Inequalities in access to the Internet internationally reflect the long-standing split between more and less developed nations. Although virtually no country lacks Internet access, the variations among and within nations (particularly between urban and rural areas) in accessibility are huge.

However, this situation is not static. The most salient feature about the Internet may be its exceedingly rapid rate of growth. Very few technologies in world history, with perhaps the exception of the mobile phone, have exhibited such explosive rates of adoption. Explosive growth is readily evident in sub-Saharan Africa and the Middle East (Figure 12.14), where growth rates between 2000 and 2017 exceeded 10,000% (and sometimes reach absurdly high

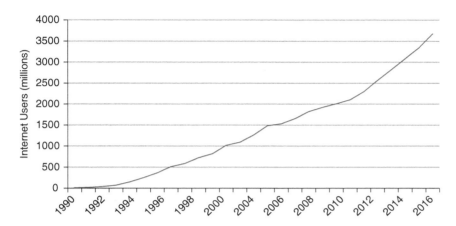

Figure 12.11: Growth in global Internet users, 1990–2016.
Source: Author, using data from Internet World Stats (internetworldstats.com).

rates such as 217 million percent, albeit from a very small base). In contrast, growth rates in the North America, Europe, Russia, Japan, and Oceania were relatively modest. Thus, although the Internet was largely confined to the developed world early in its history, it is growing the most rapidly in the developing world today, particularly in Africa and Asia. This growth brought 1.325 billion new users online

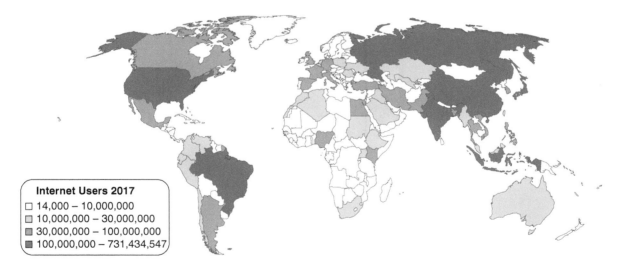

Figure 12.12: The geography of the world's "netizens" reveals that China and the United States have the largest numbers of Internet users. In China, more than 730 million people use the Internet, more than twice the entire U.S. population.

Figure 12.13: Internet penetration rates, or the proportion of people using it, reflect global distributions of incomes and the price of computers and access at cybercafés. Despite myths that cyberspace makes geography irrelevant, the truth is that where people live greatly shapes their likelihood of access.

during this period. The enormous growth rates of the Internet mean that digital divides are rapidly changing, and as access improves for many hitherto marginalized groups, may slowly decline over time.

These discrepancies in Internet access among and within countries largely follow the lines of wealth, gender, and race, a phenomenon

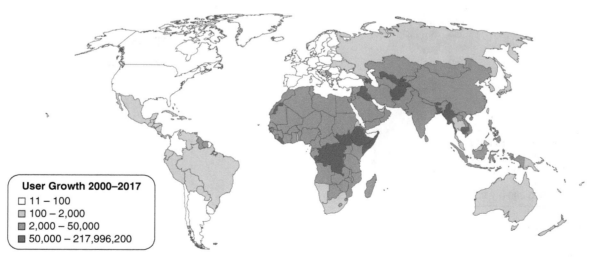

Figure 12.14: The extremely rapid growth of the Internet in the developing world is bringing hundreds of millions of new users online every year.

known as the **digital divide**. Access to the Internet, either at home or at work, correlates highly with income. Wealthier households are far more likely than the poor to have a personal computer and modem at home. In the United States, white households use networked computers more frequently than do African American or Latino ones. The elderly likewise often find the Internet to be intimidating and unaffordable, although they comprise the fastest growing demographic group of users. American Internet users thus tend to be predominantly white, middle class, well educated, younger than average, and employed in professional occupations demanding college degrees.

Social and spatial differences in access to the skills, equipment, and software necessary to get online threaten to create a large, predominantly minority underclass deprived of the benefits of cyberspace (e.g., looking for jobs). Modern economies are increasingly divided between those workers who are comfortable and proficient with digital technology and those who neither understand nor trust it. Despite the falling prices for hardware and software, basic entry-level machines for Internet access cost as little as $500, a significant sum for low-income households. Internet access at work is also unavailable to employees in poorly paying service jobs (the most rapidly growing category of employment).

Internationally, access to the Internet depends on the density, reliability, and affordability of national telephone systems. Most Internet communications occurs along fiber optic lines leased from telephone companies. Some of these companies are state regulated (in contrast to the largely unregulated state of the Internet itself) although the global wave of privatization is ending government ownership in this sector. Nations with telecommunications monopolies have higher prices than those with deregulated systems, and hence fewer people use the services. The global move toward deregulation in telecommunications has led to more user-based pricing, in which users must bear the full costs of their usage, and fewer cross-subsidies among different groups of users (e.g., commercial users subsidizing residential ones), a trend that will likely make access to cyberspace even less affordable to low-income users.

Many of the Internet's uses revolve around entertainment, personal communication, and research. However, the Internet can also be a platform for relatively powerless, marginalized individuals to challenges repressive political and social systems. The Internet has given voice to countless groups with a multiplicity of political interests and agendas, including civil and human rights advocates, sustainable development activists, antiracist and antisexist organizations, gay and lesbian rights groups, religious movements, those espousing ethnic identities and causes, youth movements, peace and disarmament parties, nonviolent action and pacifists, animal rights groups, and gays living in homophobic local environments. By facilitating the expression of political positions that otherwise may be difficult or impossible to communicate, the Internet allows for a dramatic expansion in the range of voices heard about many issues.

For example, Internet-based services such as Twitter and YouTube played an important role in the uprisings known as the "Arab Spring" that shook Tunisia, Egypt, Libya, and other countries in 2011. In this sense, cyberspace permits the local to become global, bringing local voices to a worldwide audience. Indeed, the Internet enables users to form communities based on common interests but not physical proximity.

The Internet also has a "dark side," of illegal or immoral uses. Hackers, for example, have often wreaked havoc with computer security systems. Such individuals are typically young men working for their own amusement, with varying degrees of malice. Regardless of motive, hackers often unleash dangerous computer viruses and worms. Most hacks—by some estimates as much as 95%—go unreported, but their presence has driven up the cost of computer firewalls. Cyberwarfare has emerged as an important part of military strategy: A successful cyberattack, for example, could cause airports, electricity, and communications systems to shut down, producing chaos in a country. The dark side also includes cybercrime such as identity theft; sale of counterfeit drivers' licenses, passports, and Social Security cards; securities swindles; and adoption scams. Credit card fraud is a mounting problem: 0.25% of Internet credit card transactions are fraudulent, compared with 0.08% for non-Internet transactions. Some unsavory Internet sites even offer credit card "marketplaces," where hackers sell wholesale the credit card numbers that they have stolen from merchant sites.

SOCIAL MEDIA

Another dramatic example of Internet use is exemplified through digital social networks, which greatly expedite the creation of an online persona. Networking sites such as Friendster, LinkedIn, MySpace, and Facebook began in the early 2000s. Some are aimed at finding romantic partners (e.g., Match.com), whereas others such as YouTube allow sharing of digital content with everyone, including like-minded strangers, forming classic "communities without propinquity." More than one-half of the U.S. population used social media sites in 2011, and three-quarters of those under age 30 do so.

Facebook is by far the most popular networking site in the world, with more than 1.6 billion users in 2015 (Figure 12.15), or 24% of the planet's population: If Facebook were a country, it would be the world's largest. Its users are, naturally, unevenly distributed across the planet (Figure 12.16). The United States, with 201 million users or 65% of its population, forms the largest single national group of Facebook fans. The network is also popular throughout the Americas, Europe, South Asia, parts of Southeast Asia, and Australia and New Zealand. Facebook has decisively trounced competing networking services such as Myspace, which in

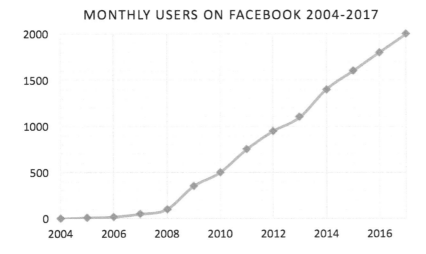

Figure 12.15: Growth of Facebook users worldwide, 2004–2016.
Source: https://commons.wikimedia.org/wiki/Category:Facebook#/media/File:Facebook_popularity.PNG.

the eyes of many users has become relegated to ethnic minorities and low-income users.

Finally, Twitter has also emerged as an important social medium. Founded in 2006, by 2016 there were worldwide roughly 350 million active Twitter accounts, which sent more than one billion tweets (short messages less than 140 characters) weekly. The vast appeal of (mobile)

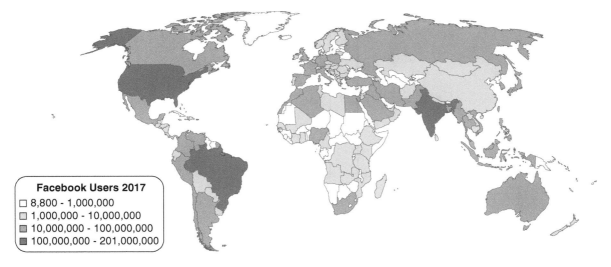

Figure 12.16: Facebook users worldwide, 2017.

social media applications proves that they are not only of relevance for younger generations, but that they impact all layers of society. Individual users, businesses, organizations, and governments alike work to determine how best to apply the various social media technologies available on the virtual horizon. Many have triumphed in doing so. Within the music sector, several stars have built their communication and marketing strategies entirely around social media. Likewise, several governments and public administration entities have discovered the usefulness of social media, including the European Union, which, via Twitter and other channels, seeks to establish a feeling of European identity among its one-half billion citizens.

■ KEY TERMS

Business services: a form of producer services that involves the creation, analysis, and sale of specialized expertise, such as law firms, engineering, marketing and public relations, accounting, and computer software and are sold mostly to firms

Consumer services: services that are sold to households and individuals, including retail trade and personal services

Digital divide: social and spatial discrepancies in access to the Internet, typically reflecting existing inequalities in income and education

Financial services: services that involve the creation, storage, lending, and manipulation of money, including banking, mortgages, and insurance

F.I.R.E. sector: finance, insurance, and real estate industries

Nondirect production: positions in manufacturing firms that are in fact service-related, including management, marketing, accounting, and clerical positions

Nonprofit sector: services that are not organized around profit maximization, including charities, zoos, blood banks, and membership organizations

Producer services: services that are sold to firms rather than individuals, including most financial and business services

Working poor: people with jobs that pay so little that they remain near or below the poverty line

■ STUDY QUESTIONS

1. What proportion of the labor force in North America, Japan, and Europe works in services?

2. Why is it difficult to measure the productivity of services?

3. What major industries comprise the service sector?

4. What are five reasons services have grown so quickly?

5. What are five ways in which labor markets in services differ from those in manufacturing?

6. Where do producer and consumer services tend to locate, and why?

7. Do telecommunications eliminate the importance of geography? Why not?

8. How did the Internet begin?

9. How much of the world uses the Internet? Where has the most rapid growth occurred?

BIBLIOGRAPHY

Beaverstock, J., Smith, R., & Taylor, P. (2000). World-city network: A new metageography? Annals of the Association of American Geographers, 90*(1), 123–134.*

Bryson, J., Daniels, P., & Warf, B. (2004). Service Worlds: People, Organizations, and Technologies. *London: Routledge.*

Corbridge, S., Martin, R., & Thrift, N. (Eds.). (1994). Money, Power, and Space. *Oxford: Basil Blackwell.*

Fischer, C. (1992). America Calling: A Social History of the Telephone to 1940. *Berkeley: University of California Press.*

Florida, R. (2014). The Rise of the Creative Class—Revisited: Revised and Expanded. *New York: Basic Books.*

Hall, M., & Page, S. (2014). The Geography of Tourism and Recreation: Environment, Place and Space. *London: Routledge.*

Schiller, D. (1999). Digital Capitalism: Networking the Global Market System. *Cambridge, MA: MIT Press.*

Schlosser, E. (2001). Fast Food Nation. *New York: Perennial.*

Standage, T. (1998). The Victorian Internet. *New York: Walker and Company.*

Warf, B. (2012). Global Geographies of the Internet. *New York: Springer.*

IMAGE CREDITS

- Figure 12.1: Stephen Morely, http://commons.wikimedia.org/wiki/File:Gdp-and-labour-force-by-sector.png. Copyright in the Public Domain.
- Figure 12.4: Wikibasti, http://commons.wikimedia.org/wiki/File:Health_care_cost_rise.svg. Copyright in the Public Domain.
- Figure 12.8: Copyright © BrendelSignature (CC BY-SA 3.0) at http://commons.wikimedia.org/wiki/File:Education_median_income_1.png.
- Figure 12.15: Copyright © Tatiraju.rishabh (CC BY-SA 3.0) at https://commons.wikimedia.org/wiki/Category:Facebook#/media/File:Facebook_popularity.PNG.

Chapter Thirteen

CULTURE, IDENTITY, AND EVERYDAY LIFE

MAIN POINTS

1. Culture is one of the most important themes in the study of human beings (Chapter 1). It refers to the webs of meaning that humans use to interpret the world and to know how to act. Culture is learned through socialization, not biologically given, and includes an enormous number of material and intangible traits.

2. A common distinction is between folk and popular culture. Essentially, the traditional lifestyles of premodern societies, folk culture changes very slowly. Popular culture refers to the rapidly changing, typically urban sets of norms, outlooks, and practices associated with modernity and advanced capitalism.

3. The cultural landscape reflects the imprint of cultures on Earth's surface, including the various ways that culture shapes land-use patterns and architecture. Landscapes are as much a part of culture as are elements such as language and religion. Culture gives meaning to place and includes all the symbolic associations that make individual locations unique.

4. Culture is deeply intertwined with social relations and can never be separated from politics. Unlike traditional views of culture, which viewed it as given and fixed, more recent ideas focus on its social and historical roots, how culture shapes individual and collective notions of identity, and the power relations that accompany this process.

5. Ethnicity is a major dimension of culture. Unlike race, which is defined as the biological differences between groups of people and has no scientific validity, ethnicity refers to

6. Culture includes different norms of behavior for males and females. Gender is an ideological and power relation that often favors men over women. Gender differences shape access to resources and the rhythms of everyday life.

7. Sexuality, or sexual preference, is often confused with gender. Although they are related, these phenomena are not identical. Like gender, sexuality shapes perceptions, behavior, and spaces.

8. Much contemporary analysis of culture is concerned with how social differences are produced and sustained, and how and why some cultural groups are politically privileged over others. How and why some groups are viewed as different—or categorized as the Other—reveals the power dynamics within and among groups.

the collective cultural traits shared by a group, including its language, religion, historical experience, identity, and territory.

Think about the various kinds of people you have met throughout your life: men and women; gays and straights; old people and children; whites, African Americans, Latinos, and Asians; tourists or immigrants from different parts of the world. What makes people think and behave differently? In large part, their backgrounds, social positions, lifestyles, and worldviews are the products of living in a certain place. Increasingly, with globalization, widely different lifestyles intersect, collide, and transform one another. In this context, understanding the social, cultural, and cognitive dimensions of human behavior, and of how differences among groups are produced and sustained, is a vital matter.

Culture is a central concept in the social sciences, including geography, and cultural geography is a vital and vibrant branch of the discipline. The study of culture and the ways in which it plays out across space takes multiple forms. We have encountered cultural geographies several times already. Chapter 1 introduced culture as a fundamental analytical theme in the social sciences made necessary by the fact that human beings are conscious and thus capable of intelligent thought. In Chapter 3, we examined the broad contours of the world's major language families, linguistic conflict, and language death. Chapter 4 focused on religions, which defined much of the cultural geography of the preindustrial world and still exert significant influence over the lives of billions of people today. Chapter 6 examined the cultural dimensions of globalization, including the invasion of Western culture into other societies. Chapter 9 focused on political geography, which is also profoundly shaped by culture, including nationalism.

This chapter examines a variety of issues concerning culture and how it shapes everyday life, group and individual identity, and people's life opportunities. It begins with an extended overview of the concept of culture, which is both complex and hugely important. In this vein, the chapter examines folk and popular culture, cultural landscapes, and the cultural construction of place. Next, it focuses on race and ethnicity, arguing that there is no biological foundation for "race," but that ethnicity is vitally important in understanding social relations and human geography. From there, it proceeds to examine gender, which intersects with other types of power relations and identity. Finally, it summarizes the notion of Orientalism. It concludes with an observation about the geography of happiness.

13.1 INTRODUCING CULTURE

Humans differ from all other animals in that they have developed not only biologically but culturally. Unlike other animals, humans can form ideas out of experiences and then act on the basis of these ideas, a process that over time led to the development of distinct ways of life. The term culture has a variety of meanings, usually associated with a way of life; the cumulative knowledge, beliefs, experiences, and values of a particular group. Although all cultures share certain basic similarities—all need to reproduce and to obtain food and shelter—they differ in the methods they use to achieve those goals. As humans settled the world, culture initially evolved in close association with the physical environment. Over time, however, cultural adaptations brought increasing freedom from environmental constraints; gradually, as human ties to the physical environment loosened, the ties to their culture increased.

Culture is a complex concept, and includes a variety of **culture traits,** or different aspects and dimensions. Often culture is divided into nonmaterial and material categories. **Nonmaterial culture** includes ideologies and beliefs, such as language and religion; traditions; values toward gender and sexuality; age-related roles; and a variety of other symbolic aspects of human life such as the norms involved in group formation and family structure. **Material culture** comprises all the human-made physical objects and elements related to people's lives and livelihood, including objects, buildings, and landscapes.

Culture is produced and reproduced both collectively, that is, as a social phenomenon, and psychologically, that is, within the minds of individuals. Early definitions of culture posited that it was simply the sum total of learned behavior. Culture is learned, not biologically inherited. A more nuanced view holds that culture is the knowledge we take for granted, what we know without knowing it, that is, common sense. Culture allows people to negotiate their way through everyday life. Everyone must internalize a culture to survive in it, to know how to act, for example, male or female, young or old, wealthy or poor. Culture defines what is normal and what is not, what is important and what is not, what is acceptable and what is not, within each social context. Culture frames our perceptions of the world, how it works, and who we are—that is, culture and identity are closely linked.

Culture is acquired through a lifelong process of **socialization**, that is, the impact of families, friendships, role models, educational systems, the media, and other institutions that inform people about how to behave. Individuals never live in a social vacuum but are continually socially produced from cradle to grave, so socialization is a lifelong process. Many people, particularly in the individualistic West, tend to think of themselves as independent beings. However in reality we are all produced through our interactions with others. The socialization of the individual and the reproduction of society and place are, therefore, two facets of one process. Such a view asserts that cultures are always intertwined with political relations and are continually contested. Representations and explanations for prevailing class, gender, and ethnic dominance are often challenged by marginalized voices from the social periphery. **Cultural assimilation** refers to the process by which minority groups incorporate the culture of a dominant majority; for example, immigrants are often assimilated into American society. When individuals or groups borrow from another group and internalize parts of its culture, the process of **acculturation** occurs. Thus, the meaning and the power of culture are closely tied to how a society is organized and how it changes.

As humans settled the world, different cultures evolved in different locations. Cultural geographers thus often speak of **cultural regions** that share numerous cultural characteristics, such as the Arab world or Latin America. Despite their many underlying similarities, each culture developed its own variations on the basic elements: language, religion, political system, kinship ties, and economic organization. Enormous variations exist within cultural regions. Each contains distinct patterns of class and gender relations and often cultural minorities.

13.2 FOLK CULTURE AND POPULAR CULTURE

Traditional cultural analysis distinguishes between "folk" and "popular" cultures. **Folk culture** refers to traditional, often rural social and symbolic practices and cultural artifacts. Folk cultures are less subject to change over time, are frequently oral and preliterate, and exhibit more local variations geographically than does popular culture. Folk cultures have preindustrial origins, and are often quite old historically. Religion and ethnicity are often key unifying variables, and traditional family and kin are paramount. Today, traditional folk cultures survive mainly in rural areas among groups linked by a distinctive ethnic background and/or religion and language. Folk cultures tend to resist change and to remain attached to long-standing attitudes and behaviors.

In contrast, **popular culture** is largely urban and tends to embrace change—even to the point of seeking change for its own sake, as in the case of fashion. Popular cultures diffuse quickly over space and time, and often homogenize peoples and landscapes in the process. Modern popular cultures began with the Industrial Revolution (Chapter 3), and often lack the historical depth of folk cultures. Popular culture is typically mass-produced and commodified within the context of commercialized culture. Finally, popular cultures are typically very urban in nature, exhibited in habits, outlooks, and practices that are predominantly concentrated in cities.

Architecture is a means of distinguishing between folk and popular culture. Building types and fences vary from one folk culture to another (Figure 13.1). Folk diets, too, reflect a preference for long-established habits, as well as the rootedness in an agrarian or maritime economy and lifestyle. The same is true of musical preferences; traditional styles are often associated with a specific region, such as for example, western swing in Texas and Oklahoma.

Figure 13.1: Traditional forms of folk culture include food, music, clothing, marriages, and architecture, such as this Amish barn. Their continuing existence given the threat of uniformity in the face of Western influence has been tenuous, and many folk cultures today are disappearing.

13.3 CULTURAL LANDSCAPES

The **cultural landscape** is the part of the planet's surface that has been directly and indirectly influenced by human interaction with the environment. This form of landscape is differentiated from the "natural" landscape, which consists of areas that have not been modified by human activity (although few such places exist today). Cultural landscapes, thus, fuse social and physical characteristics; for early cultural geographers, each cultural group adapted to its natural environment in its own way and created unique regions. The cultural landscape was thus seen as a reaction to the natural landscape. Additionally, the cultural landscape is an expression of the material features of cultural groups (Figure 13.2). As cultural geographers came to emphasize the social origins of culture, they emphasized the historical processes that led to the creation of landscapes; that is, cultural landscapes are human products—made, not given.

More recent interpretations of cultural landscapes emphasize the power relations that give rise to some landscapes and not others.

Figure 13.2: Cultural landscapes are the result of complex social, economic, and political forces interacting with the natural environment over long periods of historical time. The landscape embodies social relations, giving spatial form to culture and class.

Because there exist inequalities in terms of class, ethnicity, and gender, these are inscribed on the landscape. Landscapes are thus reflections of dominant power interests and attempts to challenge them. Such a perspective emphasizes the deeply political nature of landscapes. In this view, there is no single, objective view of landscapes, for we should "read" them like a text, decipher and interpret them, and link their symbolism to the interests of the dominant and resistant groups. Cultural landscapes that depict a historical event typically do so from a specific point of view—most often that of the elites. For example, in the case of monuments that depict historical events, geographers investigate the question of "whose" history is being depicted, and whose is not. Frequently, cultural landscapes promote the dominant social class's version of events.

13.4 RACE AND ETHNICITY

Race and ethnicity—which are related, but not identical—are major dimensions in all societies today and hugely important in human geography. Race and ethnicity are social, cultural, political, and spatial constructions that demarcate differences among people and make those differences appear natural. As such, race and ethnicity are important foundations of identity.

Race is typically defined as a biologically defined marker of difference among humans based on physical features such as skin color, facial features, or hair texture. Definitions of race typically focus on one feature, often skin color, over others. **Racism** is the belief that distinct races exist and that some are superior to others. Race played a minor role in the politics of the West until the fifteenth century, when Europeans began to encounter other cultures and conquer them. Indeed, Western racism arose in tandem with the trans-Atlantic slave trade. Non-Western examples of racism also exist, such as Arabs' treatment of Africans (including slavery) and Chinese and Japanese condescension to cultures other than their own. Racism is thus closely tied to the historical process of slavery, colonialism, and military expansion. Racism has often been used by elites to manipulate the masses. In the southern United States, for example, widespread racism convinced many poor whites that they were superior to blacks. The southern system of racial segregation began with slavery and continued throughout the long years of discrimination known as Jim Crow until the civil rights movement of the 1950s and 1960s. The Civil Rights Act of 1964 and the Voting Rights Act of 1965, for example, helped to end formalized, legal discrimination in the South aimed at minimizing political participation by African Americans.

Historically, race was treated as a biological or "scientific" method of human classification, in which racial "groups" were defined by the presence or absence of specific inheritable traits. This argument was tied to failed theories such as **Social Darwinism**, which erroneously attempted to apply Darwinian notions of competition among species to different racial groups, largely in the service of colonialism and the legitimation of class and racial inequality within Britain and the United States Another such discredited theory, **environmental determinism**, held that specific social traits were inherent to people living in specific places and were due to the environmental and climactic characteristics of each place, leaving little room for culture as a causal agent. Those in power used these notions to rationalize the exploitation or extermination of certain "racial" groups. The scientific basis for race has since been refuted, as multiple studies have shown that there is far greater genetic diversity within any single racially defined group than exists among groups. For example, people with black skin have much more variation among themselves in terms of height, weight, or intelligence than they do with groups with different skin tones.

Despite its failings as a scientific concept, race today exists as an extremely important socially constructed idea of differences among humans. Race is such an important concept because of racism. Racism assumes that specific, usually negative attributes regarding behavior, intelligence, or culture belong to specific racial groups. This ideology is then used to justify arguments for the superiority of one racial group relative to another, which leads to the production of inequality between racial groups. Thus, the term "race" has been applied almost exclusively for the purpose of rationalizing the subjugation of one group by another, regardless of the characteristics by which the supposedly inferior group was defined. For example, the British comfortably interchanged the concept of race with political nationality in their subjugation of the Irish "race." The Nazi Party in World War II sought to exterminate the Jewish "race," despite the fact that religion is not a biologically inheritable trait. In short, there is no race without racism.

Today the term "race" is commonly defined by its most visible trait: skin color. The fact that races do not have a scientific basis does not prevent people from applying the label, often with tragic consequences. For example, racism was used to justify slavery in the United States and Britain. Institutionalized racism in South Africa under the regime of apartheid (Afrikaans for "separate") relegated black South Africans to marginal economic and political roles. The argument that race is a socially constructed ideology rather than a biological or "natural" method of human classification does not deny that racism has powerful ideological and social consequences. This perspective challenges the "naturalness" of human organizational systems and the ideologies upon which they are founded. People who view race as a social construction reject outright the argument for the "naturalness" or "common sense" use of certain methods of social categorization. Critical examination of these ideologies exposes the power relations that make clear the basis for defining the concept in terms of its specific historical and geographic context. In the case of race, a critical examination of the way that racial groups are described makes clear that definitions have varied significantly both geographically and throughout history. For example, who is "black" varies over time and space; in Britain, it often refers to people from South Asia; in Russia, it includes the Caucasians of Chechnya. These variations and historical redefinitions invariably serve to rationalize the exploitation of one population by another.

The fact that people believe and behave as if race does exist makes it a social reality with material effects. Racialization refers to the process by which one comes to be viewed and/or view their self as belonging to a particular racial group. The way that people are racialized is highly dependent on where this process takes place, as noted above. For example, the Nazis stigmatized the Jews as an "inferior race" in the 1930s and 1940s, inventing a fictitious biological origin to their culture to demonize them. Similarly, American and South African racists long depicted black Africans and African Americans as an inferior race. Socialization processes are both embedded in place and connected through space such that socialization processes and geography are interdependent, not simply complementary. Geographers understand places as socially produced as well (see Chapter 1). Racialization processes are themselves shaped by and affect the social construction of places. This relationship between social process and place creates a larger racialized landscape, which provides context for how racial categories are defined.

Theorists critical of racialization argue for a relational understanding of race and racism. They maintain that one can understand racial categories, and the meanings attached to a specific racial group, only as they relate to other races. This relational understanding allows for a movement away from understanding race as a binary relationship (civilized/uncivilized, white/nonwhite) to an understanding of racism as a set of power relations acting in a specific time and place. For example, the meaning of "blackness" has varied over time and space; in the American South, it meant anyone with one drop of black blood (until they could "pass" as white); in much of Latin America, it meant poor people; to the French, the Spanish were often considered "black," and some Northern Europeans regard Mediterranean cultures as black. In each case, blackness is defined as being politically, culturally, and economically inferior. These power relations restrict individual and group access to power and resources relative to the dominant group, which in Western societies is most often defined as white. The severity of this disempowerment depends on the specific racial identity of the group in question, as well as the geographic context.

Relational perspectives on race view racism as the result of racial hierarchies within a social system of white privilege. In a racial hierarchy, racial (or ethnic) groups are positioned in a hierarchical manner based on their disempowerment relative to all other racial groups. Within the social system of white privilege, whites are at the top of the racial hierarchy with all other racial groups somewhere below. White privilege refers to the set of power relations that supports the status quo position of whites as the recipients of any number of advantages simply due to their being white. In this system "whiteness" is the norm against which all other races are understood as "different-from." White privileges are wide ranging and include access to better educational and economic opportunities, access to better housing than equally qualified minorities, and being underexposed to pollution relative to minority groups, to name just a few. White privilege does not necessarily exist through overt racism, such as the violent repression of minorities. Instead, it can be more subtle, widespread, and unintended.

Because racism is a dominant ideology in contemporary society, it affects all elements of social life. Processes of racialization attach racial understandings to places and landscapes at every scale: continents, nations, states, cities, neighborhoods, and even within the home. This means that racism unfolds at levels ranging from the local to the global. Just as people are racialized, so too are places. Racialized landscapes are an especially important part of geographical understandings of race and racism because popular understandings of landscape provide the context for people's life experiences. Racialized understandings of landscape shape the way that individuals interpret their own life experiences, and simultaneously serve to reinforce existing dominant understandings of racial categories. For example, some whites may avoid the "black part" of town, or entire cities altogether. A few common examples of racialized landscapes in the United States are "the ghetto" and "the border," which are minority landscapes, and suburbia, which is most often a "white" landscape. These racialized understandings of landscape police who or what "belongs" in a specific place. They affect a wide range of social issues, from property values to law enforcement practices.

Ethnicity is a term preferable to race because it highlights the social construction of culture and identity. Ethnicity derives from both an internal sense of distinctiveness and an external perception of difference. For example, an individual who identifies as a Muslim both thinks of him/herself or themselves as Muslim and is perceived as such by others. An ethnic category exists to classify various groups of people based on specific social and cultural characteristics, with the most typical identifier being ancestry. Unlike race, which points to biological markers, ethnicity is explicitly a social construction.

Many individual characteristics make up the building blocks of ethnic identity, including language, dialects, religious faith, literature, folklore, music, food preferences, social and political ties, traditions, values, and symbols, kinship, neighborhood, and community links, and/or migratory status. External attributes are also significant in the construction of ethnic identities. These include the role of governmental policies and social measures, racial discrimination, residential segregation, occupational concentration, and economic isolation. Ultimately, individuals conceive, maintain, and make manifest their ethnic identities in numerous ways.

Ethnicity is situational and dynamic. Individuals sustain and assert their ethnic identities in different ways, depending on the surrounding social and political environment. So, even though individuals may use the same ethnic label, they may construct their ethnicity differently. For example, among Jews, some may self-identify on the basis of a common religion; other, more secular Jews may ignore their religion and point to their cultural history and customs, such as Jewish foods. Ethnicity does not remain static over time. Instead, identities can be altered, manipulated, and transformed based on broader spatial, political, social, and economic dynamics. For example, many Europeans who migrated to the United States gradually rejected their ethnic heritage in the late nineteenth and early twentieth centuries and came to self-identify as "white," particularly as this racialized category stood in contrast to African Americans. Ethnicity, then, is a creative and complex response to both individual and social forces. The formation and expression of ethnic identity come from both historical circumstances and individual negotiations to endow ethnicity and ethnic symbols with meaning.

The United States today is an increasingly multiethnic society. In 2013, whites made up 66% of the population, whereas Latinos made up 15% (the largest ethnic minority, who may be of any race), African Americans another 13%, and Asian Americans an additional 6%. Some states are already minority-dominant, such as Hawaii, California, New Mexico, and Texas. The relatively higher rate of growth of minorities implies that the United States could become a predominantly minority- populated country around the year 2040. Notably, different minorities have distinct geographies. American Latinos, for example, are mostly concentrated in the southwestern United States (Figure 13.3), whereas African Americans are found largely in the Southeast and in the cores of large cities (Figure 13.4).

The changing ethnic composition of the United States has important implications for politics, consumption, and crime. Ethnic minorities in the United States tend to have significantly lower economic and political opportunities than do whites. For example, high school dropout rates are higher for Latinos and African Americans than they are for whites. Public schools in minority-dominant communities tend to be less well funded than those in white areas, with larger class sizes. Unemployment rates for minorities are often double or even triple that of whites, incomes are lower, and minorities are far more likely to be poor. Minorities are less likely to own homes and more likely to lack health insurance; and African American men are disproportionately concentrated in prisons. All of these factors, and others, combine to generate a lower quality of life for the minority population.

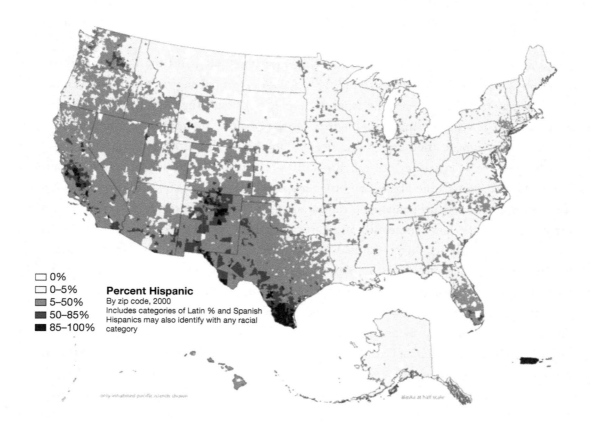

Figure 13.3: Latinos or Hispanics, the largest minority in the United States, can be of any "race" and consist of a diverse group of cultures, including people whose ancestry can be traced to Mexico, Central and South America, Cuba, Puerto Ricans, and the Dominican Republic. The heaviest concentration of Latino Americans is in the Southwest, notably along the border with Mexico.

Ethnic segregation is a significant fact of life in many large U.S. cities, in which minorities often populate large inner-city neighborhoods surrounded by white suburbs (Figure 13.5). The degree of ethnic segregation varies, of course, over time. In some cities it has gradually improved. Nonetheless, the existence of large, impoverished, minority-populated communities testifies to how class and ethnicity are closely intertwined in American society.

Most countries in the world have populations of ethnic minorities. Nation states are never ethnically homogeneous (Chapter 9), even

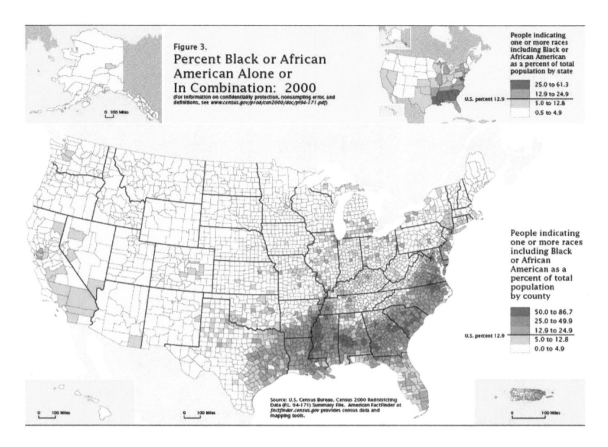

Figure 13.4: The geography of African Americans consists of a large swath of counties in the South, which reflects the historical processes of slavery and the plantation system, as well as large pools in central cities in the Northeast, Midwest, and southern California.

though nationalists often celebrate countries as such. For example, French nationalists often view France as consisting of one group of French people, ignoring ethnic minorities such as Basques, Bretons, Corsicans, and the German-speaking minority in Alsace. Japan is a relatively homogeneous nation, but Japanese nationalists overlook that country's minorities of Ainu, Koreans, Chinese, and *burakumin* (an indigenous underclass). Thus, even the most ostensibly ethnically homogeneous countries include important internal variations.

All over the world, ethnic minorities are often second-class citizens, with less access to economic and political opportunities. As noted in

Figure 13.5: The urban "ghetto" is a notable feature in the cores of most U.S. metropolitan areas, such as this map of ethnicity in Philadelphia, which illustrates the classic concentration of African Americans (shown in blue) surrounded by archipelagos of whites (in red). The map reflects of how ethnicity intersects with labor markets, housing, and public education to segregate low-income minorities into impoverished enclaves in the inner city.

Chapter 9, nation-states are never ethnically homogenous, despite the pretensions of ethnic nationalists (Figure 13.6).

In Europe, for example, a multitude of ethnic groups often cross national borders; some of these are fueled by waves of recent immigration. In France, for example, 10% of the population consists of Muslims, primarily from North Africa. In some cities such as Marseilles, this share rises to 20%. In Canada, large ethnic minorities include the French-speaking population of Québec and many indigenous peoples (Figure 13.7). In China, non-Han ethnic minorities make up only 6%

Figure 13.6: States are never politically homogeneous, but because colonial powers devised the boundaries of their former colonies, states that were formerly European colonies tend to be more diverse. Artificial boundaries set the stage for numerous horrific ethnic conflicts, civil wars, and secessionist movements.

to 8% of the population, including Mongols, Tibetans, and Uighurs, among others.

ETHNIC CONFLICTS

Occasionally ethnic-fueled violence can lead to horrific results. The disintegration of Yugoslavia in the 1990s into seven different republics, for example, resulted in the murder of 250,000 people and occurred largely along ethnic lines (Figure 13.8). In Sri Lanka, a war between the dominant Buddhist, Sinhalese-speaking majority and the Hindu, Tamil-speaking minority killed more than 80,000. In Indonesia, Muslim extremists have occasionally attacked the country's Chinese

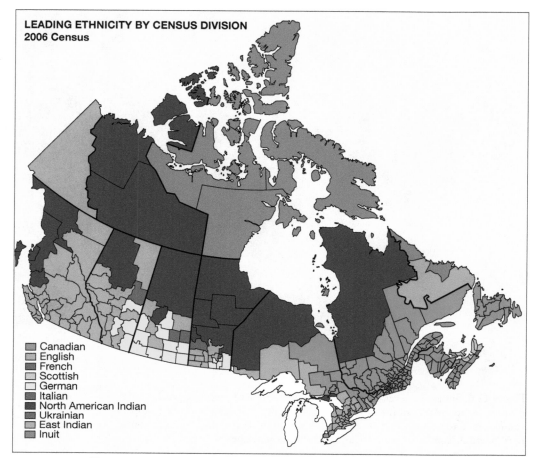

Figure 13.7: Canada, like many countries, is a patchwork of different ethnicities, including the Francophone population of Québec and the various First Nations peoples, as native Canadians are known in that country.

minority. In Iraq, Arab attacks against the Kurds have been frequent. The military government of Guatemala launched a 36-year long war (1960–1996) against the Mayan population, even though it was the numerical majority. In Rwanda, one of many ethnic conflicts on the African continent, the attempts by the Hutu majority to exterminate the Tutsi minority in 1994 led to the deaths of more than 800,000 people (Figure 13.9). More recently, as Sudan and South Sudan separated in 2011, the Arabs of Sudan have reached the brink of war with the black, largely Christian and animist population of South Sudan.

One extreme consequence of ethnic violence is the targeting of a group for genocide. The term arose from Turkish attempts to annihilate

Former Yugoslavia

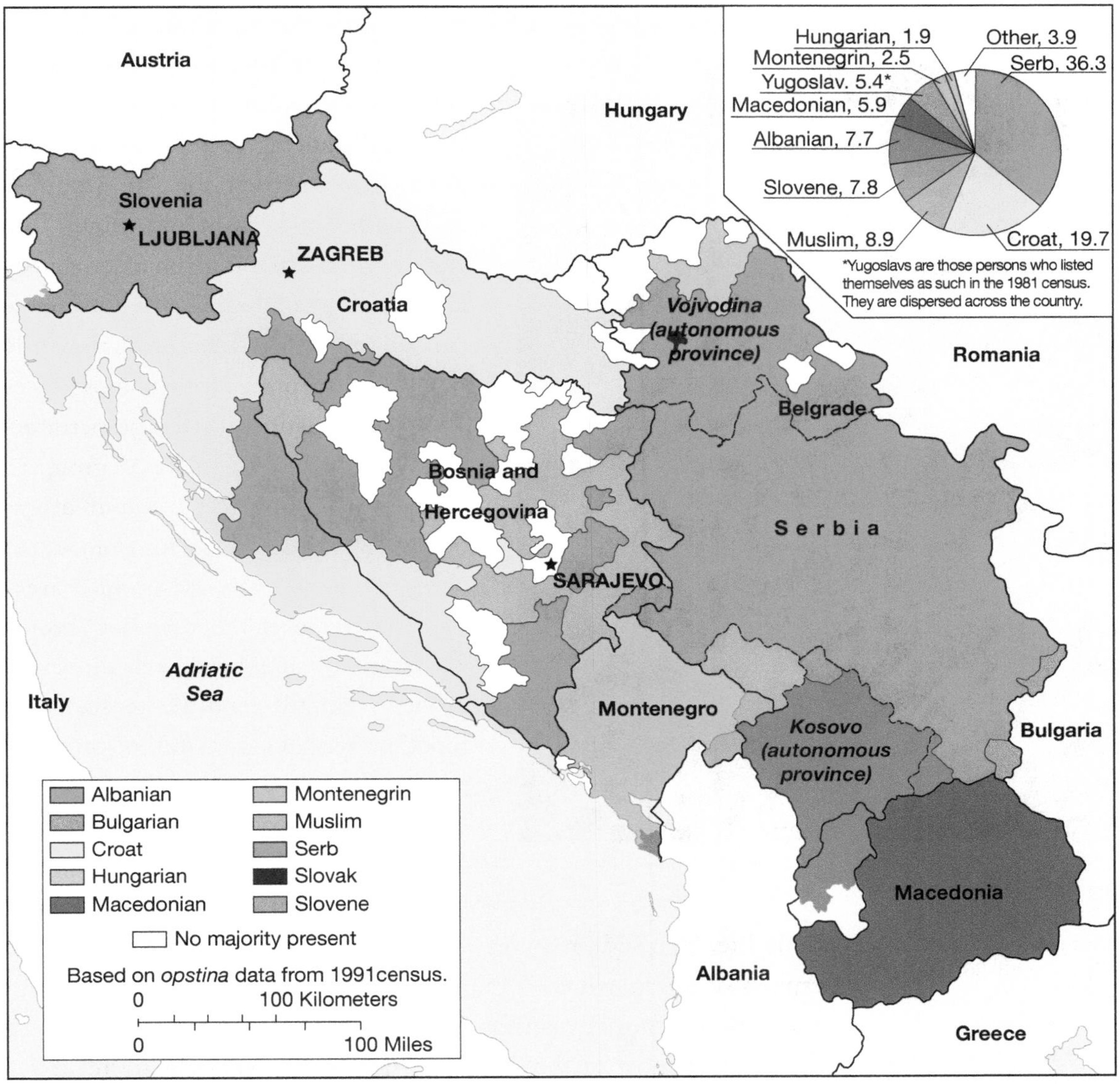

Figure 13.8: Yugoslavia was a complex quilt of different ethnicities, with different religions and even alphabets, before it disintegrated in a war in the 1990s that killed more than 250,000 people.

the Armenians during World War I, when more than 1 million people were killed. The United Nations Convention on Genocide, written following the Holocaust, defines victim groups according to national, ethnic, racial, or religious terms. Genocide is a form of one-sided mass slaughter in which a state or other authority intends to destroy the

Figure 13.9: In spring 1994, the majority Hutu tribe of Rwanda killed roughly 800,000 members of the minority Tutsi tribe. This act of genocide reflects how the colonial legacy, which pitted groups against one another, and the contemporary politics of hate play out with horrific results.

membership of an entire group, as defined by the perpetrator.

A necessary precondition for genocide is a symbolic, and sometimes spatial, distancing or separation of one group—the perpetrator "in-group"—and another, the victim "out-group." The victim group often is given a derogatory label, further emphasizing that its members do not belong. Identities frequently are linked to place, and in most cases genocide is justified by the perpetrators in terms of their "right" as a group to occupy a specific place without others being present. Of course, for genocide to occur, the perpetrator group must exercise power over the victim group. Totalitarian states have been especially prone to commit genocide because they function without constraints on their power. The worst mass killings of the twentieth century were committed by dictatorial states that controlled most aspects of cultural, political, and economic life. Genocide requires the participation of many people in the perpetrator group, which is more likely to occur when participation in mass killing is formally authorized by the state. Further, much evidence suggests that most of those who participate in genocide are relatively ordinary people engaged in extraordinary behavior that they are somehow able to define as acceptable, necessary, and even praiseworthy. Most cases of genocide occur at times of severe national tension.

Twentieth-century examples of genocide include forced movements and murder of Armenians in Turkey during World War I; population purges and deportations in the Soviet Union from the 1920s to the 1950s, in which tens of millions of people perished; the Holocaust perpetrated by Nazi Germany against the Jews, as well as against the

Roma (gypsies), the disabled, and homosexuals; mass executions of more than one million people in Cambodia by the Communist Khmer Rouge between 1975 and 1979; Saddam Hussein's extermination of many Kurds in Iraq in the 1980s; the murder and ethnic cleansing of Muslims by Serbs in the former Yugoslavia in the 1990s; and Hutu killing of roughly one million Tutsis in Rwanda in 1994. In all of these cases, genocide has been one part of larger efforts by one group to dramatically reshape the geography of peoples and places. Racist and nationalist ideas typically form the basis for stigmatizing ethnic minorities, who are blamed for any and all social and economic problems and are portrayed as impediments to progress and prosperity. The perpetrators, using powerful metaphors of "purifying" and "cleansing," justify genocide on the grounds that victim groups are less than human.

13.5 GENDER, IDENTITY, AND PLACE

In addition to race and ethnicity, **gender** is a fundamental aspect of human existence that plays a powerful role in human geography. First, we must differentiate between sex and gender. Sex includes the biological differences between males and females, largely the result of different chromosomes (XX for females, XY for males). In contrast, gender is the set of cultural norms and behaviors associated with sex. In contrast to sex, gender is a social construction: as the saying goes, "sex is what is between your legs, gender is between your ears." Gender roles are defined by the webs of masculinity and femininity associated with males and females, respectively. Historically, masculinity has been associated with power, strength, bravery, confidence, and assertiveness, whereas femininity has been associated with beauty, empathy, caring, and passivity. However, these simple stereotypes do not do justice to the complexity of gender roles, which often overlap.

If sex is given through biology, gender is socially constructed, and thus highly variable. Thus, gender roles are produced in different ways by different societies. Gender is formed initially through the different treatment of girls and boys and continues to be reinforced through the life cycle. This treatment is accompanied by different societal expectations of the values, attitudes, and behaviors of boys and girls. But it is not sufficient to note that in most societies girls and boys are raised differently and are expected to be different throughout their lives. They are also raised unequally, with boys socialized to be aggressive and to assume leadership roles, whereas girls are socialized to be passive and to be compliant followers. Although it remains controversial whether such characteristics have some initiating biological cause, the socialization process clearly emphasizes and increases any natural differences and minimizes movement across the categories of male and female. Thus, what it means to be a man or a woman varies widely historically and geographically. Because sex is relatively invariant, it cannot explain variations in gender roles. To ignore gender is to ignore all the ways in which, around the world, the lives and experiences of women and men differ from one another. Finally, gender should be distinguished from sexuality, or sexual preference. Although most people are heterosexual, some are not. Confusing gender and sexuality leads to false stereotypes—for example, that all gays are feminine or all lesbians are masculine.

Gender roles are as old as humanity itself. During the Paleolithic era (Chapter 2), for example, men hunted and women were responsible for the bulk of gathering. In all societies, women are responsible for the vast bulk of child care, an enormously time-consuming activity that often infringes on women's ability to work outside the home. In some cultures, a woman is not considered to be a woman unless she has borne children. And, although the vast bulk of unpaid housework and child care is done by women, it does not enter into national economic accounts, which privilege male forms of paid labor. Today, gender enters into social and spatial relations in an enormous number of ways, including people's self-perception and perceptions of others, their use of language (e.g., "mankind" or "fireman"), their success or lack of success at school, different friendship networks and activities, consumption patterns, propensity to migrate, types of work and occupations, voting patterns, likelihood of committing a crime and going to jail, and preferred method of suicide. Thus, in Western cultures, female friends tend to talk with one another whereas male friends do things together such as sports. Men and women prefer different movies, magazines, and cars. Men are far more likely to migrate and commit crimes. Men also tend to be more politically conservative than women.

Central to understanding gender is the concept that it is a power relation. Gender roles typically (but not always) work to the advantage of men and the disadvantage of women. Almost everywhere, today and throughout history, women live in a subordinate social position to men. More often than not, they work more, earn less, enjoy less social autonomy and fewer political rights, and culturally garner less respect. Thus, gender, like class, entails a distinction between dominant males and subordinate females. Women have made significant strides, particularly in Western societies, over the past century; they earned the right to vote in the United States in 1920, and the women's rights movement that began in the 1960s opened the door for women into higher education, many professional fields, and public office as well as reducing legal obstacles to their advancement.

Societies in which men enjoy advantages over women, simply for being men, are called **patriarchal**. Most societies in human history have been patriarchal in one form or another, and the reasons for this have been hotly debated. Some arguments propose that patriarchy is grounded in biology—that is, women's child-bearing capacities render them less powerful than men. One school of feminist thought called ecofeminism, focuses on how gender relations intersect with the natural environment. Ecofeminists argue that patriarchy arose with the Neolithic Revolution—that is, the domination of nature through agriculture was accompanied by the domination of women by men.

In traditional Hindu societies in India, widows were expected to throw themselves on their dead husband's funeral pyre in the practice of *suttee*. In Confucian China, foot binding among upper-class women left them almost unable to walk. Today, many forms of patriarchy continue to exist. Certain deeply religious and traditional societies, for example, strictly limit the display of a woman's body, especially any indication of sexuality. In many Muslim countries, women are discouraged from appearing in public, especially without a male companion (Chapter 4), and their hair, and sometimes face, must be covered (Figure 13.10).

In some African countries, as well as in Indonesia and the Middle East, the practice of female genital mutilation is common (Figure 13.11). Genital mutilation, practiced by women on girls, is justified on the grounds that it prevents women from "straying" during marriage. This practice creates enormous suffering and even death. In many countries, especially China and India, the use of ultrasound to detect the sex of a fetus has led to widespread abortion of female babies. Hundreds of millions of female fetuses have been destroyed this way, such that the usual ratio of men to women has become skewed to reflect an overabundance of males (Figure 13.12). In China, for example, there are 10% more men than women.

Figure 13.10: Many religions exert significant control over women's bodies, especially their sexuality. In traditional Islamic cultures, women are not seen in public without completely covering their bodies in a garment known as an *abaya* or *burkha* in Arabic and a *chador* in Farsi, in Iran.

In the industrialized West, women are often subjected to domestic abuse and rape. Sexually laden advertising focuses on women's bodies at the expense of other dimensions of their lives; women become valued for their looks and sexuality, not for their intelligence, bravery, persistence, or other qualities. And, on average, around the world, women are underrepresented in high political offices. Although there are some female governors, senators, and congressional representatives in the United States, they are a relatively small minority. However, in some European countries, notably in Scandinavian states such as Norway, women enjoy considerably more political influence. Several

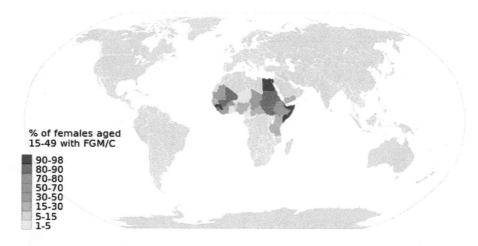

Figure 13.11: The geography of female genital mutilation. Some cultures engage in this practice to keep wives from straying from their husbands.

countries around the world have recently elected female heads of state, including Germany, Britain, Chile, India, Israel, Australia, Canada, Argentina, Sierra Leone, Liberia, and Brazil.

Despite women's recent political advances, the world of work remains deeply gendered. Manufacturing, with the exception of garments and textiles (Chapter 10), has overwhelmingly been a male

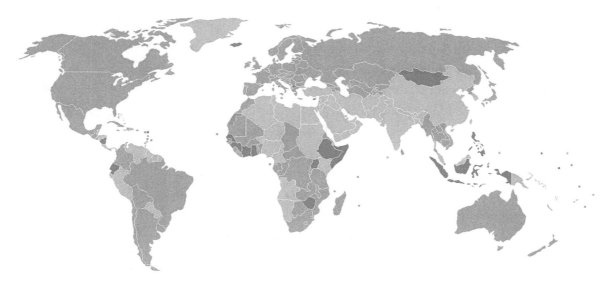

Figure 13.12: In some countries, notably China and India, the abortion of female fetuses has generated a surplus of men in the population (countries shown in blue).

world. Deindustrialization has therefore affected men's jobs much more than women's. The rise of service industries has seen a dramatic increase in the labor force participation rate of females. As women have entered the workplace, their changing roles have altered gender relations at home, and have raised issues for two-income families concerning child care. Women are usually relegated to the private domain of the home or to limited opportunities in the public domain. For example, it is still women who typically perform domestic work, which is unpaid, repetitious, and often boring. The jobs available to women outside the home are often low-paid and low-skilled, or "pink collar" jobs (Chapter 12). Thus women dominate in minimum-wage jobs in retail trade, restaurants, clerical work, personal services, child care, and nonphysician health care positions. Even professionally successful women often complain of a "glass ceiling" in large corporations or public bureaucracies, in which the positions of power and responsibility are overwhelmingly filled by men. Some corporations, such as Pepsi, eBay, and Yahoo, have female heads, but women remain underrepresented in top corporate positions. Women in the United States earn, on average, only 80% of male incomes and are disproportionately represented among low-income families, leading to the feminization of poverty. In short, women and men tend to do different work in different places, and to lead different lives, which include different visions of the world and of themselves.

Feminism is a political and social movement that asserts the equality of the genders, especially with regard to the underprivileged status of women compared with men. Thus, contrary to much popular opinion, feminism is not simply concerned with women, but with gender relations. And not only women are feminists: Because it is a political position, men can be feminists too. Feminists operate from within a wide spectrum of perspectives, but concur that social reality is gendered—that is, that gender cuts across, intersects, shapes, and is in turn shaped by other lines of social organization such as class, age, ethnicity, sexuality, and geographic location. The gendered nature of social relations typically works to the advantage of men and the disadvantage of women. Feminism seeks not to end gender differences per se (the stereotype advanced by many antifeminists) but to end the *power relations* that accompany gender, i.e., male advantage. In this respect, being a feminist is not the same as being feminine; feminist men, for example, are not necessarily effeminate. Moreover, many women are not feminists, and subscribe to traditional, sexist gender roles.

GENDER IN THE LANDSCAPE

Genders shapes landscapes, and the landscapes then provide the contexts for the reproduction of gender roles and relationships. Feminist geographers explore the many ways in which space is gendered, including how places and gender shape one another. Numerous examples of landscapes, both visible and symbolic, reflect the power inequalities between women and men. Such landscapes embody patriarchal cultural values and related **sexism**. In urban areas, statues and monuments often reinforce the idea of male power by commemorating male military and political leaders, a celebration of masculine power, accomplishment, and heroism.

Feminist geographers have explored the intersection of gender and space in many ways. For example, some have focused on the body, "the most intimate of geographies," that "closest in" to the self as it plays out over time and space. For example, there are geographies of access to abortion and contraception at several spatial scales. This microlevel geography extends to the psychic landscapes that women inhabit. For most women, the landscapes of fear, especially fear of violent sexual assault, loom large; sadly, half the human race lives in terror of the other half.

The design of domestic space is strongly influenced by ideas about gender roles. The domestic sphere is thus usually coded as feminine, whereas public spaces are more often coded masculine. In many cases, including traditional Chinese and Orthodox Jewish societies, areas for women and for men are strictly separated. Those designed for women are often isolated from the larger world. In Western societies, the most common domestic design since the 1950s has centered on the home as the domain of women and as a retreat from the larger world. Certainly, it has been usual in the Western world to locate the kitchen, designed as a separate area for the unpaid work of women, at the rear of the home. Typically women do the majority of domestic work inside the house whereas men do the bulk of outside chores.

Again, according to some feminist arguments, this same patriarchal logic influenced the expansion and shape of suburbs. Men were seen as commuters and women as homemakers/consumers requiring ready access to shops and schools. During the era when most women stayed at home, this arrangement disadvantaged women, who were isolated because of distance from opportunities for paid employment in the suburbs.

GENDER AND HUMAN DEVELOPMENT

All over the world, women work harder than men but are typically paid less and have less power. In traditional agricultural societies, women often do the bulk of farm work, including carrying water or firewood (Figure 13.13). The global diffusion of the Industrial Revolution, including sweatshops and assembly plants in many developing countries, has created a new, feminized labor force to assemble the world's clothes, toys, and electronics goods. Women also still suffer from a host of economic and political disadvantages. In some countries, such as Saudi Arabia, women enjoy very few political rights such as voting. Women's literacy rates are often lower

Figure 13.13: Women in the developing world do most of the agricultural labor. Their subjugation and oppression reflects entrenched gender norms and roles.

than men's, especially the developing world. This reduces their ability to obtain information and cultivate job skills. Women make up the bulk of the world's 1.5 billion people living in extreme poverty, including those who earn less than one dollar per day. Women's opportunities to migrate tend to be lower than that of men, which inhibits their long-run earning power and life chances. They tend to be receivers of remittances rather than senders. Malnutrition strikes women harder than men, often because they have less power in the household. In some countries, men may eat while their wives and children go hungry.

Many observers see eliminating gender inequality as key to winning the war on global poverty. For example, part of the difference in the economic gains of China relative to those of India is attributable to the fact that Chinese society has empowered women much more than has Indian society. This point was central to the discussion in Chapter 8 concerning population growth. The empowerment of women, especially improved education and job opportunities outside the home, is critical for reducing fertility.

13.6 SEXUALITY AND SPACE

Sexuality refers to a person's sexual orientation and is not the same as gender. Most people are heterosexual, but significant minorities are not. Whereas gender refers to the roles and outlooks associated with being male or female, sexuality, although related, may depart from these. Most people are heterosexual, but it is misleading to confuse alternative sexualities (i.e., being gay or lesbian) with gender; gay men, for example, may or may not be effeminate. Geographies of sexuality focus on the interrelations between sexual preference, behavior, knowledge, power, and place. Rather than being biologically given, sexuality should be seen as a social construction, produced and expressed in different ways in varying historical and geographical contexts. Geographers study sexuality as an expression of identity and as one way in which the dominant heterosexual culture is challenged. Work on diverse sexual identities has been conducted mostly within the larger framework of feminist geography, but researchers also employ **queer theory,** which refers to views of the world from the perspective of gays and lesbians. Queer theory also emphasizes the fluidity and even merging of identities, and is concerned with empowering those who lack power.

During most of Western history, the dominant view of homosexuality was very negative, a view reinforced by both the legal and medical establishments. Over the past several decades, however, attitudes have been steadily changing, particularly among the young and in part because of the gay rights movement. For example, homosexuality is no longer classified as a disorder by the American Psychiatric Association. Relationships involving same-sex partners are being legalized as legitimate and socially acceptable lifestyle choices. Countries in northern Europe were the first to recognize same-sex unions, and numerous U.S. states have followed suit although gay marriage remains a contentious political issue. Contrary views are evident, however, especially those of religious fundamentalists. In many parts of the world, lesbians and gays continue to experience homophobia, homosexual acts are criminalized, prejudice and discrimination are routine in many workplaces, and people in same-sex relationships have reduced rights to pensions and inheritance. Some of the least tolerant places in the world include Nigeria, Uganda, Saudi Arabia, and Iran, where homosexuality can be punishable by death, and Jamaica, where male homosexuality carries a punishment of 10 years' hard labor. Even in relatively liberal countries, large numbers of gay men and lesbians do not publicly identify as such for fear of recrimination. There is, thus, a geography of the closet that ranges from everyday life to the world system.

One of the ways in which sexuality is expressed in landscape, especially in large cities in the Western world, is through the gradual identification of a part of the city where lesbians and gays dominate. Early examples include Greenwich Village in New York (where the Stonewall riots of 1968 launched the gay rights movement), the Castro district in San Francisco, West Hollywood in Los Angeles, and Soho in London. Another expression of sexuality in landscape is gay pride parades or demonstrations (Figure 13.14). These are now routine in many cities and are often major celebratory occasions attracting large tourist audiences.

13.7 ORIENTALISM

Social difference and identity are closely related; we only know who we are by knowing who we are not. Inspired by the work of the famed Palestinian scholar Edward Said, many social sciences focus on **Orientalism,** the sets of beliefs that Westerners developed about non-Western parts of the world. As Europeans colonized Asia, they constructed stereotypes that were simplistic and inaccurate, but nonetheless served the colonial imperative of representing non-Western peoples as primitive and in dire need of being civilized. In the emerging Western worldview, Western culture was inherently dynamic and innovative, and non-Western peoples were trapped in stagnant,

Figure 13.14: A gay pride march testifies to how sexuality is the defining feature of identity for some people.

Figure 13.15: Orientalism refers to the set of stereotypes that Europe, or more broadly the West, developed about non-Western societies. These were never accurate images; rather, they portray the "Orient" as inscrutable, corrupt, despotic, child-like, and feminized, in need of conquering by Western white males.

unchanging, corrupt, and despotic cultures (Figure 13.15). According to this logic, the indigenous peoples inhabiting areas that Europeans wanted to colonize were "others"— not only different from but less than the Europeans, who as "superior beings had a natural right to use them and their lands as they chose. European views of the Orient offered little that was accurate about the societies they took over, but revealed much about Western biases and myths of superiority. Orientalism is thus an expression of the construction of difference on a global scale.

It is not surprising that places often become sites of conflict between different groups of people, as in the example of the apartheid landscape or gay pride marches. Throughout history, people have endowed parts of the planet's surface with meaning—that is, they have created places. But the understanding of place held by one group may be quite different from that held by another group. For example, European views of North America as a vast wilderness waiting to be tamed were at odds with the views held by Native Americans. Colonial descriptions of the tropics as either gardens of Eden or malaria-filled hells were very different from the views of indigenous peoples in tropical area of the Americas, Africa, and Asia. British views of China as a moribund, decadent society in the nineteenth century (Chapter 5) were in stark contrast with the traditional Confucian Chinese views of the Middle Kingdom (Chapter 4). In each case, the views put forth were closely associated with economic and political interests. Hence the identity of places is often contested as rival visions of a place contend for dominance. Many recent examples of these landscapes of resistance are associated with new social movements—in support of the environment, social justice, ethnic separatism, and so on—and they are best interpreted as expressions of opposition to power. Other subcultures, especially those associated with youthful populations, often express opposition to mainstream lifestyles and mainstream places. Such groups may choose to display their difference through the adoption of alternative lifestyles, including musical and clothing preferences, and through their attachment to places such as inner-city neighborhoods.

Ethnicity, gender, and sexuality are not the only grounds for exclusion from the landscape created by dominant groups. The homeless, the unemployed, the disabled, and the elderly all find themselves in some way incapable of fitting into the landscapes constructed to serve the interests of those holding power. Some disadvantaged groups are more visible than others, and some are more controversial. All, however, are in some way oppressed because the landscapes created by dominant groups presume a set of identity characteristics that they do not possess.

■ KEY TERMS

Acculturation: the process by which a person from one culture learns about and internalizes aspects of another culture, such as occurs when immigrants absorb culture traits from their new homelands

Cultural assimilation: the incorporation of elements from one culture into another

Cultural landscape: Earth's surface as shaped by local and regional cultural practices

Cultural regions: broad parts of the planet that share roughly similar cultures based largely on language, religion, and similar attributes

Culture: the stock of knowledge, beliefs, assumptions, values, mores, priorities, and ideologies that shape an individual's or group's outlook and behavior

Culture traits: specific aspects of a culture such as language, religion, dress, food, gender roles, customs, traditions, values, architecture, and land-use patterns

Environmental determinism: an obsolete, discredited ideology once popular in geography in the early twentieth century that held that cultural variations could be explained by reference to the physical environment

Ethnicity: the set of culture traits that define a particular group, largely consisting of language, religion, a common history and geography, and a shared identity as a group with a common culture

Feminism: the ideology that recognizes gender differences as a power relation and advocates for equal rights for women

Folk culture: traditional forms of culture that largely arose before capitalism, are predominantly defined by tradition, change slowly over time but vary considerably among places, and are mostly found today in rural areas

Gender: the social construction of sex as a set of roles and behaviors

Material culture: cultural traits that are tangible expressions of values and beliefs, including objects, artifacts, and landscapes

Nonmaterial culture: cultural traits that are not tangible, including mores, assumptions, beliefs, and social roles

Orientalism: the worldview and stereotypes constructed by Westerners during the era of colonialism that erroneously portrayed non-Western societies as inferior and stagnant

Patriarchy: a type of social organization in which men enjoy more power and opportunities than do women simply because they are men

Popular culture: modern forms of culture, primarily commodified through the market, that change rapidly over time (e.g., fads and fashions, musical styles), and are largely associated with the lifestyle of urban areas

Queer theory: a theory of social relations from the standpoint of gays and lesbians in which their sexuality is not considered deviant

Race: a group of people who ostensibly share a common biological heritage, usually defined through skin color, although the term has been used loosely in different ways. What were once taken as scientific categories such as "Caucasian," "Negro," or "Mongoloid" have been discredited as lacking a scientific basis

Racism: the ideology that biological differences make some groups socially superior to others.

Sexism: the worldview that holds, implicitly or explicitly, that males are more worthy, important, intelligent, or deserving than are females

Sexuality: sexual orientation (i.e., desire for one gender or another to varying degrees)

Social Darwinism: a discredited nineteenth-century attempt to apply Darwinian notions of natural selection to social relations, typically used to legitimize racism and the social status of the wealthy

Socialization: the process of learning the culture into which one is born and raised; although very powerful in childhood, socialization is a lifelong process

■ STUDY QUESTIONS

1. Define culture. Can culture exist apart from the social and political structure of a society? Why or why not?

2. What is the difference between socialization and acculturation?

3. How is culture spatialized, or manifested geographically, in the landscape?

4. Are there spatial variations to popular culture, such as music preferences? Give an example.

5. What is the difference between space and place?

6. Why is race a dubious idea scientifically?

7. How does ethnicity differ from race?

8. What is the difference between sex and gender?

9. What is patriarchy? How has it changed?

10. How do gender roles shape things such as household chores, commuting, or access to high paying jobs?

11. Why is sexuality often confused with gender? For example, gay men are often viewed as effeminate.

12. What is Orientalism?

BIBLIOGRAPHY

Berg, L., & Longhurst, R. (2003). *Placing masculinities and geography.* Gender, Place and Culture, 10, 351–360.

Bondi, L., & Davidson, J. (2003). *Troubling the place of gender. In K. Anderson, M. Domosh, S. Pile, and N. Thrift (Eds.),* Handbook of Cultural Geography. *London: Sage.*

Brown, K., Lim, J., & Brown, G. (2009). Geographies of Sexualities. *London: Ashgate.*

Brown, M., & Knopp, L. (2003). *Queer cultural geographies—We're here! We're queer! We're over there, too! In K. Anderson, M. Domosh, S. Pile, and N. Thrift (Eds.),* Handbook of Cultural Geography. *London: Sage.*

Flint, C. (2003). Spaces of Hate: Geographies of Discrimination and Intolerance in the U.S.A. *London: Routledge.*

Foster, G. (2003). Performing Whiteness. *New York: State University of New York Press.*

Frazier, J. (2010). Multicultural Geographies: The Changing Racial/Ethnic Patterns of the United States. *Cambridge: Oxford University Press.*

Gergen, K. (1991). The Saturated Self: Dilemmas of Identity in Contemporary Life. *New York: Basic Books.*

Hanson, S. (1992). *Geography and feminism: Worlds in collision?* Annals of the Association of American Geographers, 82, 569–586.

Johnston, L., & Longhurst, R. (2009). Space, Place, and Sex: Geographies of Sexualities. *Summit, PA: Rowman and Littlefield.*

Mitchell D. (2000), Cultural Geography: A Critical Introduction. *Malden, MA: Oxford-Malden, Blackwell.*

Nelson, L., & Seager, J. (Eds.). (2004). A Companion to Feminist Geography. *Hoboken, NJ: Wiley-Blackwell.*

Rose, G. (2013). Feminism and Geography: The Limits of Geographical Knowledge. *Cambridge: Polity Press.*

Said, E. (1978). Orientalism. *New York: Vintage.*

Valentine, G. (2008). *Living with difference: Reflections on geographies of encounter.* Progress in Human Geography, 32(3), 323–337.

IMAGE CREDITS

- Figure 13.1: Copyright © Depositphotos/Kathyclark.
- Figure 13.2: Copyright © Depositphotos/hecke06.
- Figure 13.3: Copyright © Fredlyfish4 (CC BY-SA 2.5) at https://commons.wikimedia.org/wiki/File:New_2000_hispanic_percent.gif.
- Figure 13.4: U.S. Census Bureau, "Census2000 Percent Black Map," https://commons.wikimedia.org/wiki/File:Census2000_Percent_Black_Map.jpg. Copyright in the Public Domain.
- Figure 13.5: Copyright © Eric Fischer (CC BY-SA 2.0) at http://commons.wikimedia.org/wiki/File:Race_and_ethnicity_Philadelphia.png.
- Figure 13.6: Copyright © Jroehl (CC BY-SA 4.0) at https://commons.wikimedia.org/wiki/File:List_of_countries_ranked_by_ethnic_and_cultural_diversity_level,_List_based_on_Fearon%27s_analysis.png.
- Figure 13.7: Earl Andrew, http://commons.wikimedia.org/wiki/File:Censusdivisions-ethnic.png. Copyright in the Public Domain.
- Figure 13.8: U.S. Central Intelligence Agency, "Yugoslavia Ethnic Map," http://commons.wikimedia.org/wiki/File:Yugoslavia_ethnic_map.jpg. Copyright in the Public Domain.
- Figure 13.9: http://commons.wikimedia.org/wiki/File:Bodies_of_Rwandan_refugees_DF-ST-02-03035.jpg, U.S. Department of Defense, 1994. Copyright in the Public Domain.
- Figure 13.10: Copyright © Depositphotos/londondeposit.
- Figure 13.11: Copyright © M Tracy Hunter (CC BY-SA 3.0) at http://commons.wikimedia.org/wiki/File:2013_Female_Genital_Mutilation_Cutting_Circumcision_FGM_World_Map_UNICEF.SVG.
- Figure 13.12: Copyright © Nay T. Diniz (CC BY-SA 3.0) at http://commons.wikimedia.org/wiki/File:Sex_ratio_total_population_2.png.
- Figure 13.13: Copyright © Depositphotos/smithore.
- Figure 13.14: Copyright © JMazzolaa (CC BY-SA 2.0) at https://commons.wikimedia.org/wiki/File:NYC_Pride_2015_(19064384758).jpg.
- Figure 13.5: Léon Comerre, "A Bejeweled Harem Beauty," https://commons.wikimedia.org/wiki/File:L%C3%A9on-Fran%C3%A7ois_Comerre_-_A_Bejeweled_Harem_Beauty.jpg. Copyright in the Public Domain.

INDEX

A

accents, 99
acculturation, 502
acquired immune deficiency syndrome (AIDS), 303, 356–357
Afro-Asiatic language, 114–115
Agency for International Development, 351
agribusiness, 421–422, 431
Agricultural Adjustment Act, 430
agroecosystems, 432–433
ahimsa, 163
Alexander the Great, 62, 145
Alfred the Great, 147
Al-Qaeda, 285, 286, 406
Al-Shabaab, 406
American commercialism, 281
American corporate farming, 423–424
American Federation of Labor, 215
American Federation of State, County, and Municipal Employees and Service Employees International, 481
American Federation of Teachers, 481
American "War on Terror", 406
Analects of Confucius, 169
Anasazi people, 77
Anglo-Norman, 110
animate sources of energy, 189
anti-Americanism, and anti-globalization, 280–281
anti-globalization, 280–287
  anti-Americanism and, 280–281
  globalization and religious fundamentalism, 284–286
  grassroots, 282–285
Appalachian Coal Belt, 439
Arab Empire, 306
Arabic language, 116
Arabs
  defined, 154
  Muslims and, 156–157
"Arab Spring", 286, 328, 492
Arab world, 227–228
Archer Daniels Midland, 421, 423
aristocracy, 189
artificial selection, 42
Aryans, 103
Ashkenazi Judaism, 144–145
Asian rice paddy cultivation, 53–54
Association of Southeast Asian Nations, 247
astronomy, 45
atheism, 175–176
atheists, 175
Atlantic slave trade, 374
Audi, 443
Augustus Caesar, 64
Australopithecus, 28
Austro-Asiatic languages, 124, 126
Austro-Hungarian Empire, 385, 393
Austronesian languages, 123–124
automobile industry, 456–458
Aztecs, 70–73

mythology, 70
warlike culture, 71

B

baby boom, 368
Bacon, Francis, 206
Balochi language, 105
banks/banking
  internationalization of, 253
  modern, 254
Bantus, 118
Barber, Benjamin, 280
Basque language, 128, 130–131
Battle of Dien Bien Phu, 231
Bell Telephone Laboratories, 458
Berber, 116
Berlin Conference, 226, 237, 393
Bhagavad-Gita, 159
Bharatiya Janata Party, 284
bin Laden, Osama, 286, 406
bipedalism, 28
birth rate, 345
Black Death, 194–195
Boko Haram, 406
Bolivar, Simon, 219
Bolshevik revolution, 420
Borlaug, Norman, 352
Boserup, Esther, 55
bourgeoisie, 199
Bové, José, 281
Boxer Rebellion, 218, 231
brain drain, 319
Brexit, 282, 283
British East India Company, 228
British Empire, 400. *See also* Great Britain
Bronze Age, 49–50
Bronze Age Greece, 83
Brown, Lester, 350
bubonic plague, 194–195
Buddhism, 161–166
  ahimsa in, 163
  described, 164
  Hinduism and, 162
  Mahayana, 164–165
  Siddhartha Gautama, 164
  Theravada, 166–167
  Vajrayana, 165
Bunge, 420
business services, 469

C

campus, 2–3
Cantonese, 121
capital, 199
  defined, 253
  forms of, 253
capital flight, 320
capitalism, 384
  class relations and, 199–200
  colonialism and, 217–240
  emergence of, 196–209, 384
  expansion of, 385
  feudalism and, 187
  finance and, 200–201
  global, 398
  ideologies and, 203–206
  Industrial Revolution and, 206–219
  long-distance trade and, 202–203
  markets and, 196–197
  Marx on, 351
  nature of, 196–209
  rising productivity levels and, 352
  territorial and geographic changes, 201–203
  world-systems theorists on, 398–399
capital markets
  defined, 253
  international money and, 253–254
capital resources, 249
Cargill, 420
Carnegie, Andrew, 212, 455
cartographers, 10–11
cartography, defined, 10
castes, in Hinduism, 161–162
Catalan language, 105
Caxton, William, 110
Celtic language, 107
chaebols, 447
Champollion, Louis, 61
Chase Manhattan Bank, 330
Chavin culture, 74
cheap labor, 251
chi, in Daoism, 168
child labor, 309
China, 6, 230
  as coal consumer, 268
  Bronze Age, 65
  civil wars occurred in, 401
  early civilization of, 65–68
  foreign investments in, 255
  invasion of Vietnam, 401
  inventions in, 66–67
  rice cultivation in, 65

Chin dynasty, 65
Chinese languages, 118–119, 119–121
Chinese pictographs, 47
Christianity, 89, 145–150, 406
   collpase of Roman Empire and, 146
   Eastern Orthodox forms of, 148–149
   medieval Europe and, 147
   origin, 145
   Roman Catholic, 148–149
Chrysler, 457
Citicorp, 330
civic nationalism, 391
civilizations, 55–87
   Chinese, 65–68
   early urbanization, 57
   Egyptian, 59–62
   European, 83–87
   Indus River Valley, 66–68
   in South America, 74–75
   Mesoamerica, 67–73
   Mesopotamia, 57–58
   North American, 76–77
   sub-Saharan African, 78–80
civil liberties, 328
Civil Rights Act of 1964, 505
civil wars, 401–402
civitas, 56
classical geopolitics, 394–397
Classical or Hellenic Greece, 85–87
class relations
   bourgeoisie, 199
   capitalism and, 198–199
Club of Rome, 350
coal, 268
Cold War, 280, 284, 294, 400, 403
Colombian drug cartels, 404
colonialism, 306
   annihilation of indigenous peoples and, 235
   British Empire and, 220
   capitalism and, 217–240
   cultural westernization and, 237–238
   dual society and, 236
   effects of, 235–239
   end of, 238–239
   French and, 220–221
   impact on colonizers and colonized, 218
   polarized geographies and, 236–237
   political geographies and, 237
   Portugal and, 220
   waves of, 218–219
Columbian Exchange, 222–223
Columbus, Christopher, 76

commercial energy
   consumption in developed countries, 265
   consumption in developing countries, 265
commercial grain farms, 426
commercialized agriculture, 420–431
   types of, 425–430
commodities, 196
Common Agricultural Policy, 431
common market, 259
Communist Khmer Rouge, 517
comparative advantage, 203, 249–250
ConAgra, 420
Confucianism, 169–172
   emperor and his subjects' relations under, 170–171
   filial piety in, 170
   males and females relations under, 170
   older and younger people relations under, 170
Confucius, 169–170
Congress of Industrial Organizations, 215
consumer goods, 299–300
consumer services, 470
contraceptives, 349
Copernican revolution, 206
corrupt and oppressive governments, 328–329
Cortes, Hernando, 72
Counter-Reformation, 386
"cradle of civilization", 56
credit card fraud, 492
creoles, 99
critical geopolitics, 397–398
crude birth rate (CBR), 345
cultural assimilation, 502
cultural globalization, 272–276
cultural homogenization, 275
cultural landscapes, 504–505
cultural regions, 502
cultural westernization, 237–238
culture
   acculturation and, 502
   defined, 18–19, 500
   folk, 502–503
   human geography and, 18–19
   introducing, 501–503
   material, 501
   nonmaterial, 501
   popular, 502–503
   socialization and, 502
   traits, 501
customs union, 259
cyberwarfare, 492
Cyrillic, 107

D

Daimler-Benz, 457
dairy farming, 426–427
Dao De Ching (the Book of Changes) (Lao Tzu), 166–167
Daoism, 166–169
    chi in, 168
    representation of yin and yang in, 168–170
da Vinci, Leonardo, 205
death rate (mortality), 345
debt repayments, 325
Deere, John, 212
degree of unionization, 481
deindustrialization, 449–452
Del Monte, 423
demographic transition, 353–369
    contrasting the demographic transition and Malthusianism, 367–368
    criticisms of demographic transition theory, 367
    stage 1: preindustrial society, 353–359
    stage 2: early industrial society, 359–361
    stage 3: late industrial society, 361–362
    stage 4: postindustrial society, 362–366
demographic transition theory, 353, 367
developed countries
    commercial energy consumption in, 265
    world trade and, 253
developing countries, 316–331
    commercial energy consumption in, 265
    corrupt and oppressive governments, 328–329
    foreign debt, 323–327
    IMF loans and, 257–258
    inadequate and insufficient technology, 320–322
    lack of capital and investment, 319–320
    low labor productivity, 318–319
    poor terms of trade, 323–324
    rapid population growth, 316–318
    residential patterns in, 311
    restrictive gender roles, 326–327
    tourism and, 272
    unemployment and underemployment, 318–319
    unequal land distribution, 321–322
development, defined, 295
development strategies, 329–333
    expansion of international trade, 330
    foreign aid, 331–332
    international private investment, 330–331
dialects, 99, 119
Diamond, Jared, 72, 221
diaspora, 142
Dickens, Charles, 452
diffusion (spread) of information, 15
digital divide, 491

diminishing marginal returns, 348
Dole, 420
domestication
    of animals, 28, 40–43
    of plants, 28, 40–41, 43
Donald Trump, election of, 283–284
doubling time, 346
Dravidian languages, 124
Dravidians, 66
drug production, movement, and conflict, 404–406
drug trafficking, 404
dual society, and colonialism, 236
Duke, James, 212
Dupont, Eleuthere, 212
Dust Bowl, 430
Dutch East Indies Company, 234

E

early industrial society, 359–361
East Asia, 229–232
East India Company, 228
eBay, 521
economic and legal environment, 474–475
economic development, 295–304
    economic structure of the labor force, 296–297
    education and literacy of a population, 299–300
    health of a population, 301–304
    per capita gross domestic product (GDP), 296–298
    production of consumer goods, 299–300
economic sector
    primary, 296
    secondary, 298
    tertiary, 298
economic structure of the labor force, 296–297
economic union, 260
economy
    defined, 20
    informal, 309
    political. *See* political economy
education
    demand for, 472–473
    of a population, 299–300
    services labor markets and, 482–484
Egyptian civilization, 59–62
    gods and goddesses, 61
    hieroglyphs, 60–61
    Nile river and, 59
    pharaoh, 61
Egyptian hieroglyphics, 47, 60–61
electoral geography, 408–412
    redistricting, 409–411

electronics industry, 457–459
emigration, 346
employment. *See also* unemployment
   informal, 309–310
   tourism and, 269
encomiendas, 223
Energy resources, 265–266
English language, 108–115
   evolution of, 110–111
   French influence and, 110
   Germanic grammar and, 109
   Latin words and, 110
   loan words, acquired from other languages, 110, 112
   origin of, 108
Enlightenment, 174
entrepreneurship, 249
environmental determinism, 16, 506
epidemiological transition, 356
Erasmus, 205
Essay on the Principle of Population Growth (Malthus), 347
ETA (Euskadi ta Askatasuna, "Basque Fatherland and Liberty"), 406
ethnic conflicts, 513–516
ethnicity
   defined, 508
   dynamic, 509
   perceived, 391
   power and, 22
   situational, 509
   social relations and, 22
ethnic nationalism, 391, 391–392
ethnic religions, 140
Europe
   colonization of Arab world, 227–228
   colonization of Latin America, 222–223
   colonization of Southeast Asia, 231–233
   colonization of sub-Saharan Africa, 225–227
   colonizing of less developed countries, 218–219
   gasoline taxes in, 268
   manufacturing, 441–445
   technological advantages, 221
European civilizations, 82–87
   Classical or Hellenic Greece, 85–87
   Minoan culture, 83–84
   Mycenaean culture, 84
   Roman Republic, 86–88
European Economic Community (EEC), 260. *See also* European Union (EU)
European Union (EU), 247, 260–262, 386, 403, 431
export, defined, 475

F

Facebook, 492–493
failed states, 403–404
Farsi language, 105
Federal Interstate Highway System, 457
feedlots, 429
feminism, 521
fertility
   defined, 345
   rate, 345
feudal guilds, 192
feudalism
   bubonic plague and, 194–195
   capitalism and, 187
   characteristics of, 187–191
   defined, 187
   end of, 192–195
   farming under, 189
   ruling class of, 189
finance
   banking and, 200
   capitalism and, 200–201
   commercial credit and, 200–201
finance, insurance, and real estate (FIRE) sector, 469
financial services, 468
Finno-Ugric languages, 116
"Five Pillars of Islam", 150–151
folk culture, 502–503
Ford, Henry, 456–457
"Fordism", 456
Ford Motor Company, 457
foreign aid, as development strategy, 331–332
foreign debt, in developing countries, 323–327
foreign direct investment (FDI), 255, 318, 330
fossil fuels, 263
Franco, Francisco, 130
"free market" societies, 199
free trade area, 259
French Alsace-Lorraine, 393
French language, 113, 114
French Revolution, 347, 386, 400
Friendster, 492
full economic integration, 260

G

Gaelic, 130
Galileo, 206
Gandhi, Mahatma Mohandas, 228
garments, 446
gender, 517–526
   composition, 479–481

defined, 326, 517
human development and, 522–524
in the landscape, 521–522
power relation and, 518, 521
socially constructed, 517
gender roles, 517
in developing countries, 327–328
restrictive, 326–327
General Foods, 420
General Mills, 420
General Motors, 457
genocide, 514–515
Geographic Information Systems, 410
geographies of the Internet, 487–492
geography
defined, 2
human. *See* human geography
of Industrial Revolution, 210–212
of languages, 97–133
of nuclear weapons, 407–408
of secularism, 174–178
of services, 483–484
physical. *See* physical geography
regional. *See* regional geography
space and, 10
systematic. *See* systematic geography
telecommunications and, 484–495
geography of wars and terrorism, 400–409
civil wars, 402–403
drug production, movement, and conflict, 404–406
failed states, 403–404
geography of nuclear weapons, 407–408
terrorism, 404–406
geopolitics, 394–398
classical geopolitics, 394–397
critical geopolitics, 397–398
Georgian languages, 128
German Empire, 393
Germany
exports of heavy manufactured goods, 249
industrialization in, 217–218
gerrymandering, 410–411
Gibbon, Edward, 89, 146
global division of labor
Great Britain and, 250–251
international trade and, 250–253
globalization, 4
cultural, 272–276
defined, 246–247
internationalization of banking, 253
international trade, 246–253
regional economic integration, 259–265

religious fundamentalism and, 284–286
tourism and, 269–272
global markets
growth, and Industrial Revolution, 216–218
productivity and, 252
governments
corrupt, 328–329
oppressive, 328–329
grassroots antiglobalization, 282–285
Brexit, 282
Donald Trump, election of, 283–284
Great Britain
as industrial economy, 211
colonization of North America, 224
colonization of South Asia, 228–229
global division of labor and, 250–251
Industrial Revolution and, 210
international trade and, 216
Great Depression, 368, 430, 441
Great Wall of China, 65
Green Revolution, 352, 430
gross domestic product (GDP), 296
guilds, feudal, 192
Guns, Germs, and Steel (Diamond), 72, 221

H

hadith, 152
haj, 152
Hamas, 286
hara-kiri, 173
Harappa civilization, 66
Hasidic Judaism, 144
Hasidim, 284
Haushofer, Karl, 396
health care
demand for, 472–473
of a population, 301–304
urbanization in developing countries and, 316–317
Heartland theory, 395–396
heavy manufacturing industries, 480
Hebrew, 116
hegemony, 399
Hellenic Greece, 85–87
hieroglyphs, 60–61
Hinduism, 159–162
Bhagavad-Gita, 159
Buddhism and, 162
castes in, 161–162
cows' regard in, 160
origin of, 159
polytheistic religion, 159
samsara, 160

Hitler, Adolf, 395, 396
Hittites, 105
Holy Roman Empire, 385, 386
hominids
    Australopithecus, 28
    bipedalism and, 28
    defined, 28
    evolution of, 28–34
Homo erectus, 29
Homo habilis, 29
Homo sapiens, 30–31, 33
Hudson Bay Company, 224
Huitzilopochtl (sun god), 71
human capital, 319
human geography
    culture and space, 18–19
    described, 3–4
    historical context, 13–15
    networks and interdependence, 15–16
    people and the environment, 16–17
    physical geography and, 4
    social relations and, 20–22
    strengths of, 4
human immunodeficiency virus (HIV), 356–357
human resources, 248
Hume, David, 206
Hungarian language, 116
hunting and gathering society, 34–40
    fire and, 35
    religious beliefs, 37–38
    resource exploitation in, 34
    sexual reproduction, 37
    weapons and tools used in, 34–35
Hunt-Wesson, 421
Hussein, Saddam, 154, 517
Hyksos, 105

I

Ice Age, 43
identity, 517–526
    building blocks of ethnic, 508
    ethnicity and, 508
    sexuality as an expression of, 524
    social difference and, 525
ideologies
    capitalism and, 203–206
    printing invention, 204–205
    Protestant ethic, 206
    Protestant Reformation, 205–206
    Renaissance, 205
immigration, 345

inadequate and insufficient technology, 320–322
inanimate energy, 207–209
Inca civilization, 75–76
Inca rebellions, 218
income distribution, in services labor markets, 477–480
Indian Sepoy uprising of 1857, 218, 228
India–Pakistan wars, 401
Indo-European language family, 101–114
    Aryans and, 101–102
    Balochi, 105
    Celtic language, 106
    Cyrillic, 107
    English language, 108–115
    Farsi, 105
    Latin-based Romance languages, 105–106
    Latvian languages, 107
    Sanskrit, 103–105
    Slavic languages, 107
Indus River Valley civilization, 66–68
industrial food-production truck farms, 422
industrialization
    capitalism and, 212
    cycles of, 212–213
industrialized agricultural systems vs. preindustrial agriculture, 50
Industrial Revolution, 18, 48, 86, 306, 309, 338, 344, 347, 349, 352–353, 418, 438, 439, 441, 452, 477, 503, 522
    capitalism and, 206–219
    consequences of, 213–215
    cycles of industrialization, 212–213
    geography of, 210–212
    growth of global markets, 216–218
    inanimate energy, 207–209
    increase in productivity, 210
    international trade and, 216–218
    population effects and, 216
    technological innovation, 208–209
Industrial Workers of the World, 215
infant mortality rate, 303
informal economy, 309
informal employment, 309–310. *See also* employment
infrastructure, 199
institutionalized racism, 506
International Bank for Reconstruction and Development. *See* World Bank
international migrant labor, 375
International Monetary Fund (IMF), 247, 257–259, 326–327
    loan acquisition from, 257
    origin of, 258
    role of, 257–258

international private investment, 330–331
international trade, 247–254
   expansion of, 330
   global division of labor and, 250–253
   Great Britain and, 217
   Industrial Revolution and, 216–218
   specialization and, 247–250
Internet, 487–488
   discrepancies in access of, 490–491
   geographies of, 487–492
   growth of, 488–489
   mobile access to, 485
   National Science Foundation and, 488
   originated in, 488
   penetration rates, 490
   social media and, 492–493
   users of, 489
Irish Republican Army (IRA), 406
Iron Age, 49–50
Islam, 79, 149–158, 284, 406
   gender norms in, 153
   hadith, 152
   Muhammad and, 150
   origin of, 149
   sharia, 152
   Shiite–Sunni divide, 153
Islamic fundamentalism, 332
Islamic State (ISIS), 406
isolates, 100

J

Jainism, 161
Japanese languages, 118
Japan's Ministry of International Trade and Industry (MITI), 446
Jefferson, Thomas, 206
Jihad vs. McWorld (Barber), 280
Judaism, 141–145, 284
   Ashkenazi, 144–145
   Hasidic, 144
   origin, 141
   Sephardic, 143
   Talmud and, 141
   Torah and, 141

K

keiretsu, 446
Khoisan languages, 118
Khomeini, Ayatollah, 285
Kingdom of Axum, 80
Kingdom of Egypt, 60
Kingdom of Zimbabwe, 82
King Hammurabi, 59
Kissinger, Henry, 396
Kjellen, Rudolf, 394
Knights of Labor, 215
Koran, 116, 150, 285
Korean language, 118
Korean War, 323

L

labor force, economic structure of, 296–297
labor intensity services labor markets, 477
labor market, 309
labor productivity
   in developing countries, 318–319
   low, 318–319
lack of capital and investment, 319–320
land distribution
   in developing countries, 321–322
   unequal, 321–322
land-grant system, 223
land reform, 323
landscape, 2
   societies and, 6
language(s)
   death, 131–132
   defined, 98–100
   families, of world, 100–128
   geography of, 97–133
   linguistic differences, 129–131
Lao Tzu, 166
Lashkar e-Taiba, 406
late industrial society, 361–362
Latin America
   Columbian Exchange, 222–223
   Europe colonization of, 222–223
Latin-based Romance languages, 105–106
Latvian languages, 107
The Limits to Growth, 351
Le Pen, Marine, 284
less developed countries, 11, 294
   capital flight in, 320
   characteristics of, 298–299
   debt crisis in, 326–327
   employment opportunities in, 309–310
   governments in, 328–329
   IMF loans and, 257
   industrialization of, 266
   international private investment in, 330–331
   international trade and, 330
   investment obstacles in, 320

low levels of productivity in, 318–319
protectionism and, 330
unemployment and underemployment in, 318–319
Washington Consensus and, 258
world's energy, consumed by, 258
lingua franca, 114
linguistic conflict, 129–131
LinkedIn, 492
literacy of a population, 299–300
literacy rate, 299
Lithuanian languages, 107
location, 2
Locke, John, 206
long-distance trade, and capitalism, 202–203
Luther, Martin, 149, 205

M

Machu Picchu, 76
Mackinder, Sir Halford, 395–396
Mahayana Buddhism, 164–165
"Make America Great Again" campaign, 283
Malagasy language, 124
Malay languages, 123
Malayo-Polynesian languages, 123–124
Malthusianism, 348, 349, 350, 351, 353, 367
    contrasting the demographic transition and, 367–368
Malthusian theory, 347–350
Malthus, Thomas Robert, 216, 249, 338, 347–350
Manchu language, 118
Mandarin, 120–121
mandate of heaven, 170
manufacturing, 451–460
    automobiles, 456–458
    East Asia, 444–449
    electronics, 457–459
    Europe and Russia, 441–445
    major concentrations of, 439–451
    North America, 439–441
    steel industry, 453–456
    textiles and garments, 452–453
map projections, 10–11
maps, defined, 10
maquiladora plants, 330
markets
    capitalism and, 196–197
    defined, 196
    infrastructure, 199
market societies, 390
Marxism, 338, 351–352
Marxists, 351
Marx, Karl, 351–352

mass external migrations, 374
material culture, 501
Mayan culture, 69–70
Mayan glyphs, 47
McDonald's, 418
Medicaid, 472
Medicare, 472
Melanesians, 128
Menes, King of Egypt, 60
Meng Tzu, 169
Mercedes Benz, 443
Mesoamerica, 67–73
    Aztecs, 70–73
    Mayan culture, 69–70
Mesopotamia, 57–58
    time system, 60
    writing system/cuneiform, 57
    ziggurats, 57, 61
microelectronic technology, 458
migration, 345, 370
    barriers to, 371
    causes of, 371–373
    consequences of, 372–373
    defined, 370
    economic effects of, 373
    examples of colonizing, 376
    in highly urbanized countries, 377
    intercontinental, 374–375
    international, 373
    patterns of, 374–377
    social conflict and, 373–374
minerals
    defined, 264
    extraction, 265
    United States and, 264–265
minimills, 455
Minoan culture, 83–84
Mississippian mound-building culture, 77
mobile phones, 485–486
mobile telephony, 485
Moche people, 74
Mohenjo-Daro civilization, 66
Mongol language, 118
Mongols, 398
monocultures, 419
monotheism, 49
more developed countries, 11
Morgan, J. P., 212
Morse, Samuel, 484
Muhammad, 149
Muslim Ayodhya mosque, 284
Muslim Brotherhood, 285, 286

Muslims
  Arabs and, 156–157
  defined, 150
  population in the world, 156–158
Mycenaean culture, 84
MySpace, 492–493

N

Nabisco, 420
Napoleonic Wars, 219, 223, 388, 400
nation, 385
National Front, 284
nationalism, 390–393
  civic, 391
  defined, 390
  Japanese, 172
  Jewish, 116, 143
National Science Foundation, 488
nation-state, 384–390
  defined, 385
natural environment
  human geography and, 18–19
  people and, 18–19
natural gas, 268–269
natural growth rate (NGR), 346
natural resources, 263–264
natural selection, 42
Nazca people, 74
Nazi Party, 506
Nazis, 393, 395, 507
Neanderthals, 33
Neolithic Age, 34
Neolithic Revolution, 38, 39–49, 338, 418, 419, 518
  agricultural innovations, 45
  astronomy, 45
  different forms of writing, 46–47
  domestication of animals, 40–43
  domestication of plants, 40–41, 42
  first class-based societies in, 46
  Ice Age and, 43
  metals, use of, 49
  organized religious systems, 48–49
  shift to agriculture in, 44
neo-Malthusian family planning, 362
Neo-Malthusianism, 350–351
neo-Malthusians, 338
net migration rate (NMR), 346
networks, and human geography, 15–16
Newcomen, James, 207
"New Stone Age", 34
New Testament, 145

Newton, Isaac, 206
newtowns, 194
New World, 371, 374
Niger-Congo languages, 118–119
Nile river, 59
nirvana, 160
nomadic herding, 51–52
"nondirect production", 470
nongovernmental organizations (NGOs), 280
nonmaterial culture, 501
nonprofit sector, 470
nonrenewable resources
  defined, 263
  fossil fuels, 263
Nontonal languages, 121
North America, 224
  capital flight to, 320
  fertility levels in, 362
  manufacturing, 439–441
  urbanization rates, 305
North American civilizations, 76–77
North American Free Trade Agreement (NAFTA), 16, 247, 262–264, 283, 330
North American Manufacturing Belt, 439–440
nuclear families, 362
nuclear war, 408
nuclear weapons, geography of, 407–408
"nuclear winter", 408

O

Oceania, 234–235
"Old Stone Age", 34
Old Testament of the Bible, 142
Olmecs, 68
Opium Wars of the 1840s, 231
opportunity cost, 361
Organization of the Petroleum Exporting Countries (OPEC), 266–267, 320, 325
Orientalism, 501, 525–526
Ottoman Empire, 227–228, 393
Ottomans, 398

P

Paleolithic Age, 34
Palestine Liberation Organization, 285
pandemic, 357
Paris Accord, 283
patriarchal societies, 518
Pax Romana, 86
Peace Corps, 351
Peloponnesian Wa, 85

people, and natural environment, 16–17
Pepsi, 521
per capita gross domestic product (GDP), 296–298
petroleum, 266–268
   Middle East and, 266–267
   OPEC and, 266–267
pharaohs, 61
physical geography
   described, 3
   human geography and, 4–5
physical resources, 248
pictographic writing system, 120
Pizarro, Francisco, 76
place, 2, 517–526
polarized geographies, 236–237
political boundaries, 11–12
political economy
   defined, 20
   social relations and, 20
politics, 20
polytheistic religion, 159
popular culture, 502–503
population
   change, 345–346
   density, 342–345
   education of, 299–300
   health of a, 301–304
   literacy of, 299–300
   pyramids, 368
   structure, 368–371
population growth
   in developing countries, 316–318
   Marxism, 351–352
   technological optimism, 352–353
Porsche, 443
Portuguese language, 112
postindustrial society, 362–366
power
   ethnicity and, 22
   social relations and, 22
preindustrial agriculture, 50–54
   Asian rice paddy cultivation, 53–54
   nomadic herding, 51–52
   shifting cultivation, 51–52
   vs. industrialized agricultural systems, 50
preindustrial society, 353–359
"premature urbanization", 311
price-support programs, 431
priests, 48
primary economic sector, 296
private investment, international, 330–331
producer services, 468

production factors, 248
production of consumer goods, 299–300
productivity
   increases, and Industrial Revolution, 210
   international trade and, 251–252
protectionism, 330
Protestant ethic, 206
Protestant Reformation, 149, 205–206
public sector, growth of, 474–475

Q

quality, of products, 252
Quechua language, 128
queer theory, 524
Quran. *See* Koran

R

Ra (sun god), 61
Rabin, Yitzhak, 284
race, 31–32
   defined, 505
   ethnicity and, 505–518, 508–509
   human classification and, 506
   Irish, 506
   relational perspectives on, 507–508
   scientific basis for, 506
racialization
   defined, 507
   processes of, 507–508
   theorists critical of, 507
racism
   defined, 505
   institutionalized, 506
   relational perspectives on, 507–508
   slavery and, 506
   Western, 505
Ramadan, 151
Ramses II, pharaoh of Egypt, 62
rational secularization, 174
Ratzel, Friedrich, 394–395
redistricting, 409–411
Reformation, 386
region, 2
regional economic integration, 259–265
   common market, 259
   customs union, 259
   economic union, 260
   European Union (EU), 260–262
   free trade area, 259
   full economic integration, 260

North American Free Trade Agreement (NAFTA), 262–264
regional geography, 6
religion(s). *See also* specific religions
   ethnic, 140
   hunting and gathering society, 36–37
   major world, 140–174
   Neolithic Revolution, 48–49
   United States and, 177–178
   universalizing, 140–141
religious fundamentalism, 284
   globalization and, 284–286
Renaissance, 205
renewable resources, defined, 264
Republican Party, 283
residential patterns in the developing world, 311
Ricardo, David, 249, 348
rising incomes, 471–473
Rockefeller, John D., 212
Romance languages, 105–106
Roman latifundia, 51
Roman Republic, 86–88
Romans, 398
Ronald McDonald, 280
Russia
   coal reserves in, 268
   manufacturing, 440–444
   natural gas and, 268

S

Sadat, Anwar, 285
Said, Edward, 525
samsara, 160
Sanskrit language, 103–105, 122
   Indic branch of languages and, 103
   overview, 103
   role in northern India, 103
science, and capitalism, 206
secondary economic sector, 298
secularism, 174–178
   atheists and, 175
   Max Weber and, 174
Semitic, 116
Sephardic Judaism, 143
Sepoy Rebellion, 228, 230, 280. *See also* Indian Sepoy uprising of 1857
September 11, 2001 terrorist attacks, 16, 406
sepuku, 173
serfs, 190–191
service exports, 475–476
services
   defined, 470
   defining, 468–471
   demand for health care and education, 472–473
   geography of, 483–484
   growth of, 471–477
   growth of the public sector, 474–475
   increasingly complex economic and legal environment, 474–475
   rising incomes, 471–473
   service exports, 475–476
   services labor markets, 476–484
   telecommunications and geography, 484–495
services labor markets, 476–484
   educational requirements, 482–484
   gender composition, 479–481
   income distribution, 477–480
   labor intensity, 477
   low degree of unionization, 481
sexism, 522
sex ratio, 370
sexuality
   and space, 524–525
   defined, 524
Shah of Iran, 285
Shang dynasty, 65
sharia, 152
shifting cultivation, 51–52
Shih Huang Ti, 65
Shiites, 153–154
Shintoism, 172–174
Siddhartha Gautama, 164
Sikhism, 161
Silk Road, 193–194
Sino-Tibetan languages, 119–122, 124
slash-and-burn cultivation, 51, 53
Slavic languages, 107
slums
   growth of, 311–315
   urbanization in developing countries and, 311–315
Smith, Adam, 206
Social Darwinism, 349, 394, 506
socialization, 18, 502
social media, 492–493
social phenomena, and space, 9–10
social relations
   ethnicity and, 22
   human geography and, 20–21
   political economy and, 20
Social Security, 353, 373
Social Security cards, 492
social surplus, 55
societies, and landscape, 6

Soil Bank program, 431
South American civilizations, 74–75
   Chavin culture, 74
   Inca culture, 75–76
   Moche people, 74
   Nazca people, 74
South Asia, 228–229
Southeast Asia, 231–233
sovereignty, 385
space, 2
   described, 7–12
   geography and, 10
   human geography and, 18–19
   physical bodies and, 5
   physical surface and, 8–9
   social phenomena and, 9–10
Spain, 223
Spanish-American War of 1898, 223, 231
Spanish land grant system (encomiendas), 322
Spanish language, 112
spatial distributions, 3
spatial scales, 6
Stalin, Joseph, 175
Standard Arabic, 116
steel industry, 453–456
structural adjustment policies, 326
sub-Saharan Africa
   civilizations, 78–80
   colonization by Europe, 225–227
subsistence agriculture, 420
Sumerian cuneiform, 47
Sumerians, 57–58
Sunni, 153–154
"survival of the fittest" theory, 349
sustainable agriculture, 432, 432–434
suttee, or widow burning, 284
Swahili, 118
swidden/shifting cultivation, 51, 53
systematic geography, 6

T

Tagalog language, 123
Taiping Rebellion, 231, 233
Taliban, 285, 286
Talmud, 141
Target, 469
Tatar language, 118
Tawantinsuyu, 75
TCP/IP (Transmission Control Protocol/Internet Protocol), 488
technological innovation, 208–209

technological optimism, 352–353
technology, 320–322
telecommunications and geography, 484–495
telegraphy, 484
Tennessee Valley Authority, 441
terms of trade
   defined, 323
   in developing countries, 323–324
terrorism, 404–406
   defined, 404
   geography of, 400–409
   incidence of, 404
   Islamic, 406
   Muslim, 283
tertiary economic sector, 298
textile and garment industries, 452–453
texting, 485
Thai-Kradai languages, 124
The Decline and Fall of the Roman Empire (Gibbons), 146
Theravada Buddhism, 166–167
Thirty Years' War, 386
three-field system of farming, 192
Timbuktu, 79
time-space compression, 203, 216
Torah, 141
tourism
   described, 269
   employment and, 269
   globalization and, 269–272
   growth after World War II, 269–270
   less developed countries and, 271
   types of, 270–271
trade
   capitalism and, 202–203
   comparative advantage and, 203
transhumance, 51
transnational corporations (TNCs), 254–255
Trans-Pacific Partnership, 283
Treaty of Tordesillas, 222
Treaty of Vienna, 219
Treaty of Westphalia, 386
truck farming, 422
Trump, Donald J., 282, 283–284
Turkic languages, 116–117
Twitter, 492, 493

U

underemployment
   defined, 318–319
   in developing countries, 318–319

unemployment. *See also* employment
    defined, 318
        in developing countries, 318
unequal exchange, in international trade, 250
Unilever, 423
unionization, degree of, 481
Union of Soviet Socialist Republics (USSR), 175
United Brands, 421, 423
United Kingdom Independence Party (UKIP), 282
United Nations, 401
United States
    coal reserves in, 268
    fossil fuels consumption in, 265
    invasion of Afghanistan and, 402
    minerals and, 264–265
    oil demand in, 266
    Winter Wheat Belt, 427–428
universalizing religions, 140–141
Upper Rhine–Alsace-Lorraine Region, 443
Ural-Altaic, 116–118
urbanism, 57
urbanization, defined, 215
urbanization in developing countries, 305–317
    growth of slums, 311–315
    health and environmental concerns, 315–316
    informal employment, 309–310
    residential patterns in the developing world, 311
U.S. agricultural policy, 429–432
U.S. Defense Department's Agency Research Projects Administration (ARPA), 488
U.S. Department of Justice, 412
U.S. Steel Company, 455
U.S. Supreme Court, 411

V

Vajrayana Buddhism, 165
value added, 438
Vietnam War, 401
villanovas, 194
Voice Over Internet Protocol (VOIP) telephony, 488
Voting Rights Act, 412, 505

W

Wallerstein, Immanuel, 398

Wal-Mart, 469
wars, geography of, 400–409
Washington Consensus, 258
Watt, James, 207
Weber, Max, 174, 206
Westphalian system of nation-states, 386
women
    Confucianism and, 169
    domestic abuse and rape and, 519
    economic, political, and social status of, 327
    feminism and, 521
    incomes, 361
    labor-force participation of, 327–328
    Muslims and, 153–154
    political advances of, 520
    power relations between men and, 21–22
    twenty leading occupations of employed, 478
working class, 213–214
"working poor", 478
World Bank, 247, 257–259, 326
    origin of, 257
    role of, 257–258
world's population
    distribution of, 338–344
    growth of, 338–344
    population density, 342
world-systems theory, 398
world trade, 253
World Trade Center attacks, 286
World Trade Organization (WTO), 247, 256–257, 283
World War I, 374, 396
World War II, 370, 374, 377
World Wide Web, 488

Y

Yahoo, 521
yayesakura, 173
YouTube, 492

Z

zero population growth (ZPG), 363, 370
ziggurats, 57, 61
Zionism, 116, 143
Zulu attacks, 218

CPSIA information can be obtained
at www.ICGtesting.com
Printed in the USA
FSHW021913170719
60124FS

9 781516 529025